Male-mediated Developmental Toxicity

Issues in Toxicology

Series Editors

Professor Diana Anderson, *University of Bradford, UK*
Dr Michael D Waters, *National Institute of Environmental Health Science, N Carolina, USA*
Dr Timothy C Marrs, *Edentox Associates, Kent, UK*

This Series is devoted to coverage of modern toxicology and assessment of risk and is responding to the resurgence in interest in these areas of scientific investigation.

Ideal as a reference and guide to investigations in the biomedical, biochemical and pharmaceutical sciences at the graduate and post graduate level.

Titles in the series:

Hair in Toxicology: An Important Bio-Monitor
Edited by Desmond John Tobin, *University of Bradford*

Male-mediated Developmental Toxicity
Edited by Diana Anderson and Martin H Brinkworth, *Department of Biomedical Sciences, University of Bradford, Bradford, UK*

Visit our website at www.rsc.org/issuesintoxicology

For further information please contact:
Sales and Customer Care, Royal Society of Chemistry, Thomas Graham House, Science Park, Milton Road, Cambridge, CB4 0WF, UK
Telephone: +44 (0)1223 432360, Fax: +44 (0)1223 426017, Email: sales@rsc.org

Male-mediated Developmental Toxicity

Edited by

Diana Anderson and Martin H Brinkworth
Department of Biomedical Sciences, University of Bradford, Bradford, UK

RSC Publishing

ISBN-13: 978-0-85404-847-2

Published by The Royal Society of Chemistry,
Thomas Graham House, Science Park, Milton Road,
Cambridge CB4 0WF, UK

Registered Charity Number 207890

For further information see our web site at www.rsc.org

Foreword

By the Renaissance, Europe was ravaged by epidemics of infectious diseases and most people felt this must be divine punishment. When people travelled they brought the plague with them. Was divine punishment transportable?

Nowadays, some people may still hold much the same views. The spreading of industrialization and of man-made chemicals epitomizes human eagerness and the world deserves a conveyable calamity. Similar dilemmas loom: do exposures to chemicals produce "transportable" traits passing, in this case, from generation to generation? Opposite views consider inheritance the decisive factor for health and disease, a sort of lottery-producing winners and losers.

Fortunately, recent advances in genetics are reordering our conceptions of the respective roles of environment and genetics in diseases and reproduction, and a dichotomy between environmentalists and geneticists is no longer justified.

Whereas it is not entirely clear to anyone where the genomic revolution will lead, it is obvious that this milestone brings with it enormously compelling opportunities to understand better human health and disease, including reproductive disorders.

In that respect, this meeting on male-mediated reproductive toxicity is providing new information and understanding. However, a broad question now arises: how do we perceive a newly forming order in a developing area of science, how to insert it into the general frame of our complex intellectual constructions?

Perhaps, the answer comes from a fascinating book by Hofstadter[1] that illuminates one of the greatest mysteries of modern science: the nature of human thought processes. In his view, all constructions of human mind can be explained by the phenomenon of "strange loops", that occurs whenever, moving upwards (or downwards) through the levels of some hierarchical system, we unexpectedly find ourselves back where we started. He illustrates examples of "strange loops" in music, art and logic, as well as in ideas drawn from mathematics, biology, psychology, physics and linguistics.

In music, "strange loops" are the canons as they might be found in the Musical Offering of J. S. Bach. A canon is a single theme played against itself. This is done by having copies of the theme sung by various participating voices.

After successive modulations, all voices are exactly one octave higher than they were at the beginning and the process could go on forever.

The "strange loop" is also one of the most recurrent themes in the work of the Dutch graphic artist M. C. Escher. His "drawings hands" is one of the best visual realizations; each of the two hands draws the other.

In logic, the ancient philosophical paradox of Epimenides shows the phenomenon. Epimenides was a Cretan who made the statement "all Cretans are liars".

Several papers have been presented at this meeting, where paternal exposures to chemicals have been associated with a range of adverse outcomes on reproduction. These include infertility, miscarriage and a variety of malformations observed postnatally, some of which have arisen from stable mutations in the male germ line.

However, data have also been presented[2] where, instead, maternal exposures to certain endocrine disruptors correlate with transgenerational male-mediated adverse effects on fertility. An epigenetic mechanism was suggested because the described effects correlate with altered DNA methylation patterns in germ lines.

These results best illustrate the concept of "strange loop" as applied to male-mediated reproductive toxicity/effects. Reduced sperm count and motility, and increased apoptotic rates in spermatogenic cells were passed down to nearly every male in four subsequent generations. This was achieved with a transient exposure to vinclozolin and methoxychlor of female-gestating rats during the period of gonadal sex determination. Whereas in this study DNA mutations have not been ruled out, and epigenetic mechanisms have already been suggested for male-mediated toxicity,[3] there is very little previous evidence for toxins to cause these transgenerational effects by this mechanism.

Thus, a male-mediated reduction of fertility through generations was initiated by exposures *in utero* of first-generation male rats. An example of a "strange loop", where evidence takes us back to a teratogenic effect classically induced by exposing pregnant females to toxicants, but with the addition of a relatively new notion: changes induced in this way can be transmitted, male-mediated, through several generations.

Progress in science derives from the addition of new facts and this meeting certainly contributed to it. We may wish to consider that often new facts take us back where we started, though enriched with new knowledge and ready to start another loop. And so it goes on. I wonder whether the new order between genetic and environmental factors we are witnessing could be depicted, according to Hofstadter, as a golden braid made of newly forming "strange loops".

Marcello Lotti

References

1. D.R. Hofstadter, *Gödel, Escher, Bach: An Eternal Golden Braid*, Penguin Books, Rome, 1980.

2. M.D. Anway, A.S. Cupp, M. Uzumcu and M.K. Skinner, Epigenetic transgenerational actions of endocrine disruptors and male fertility, *Science*, 2005, **308**, 1466–1469.
3. J.M. Trasler, Translational research in male mediated developmental toxicity, in *Advances in Male Mediated Developmental Toxicity*, B. Robaire and BF. Hales (eds), Kluwer Academic/Plenum Publishers, New York, 2003.

Preface

This book was developed from the 3rd International Congress on Male-Mediated Developmental Toxicity held at the University of Bradford from 31st July 2005 to 3rd August, 2005. The International Congresses on Male-Mediated Developmental Toxicity are held every 4–5 years and serve as a focal point for discussion of male germ line mutagenesis that can lead to developmental defects in the next generation. There is considerable concern about the possibility that environmental, therapeutic or occupational exposure to certain agents could lead to the induction of heritable damage in sperm that in turn could lead to childhood abnormality or disease. This topic is of immense significance to populations around the world; some of these concerns are outlined below.

(i) Since 2001, there has been an increased awareness of epigenetic trans-generational actions of certain chemicals through the male germ line, where effects are not diluted as they are in genetic transmission.

(ii) There has also been a steadily increasing awareness of the risk of transmission of genetic defects as a result of assisted-conception techniques such as intra-cytoplasmic sperm injection (ICSI), especially if the sperm used are from men whose spermatogenesis has been impaired by a genetic defect.

(iii) A further, ever-present, clinical issue, is the damage to spermatogenesis arising from various anti-cancer regimens that use highly genotoxic agents and thus may pose a risk of inducing mutations in surviving spermatogonial stem cells.

(iv) In addition, fundamental mechanisms of germ line mutagenesis form a strong strand of discussion, as it is only by understanding these mechanisms that the toxicological and clinical problems can be resolved.

(v) The rapid advances in reproductive and biomedical technologies of the last few years (*e.g.*, ICSI and cloning) have raised significant ethical issues that are only beginning to be confronted.

This book encapsulates the characteristics of these meetings which achieve high-level attendance among researchers in the field from all over the world, cutting-edge talks from leading scientists and the generation of new ideas that drive research forward and assist in forging of international collaborations. Previously, these ideas have been developed in books of proceedings of the conferences: (1) Reproductive Biology: *Male-Mediated Developmental Toxicity*, AF Olshan and DR Mattison (eds), Plenum Press; and (2) Advances in Experimental Medicine and Biology, *Advances in Male-Mediated Developmental Toxicity*, vol 518, B Robaire and BF Hales (eds), Kluwer Academic/Plenum Publishers, which have also been influential among regulatory scientists, informing thinking among those whose duty it is to address the challenges posed above and maintain public health and safety. These sentiments are also relevant to the current book.

In the light of recent scientific and medical developments, the time was felt to be right to stage the 3rd such meeting at the University of Bradford, UK, organised under the chairmanship of Professor Diana Anderson, who has published over many years on studies of male-mediated developmental toxicity. The contents of the book reflects the meeting with introductory scientific papers by Professors M. Lotti and M. Skinner (Foreword and Chapter 1), Section 1 on Heritable Effects in Humans (Chapters 2–8) and Section 2 on Animal Models (Chapters 9–15). Section 3 Germ line Mutagenesis (Chapters 16–24) incorporates the link with the 24th annual meeting of the United Kingdom Environmental Mutagen Society (UKEMS). This is the national society for those working in genetic toxicology and has long-standing interests in germ line mutagenesis and its consequences. The meeting was organised under the Chairmanship of Dr Martin Brinkworth who has published on the topic for over 15 years.

Diana Anderson and Martin Brinkworth

Contents

Animal Models

Chapter 11 Molecular Changes in Sperm and Early Embryos after Paternal Exposure to a Chemotherapeutic Agent

Bernard Robaire, Alexis M. Codrington and Barbara F. Hales

Chapter 12 Transmissible Genetic Risk Causing Tumours in Mice and Humans

Taisei Nomura

Germline Mutagenesis

Chapter 23   **DNA Repair Capacities in Testicular Cells of Rodents
and Man**
*Gunnar Brunborg, Nur Duale, Julie Tesdal Haaland,
Christine Bjørge, Erik Søderlund, Erik Dybing,
Richard Wiger and Ann-Karin Olsen*

Chapter 24   **3rd International Congress on Male-Mediated Development
Toxicity : Closing Panel Discussion**
Jack Bishop and Barbara F. Hales

Dedication

We dedicate this book to the memory of Professor Terry G. Baker, PhD, DSc, FRSE C.Biol, FIBiol, FRCPath, FIBMS who died on February 22nd, 2006. He made some introductory remarks at the opening ceremony of this Congress, and it was the last time he spoke at a scientific meeting. He was the person who founded the Department of Biomedical Sciences at the University of Bradford in 1980 and continued as its Head until 2000 when he became Emeritus Professor. During this time, he was also Dean and pro-Vice Chancellor. His field of research was Reproductive Biology, where he did pioneering studies on öoctye development, but in later years turned his attention to the male, and created the environment for Male Reproductive Toxicology which led to the meeting being staged at Bradford.

Acknowledgements

The editors gratefully acknowledge a financial award from the UK Medical Research Council, a grant from CEFIC (the European Chemical Industry Council) and other financial contributions from the US National Institute of Environmental Health Sciences, the UK Environmental Mutagen Society, AstraZeneca, GlaxoSmithKline and the EU ChildrenGenoNetwork.

Introductory Scientific Presentations

CHAPTER 1

Epigenetic Transgenerational Actions of Endocrine Disruptors through the Male Germ-Line

MICHAEL K. SKINNER

Center for Reproductive Biology, School of Molecular Biosciences, Washington State University, Pullman WA 99164-4231

1.1 Review

Embryonic exposure to environmental factors has been shown to cause adult onset disease,[1-3] but few have looked at the second F2 generation. Examples include the late embryonic and early postnatal exposure to cyclophosphamide causing embryonic defects,[4] embryonic nutritional defects causing immune defects,[5,6] diethylstilbesterol (DES) causing female reproductive tract abnormalities,[7,8] and other endocrine disruptors causing male reproductive defects.[9,10] Studies suggest effects of environmental factors on the first generation. Any transgenerational phenotype would require transmission through the germ-line.

A recent observation demonstrated that the exposure of a pregnant rat transiently to endocrine disruptors caused a spermatogenic cell defect and subfertility in the F1 generation and all subsequent generations examined (F1–F4).[11] The endocrine disruptors used were the anti-androgenic fungicide vinclozolin used in the fruit (*e.g.*, wine) industry[12] and the pesticide methoxychlor used to replace DDT.[13] The critical exposure period was at the time of sex determination and the transgenerational phenotype was transmitted through the male germ-line.[11] The phenotype of increased spermatogenic cell apoptosis, decreased sperm numbers and sperm motility was observed in greater than 90% of all males of all the generations examined. When the animals were allowed to age up to 1 year additional diseases developed including cancer, prostate disease, kidney disease, and immune cell defects.[14] A high frequency of transmission was observed in all generations examined for all the disease states.

The frequency of the transgenerational phenotype was such that a DNA sequence mutational event could not be involved. The random nature of a DNA sequence mutation has a phenotype typically less than 1% and this often

1

declines in subsequent generations.[1,15] An epigenetic mechanism is found to be involved due to the frequency of the phenotype. To support these conclusion, two genes were identified in the sperm that had altered methylation patterns associated with the transgenerational phenotype discussed.[11] Therefore, the endocrine disruptors appear to induce an epigenetic transgenerational disease condition for four generations through the male germ-line.[11] The epigenetics appears to involve altered DNA methylation. Although most genes get re-set in early embryonic development, a subset of genes called imprinted genes maintains their DNA methylation pattern which appears to be permanently programmed. In contrast to all somatic cells the primordial germ cells undergo a de-methylation during migration and early colonization of the embryonic gonad, followed by a re-methylation starting at the time of sex determination in a sex-specific manner.[16–18] The exposure of the pregnant mother at the time of sex determination appears to have altered the re-methylation of the germ-line and permanently re-programmed the imprinted pattern of DNA methylation.[11] This provides a unique epigenetic mechanism to promote a transgenerational phenotype induced by an environmental factor.

Altered methylation of imprinted genes has been shown to promote disease states.[19] Cancer and tumor development has also been shown to be involved in epigenetic alteration of DNA methylation.[20] Therefore, the epigenetic reprogramming of the male germ-line causes numerous transgenerational disease states that can be explained by this epigenetic mechanism. The identification of the altered DNA methylation sites and associated genes will provide more insight into the proposed epigenetic transgenerational phenotype.[11]

The level of endocrine disruptors used in the recent studies[11,14] is higher than levels anticipated in the environment, such that conclusions regarding the toxicology of these endocrine disruptors are not possible. However, the important factor is the identification of this novel phenomenon, that an environmental factor can promote an epigenetic transgenerational phenotype.[11] Due to this observation the potential hazards of environmental factors need to be carefully evaluated. If the exposure of your grandmother at mid-gestation to environmental toxins can cause a disease state in you with no exposure, and you will pass it on to your grandchildren, the potential hazards of environmental toxicants must be rigorously assessed. Transgenerational studies need to be performed in evaluating the toxicology of environmental compounds.

The epigenetic transgenerational phenotype also provides critical insights into disease etiology. Since a number of common disease states were induced,[14] an epigenetic component of disease now needs to be seriously considered. In the event a major epigenetic component exists, the epigenetic background of an individual may be a major factor in susceptibility to disease development. Therefore, identification of the genes involved with altered methylation may provide essential new diagnostics to assess future onset of disease. This will allow new therapeutic targets and therapies to potentially prevent the onset of disease. This is a new paradigm in disease etiology that needs to be considered.

In a broader biological perspective, the ability of an environmental factor to cause a permanent genetic trait in all subsequent progeny of an affected individual can significantly impact our understanding of evolutionary biology. Currently, a DNA sequence mutation event that allows an adaptation and natural selection is considered the driving factor in evolutionary biology. However, the frequency of specific evolutionary events[21,22] and regional influences on evolution suggests that an additional epigenetic mechanism should be considered. Although a DNA sequence mutational event will be important for evolutionary biology, an epigenetic component influenced by an environmental factor needs to be considered as an alternate factor that will help explain some aspects of evolutionary biology.

The epigenetic transgenerational actions of endocrine disruptors observed[11] provide novel insights into several areas of biology. The ability of an environmental compound to promote a transgenerational phenotype suggests toxicology studies need to consider transgenerational elements of the actions of potential toxic agents. Future studies need to investigate the types of compounds that can induce the epigenetic effect. Currently we know anti-androgenic compounds can, but need to assess if other factors can as well. The toxicology studies need to be done to assess the minimum required dose to obtain a phenotype and compare this to potential environmental levels. This information will reveal if the levels in our environment are a problem. The epigenetic effects on the methylation state of specific genes needs to be determined to provide insights into the mechanisms of action of the environmental factors. In addition, these genes will provide potentially critical diagnostic markers and therapeutic targets for a variety of common diseases. The basic elements of disease etiology now need to consider epigenetic factors as markers and/or causal factors. In a broader context, the epigenetic transgenerational impact of environmental factors needs to be considered in the mechanisms involved in evolutionary biology. Epigenetics will likely be a much more important factor in biology than currently appreciated. Epigenetics is the next layer of complexity beyond the DNA sequence.

Acknowledgments

Thanks to Ms. Jill Griffin for assistance in the preparation of this manuscript.

References

1. R. Barber, M.A. Plumb, E. Boulton, I. Roux and Y. E. Dubrova, *Proc. Natl. Acad. Sci. U. S. A.*, 2002, **99**, 6877–6882.
2. I.D. Morris, *Int. J. Androl.*, 2002, **25**, 255–261.
3. C.M. Foran, B.N. Peterson and W.H. Benson, *Toxicol. Sci.*, 2002, **68**, 389–402.
4. T.S. Barton, B. Robaire and B.F. Hales, *Proc. Natl. Acad. Sci. U. S. A.*, 2005, **102**, 7865–7870.
5. F. Bernard, C. Picard, V. Cormier-Daire, C. Eidenschenk, G. Pinto, J.C. Bustamante, E. Jouanguy, D. Teillac-Hamel, V. Colomb, I. Funck-Brentano, V. Pascal, E. Vivier, A. Fischer, F. Le Deist and J.L. Casanova, *Pediatrics*, 2004, **113**, 136–141.

6. P.D. Gluckman and M.A. Hanson, *Semin. Fetal Neonatal Med.*, 2004, **9**, 419–425.

7. A.L. Herbst, H. Ulfelder and D.C. Poskanzer, *N. Engl. J. Med.*, 1971, **284**, 878–881.

8. J.D. Cook, B.J. Davis, S.L. Cai, J.C. Barrett, C.J. Conti and C.L. Walker, *Proc. Natl. Acad. Sci. U. S. A.*, 2005, **102**, 8644–8649.

9. A.S. Cupp, M. Uzumcu, H. Suzuki, K. Dirks, B. Phillips and M.K. Skinner, *J. Androl.*, 2003, **24**, 736–745.

10. M. Uzumcu, H. Suzuki and M.K. Skinner, *Reprod. Toxicol.*, 2004, **18**, 765–774.

11. M.D. Anway, A.S. Cupp, M. Uzumcu and M.K. Skinner, *Science*, 2005, **308**, 1466–1469.

12. W.R. Kelce, E. Monosson, M.P. Gamcsik, S.C. Laws and L.E. Gray Jr., *Toxicol. Appl. Pharmacol.*, 1994, **126**, 276–285.

13. W.R. Kelce, C.R. Lambright, L.E. Gray Jr. and K.P. Roberts, *Toxicol. Appl. Pharmacol.*, 1997, **142**, 192–200.

14. M.D. Anyway, C. Leathers and M.K. Skinner, *Endocrinology*, 2006, in press.

15. B.S. Shi, Z.N. Cai, J. Yang and Y.N. Yu, *Mutat. Res.*, 2004, **556**, 1–9.

16. P. Hajkova, S. Erhardt, N. Lane, T. Haaf, O. El-Maarri, W. Reik, J. Walter and M. A. Surani, *Mech. Dev.*, 2002, **117**, 15–23.

17. G. Durcova-Hills, J. Ainscough and A. McLaren, *Differentiation*, 2001, **68**, 220–226.

18. W. Reik and J. Walter, *Nat. Rev. Genet.*, 2001, **2**, 21–32.

19. Y.H. Jiang, J. Bressler and A.L. Beaudet, *Annu. Rev. Genomics Hum. Genet.*, 2004, **5**, 479–510.

20. A.P. Feinberg and B. Tycko, *Nat. Rev. Cancer*, 2004, **4**, 143–153.

21. D. Penny, *Nature*, 2005, **436**, 183–184.

22. M. Balter, *Science*, 2005, **309**, 234–237.

Heritable Effects in Humans

CHAPTER 2
Reproductive Outcomes among Men Treated for Cancer

JOHN J. MULVIHILL AND TIMOTHY J. GARLOW

Department of Pediatrics and General Clinical Research Center, University of Oklahoma, Oklahoma City OK 73104, USA

2.1 Why Study Cancer Survivors?

A report from the US Institute of Medicine and National Research Council showed that, as of the end of 2002, an estimated 10.1 million individuals in the US had received the diagnosis of a cancer (other than nonmelanotic skin cancer) in their lifetime; 40% of them were under age 65 years.[1] One can assume half are male and hence estimate up to two million men in the US are cancer survivors with reproductive potential. Estimated another way, close to a half million men of reproductive age are added to US population each year, with the inclusion of recipients of organ transplants (large kidneys) who tend to receive some chemotherapy-like agents and the exclusion of men who die in within 5 years of diagnosis (Table 1).

The therapies for men with cancer comprise several categories: physical agents (namely, ionizing radiation), antimetabolites, alkylating agents, antibiotics, and alkaloids. These agents are potent mutagens since they are intended to interfere with DNA metabolism. They are often given not as a single agent, but rather as combinations. So, in contrast to the pure exposure of experimental investigations

Table 1 *Males of reproductive age after cancer therapy*[a]

Disorder	Annual frequency among males	Five-year survival (%)	Calculated five-year survivors
All cancers (<45 years of age)	710,040	65	461,526
Hodgkin's disease	3980	84	3343
Testicular cancer	8010	96	7690
Childhood cancer	5350	79	4336
Transplant recipients	15647	~60	8388

[a] http://seer.cancer.gov/; and, http://www.ustransplant.org/

of male reproductive toxicity, the human exposure to cancer treatment is complex. But, unlike other human exposures of concern for mutagenicity, the timing and dosage of exposure to cancer therapy can be precisely known. Moreover, dosage can be independently corroborated by medical records and occasionally by direct *in vivo* measurements, for example, with blood levels of DNA adducts.

2.2   Endpoints

The endpoints that can be studied are those mentioned throughout this volume. Some are *clinical* endpoints that do not require laboratory collaboration such as spontaneous abortion, infertility, and genetic diseases, including birth defects and especially the so-called sentinel phenotypes. A sentinel phenotype is a clinical disorder or syndrome that occurs sporadically as a consequence of a single, highly penetrant mutant gene that is a dominant trait of some frequency and low fitness and that is uniformly expressed and accurately diagnosable, with minimal effort at or near birth.[2] In *Online Mendelian Inheritance in Man* < http://www.ncbi.nlm.nih.gov/entrez/query.fcgi?db=OMIM >, there are about 40–80 traits that meet this definition. Together they have an estimated total frequency around one in a thousand. In short, sentinel phenotypes, such as achondroplasia or aniridia, are powerful endpoints, clinically relevant, almost certainly representing new mutation, and possibly attributed to environmental exposure. But, they are rare events and large numbers of exposed men would have to be counted to observe their offspring for such defects.

Other endpoints require *laboratory* collaboration: sperm analyses and, in offspring, chromosomal abnormalities and variations in proteins or nucleic acids. With many different gene tests on single subjects, fewer exposed people are needed to achieve statistical power. However, the relevance of molecular abnormalities to the health of clinically normal offspring is problematic.

2.3   Decreased Fertility

There are many barriers to reproduction by men (and women) after cancer and its treatment. The focus of this chapter is the narrow issue of testicular impairment and germ cell mutation. However, reproductive potential of cancer survivors could be limited by the many concomitant associations of the experience of cancer in childhood, adolescence, and early adulthood: impaired education, growth, and development, limited job opportunities and income, sterility *per se*, dismemberment or deformity due to therapy, or fear and uncertainty about the future.

Many variables influence even a single measure of reproduction or reproductive capacity, even one as discrete as gonadotrophin levels. For example, young men were studied who had received cyclophosphamide, not for cancer but for immunologic renal disease in childhood.[3] In general, the frequency of gonadal dysfunction, as reflected by gonadotrophin levels in blood, increased with the dose of cyclophosphamide; but, it also varied with age at exposure or, more precisely, with the pubertal status. The post-pubertal testis is much more sensitive to sterilization by cyclophosphamide than the pre-pubertal testis, which is only slightly impaired at doses given for immune renal disease.

By contrast, there is a large male–female difference in reproductive outcomes among long-time survivors of Wilms' tumor of the kidney. In a network of seven pediatric hospitals,[4] 30% of 114 pregnancies of women who survived Wilms' tumor and who underwent abdominal radiation had adverse outcomes, defined as fetal or neonatal death or birth weights under 2.5 kg. The rates of such outcomes among female survivors without abdominal radiation and among the wives of male survivors were 0% and 3%, respectively. A postal survey of British general practitioners who treated survivors of cancer diagnosed in childhood between 1946 and 1977, identified 20 Wilms' tumor survivors.[5] The birth weight of offspring of women who received abdominal radiation was 2584 g; among the wives of male survivors and women who did not undergo radiation, birthweight was 3146 g, a highly significant difference. In short, the offspring of men with Wilms' tumor did not suffer these consequences.

2.4 The Five-Center Study

2.4.1 Design

To clarify several issues concerning late reproductive and other effects among cancer survivors, a team at the National Cancer Institute struck collaborations with five cancer registries around the US in the late 1970s.[6,7] The first goal was to assemble a cohort of long-term survivors of childhood and adolescent cancer. Cases had either histologically confirmed cancer or clinically diagnosed brain tumors. Diagnosis took place from 1945 through 1974, under the age of 20 years. Finally, survival of 5 years and attainment of 21 years were required by an arbitrary date of study cutoff.

These criteria produced a special set of 2498 eligible cancer survivors. Because three decades of cancer treatment were included, half of the cases had surgery only; these represent a control group within the study itself that could be used to distinguish the effects of therapy from any determinants of cancer risk. One-third had radiotherapy only, and another third had some chemotherapy, often far less than the multiagent chemotherapy that pediatric oncologists presently give. Twenty-one percent of study subjects were pre-pubertal boys and girls, so we were able to do some comparisons between pre- and post-pubertal exposures. The collected data included an interview about interval medical and social history, infertility, any ill health of the offspring and, importantly, consent for records so that certain medical events, such as cancer, birth defects, and infertility, could be documented by actual medical records or death certificates. Cases' permissions were also gained to interview their brothers and sisters, who were selected as controls in a ratio of two controls per one case.

2.4.2   Results

One early analysis of the large database addressed Wilms' tumor,[8] specifically
the 26 female and 21 male survivors and 77 of their siblings (as controls).
The apparent excess of fetal deaths in females was due to five miscarriages
in one woman with bicornuate uterus. Again, female, but not male survivors
had a significant excess (four-fold) of adverse, live-born outcomes. One out of
37 offspring had Wilms' tumor: a female with unilateral Wilms' tumor had a
male with bilateral tumors and also a ventricular septal defect.[6] These findings
were consistent with other studies, and in a sense, validated the study design.

A major analysis addressed infertility, measured as the time from first
marriage to first pregnancy.[7] There were big differences by type of therapy
among the 595 male and 637 female cases. By the crude and the adjusted
fertility rate, men had a greater loss of fertility than women, especially those
men exposed to alkylating agents. The mechanism of infertility seemed to be
germ cell aplasia, worse in men than women.

If fertility was maintained – and it was in many – was there any increase in
genetic disease among the offspring? One measure of genetic disease is cancer.
Of the 2308 offspring, 7 cancers occurred.[6] This number was no greater than
expected either in the controls (cousins of the offspring) or in population
estimates in a study that had about an 80% power to detect a tripling. By
inspection, most of the cases' offspring who had cancer had a known hereditary
or familial type of cancer: two with retinoblastoma, one with multiple endocrine neoplasia, and one with Wilms' tumor. There seems to be no excess of
cancer apart from the known hereditary and familial syndromes.

As another measure of genetic disease, we arbitrarily defined "genetic
disease" as a known or probable cytogenetic syndrome, a single gene trait, or
one of 15 simple malformations tracked by the US Centers for Disease Control.
To detect what could represent a mutation (in contrast to possible genetic
disorders already in the family), we established that "sporadic" occurrences of
genetic disease had to have no affected relatives with a similar genetic disease
and that the other group, "familial," had a similarly affected relative. We
scored any recessive disorder as "familial," because we assumed both parents
were carriers and neither had a new mutation from germ cell exposure to the
cancer treatments. Offspring achieved an average of 11.5 years, and 75 (3.4%)
of them had a genetic disease by our definition. In comparison, the frequency in
cousins, the offspring of sibling controls, was 2.8%, not a statistically significant
difference.[9]

A few of the so-called sentinel phenotypes occurred, but almost all of them
were already in the family. One control offspring had a sporadic genetic disease;
a congenital cataract, but we could not be sure it was a new dominant
mutation. One child of a case survivor had albinism; by a stretch of reasoning,
he could represent a new mutation, under the assumption that the disorder was
X-linked and the mother had been well examined by an ophthalmologist and
found to have normal retinas. These assumptions are unlikely; more credible is
the interpretation that the child's disorder is autosomal recessive and that both

parents inherited and passed on the mutant allele. In short, there was no clear instance of sentinel phenotype in the offspring of our cancer survivors. As a nested case-control study, the parents of individuals with sporadic genetic disease had "potentially mutagenic therapy" no more often than the parents of the larger number of offspring without genetic disease. By χ^2 test, there is no significant difference in a study that had an 87% power to detect a doubling.

2.4.3 The Male Experience

Additional tabulations have been done that allow inspection of the data on males only. The frequency of genetic disease in the offspring of males was 2.6%, lower than the frequency in offspring of females (4.0%). The frequency of fetal deaths was 10.7% among pregnancies by male survivors and 11.0% among pregnancies of female survivors. The sex ratio in offspring of males was 1.04 (male:female), compared to 1.01 in offspring of female (and 1.00 and 1.04, respectively, in offspring of controls). The sex ratio is used to examine the possibility of X-linked mutations in human beings.

2.5 The Childhood Cancer Survivor Study

2.5.1 Design

In a larger and more recent study, also funded by the US National Cancer Institute, 25 institutions began collaboration in 1994, to establish a retrospective cohort study of a large population of childhood cancer survivors in North America, following much the same design as the above-mentioned Five-Center Study. Eligible patients had to have been diagnosed with selected cancers prior to age 21 years between 1970 and 1986 and to have survived for at least 5 years after treatment end.[10] The participating centers extracted data from the medical records of the survivors that pertained to their cancer treatment (chemotherapy, radiotherapy, and/or surgery). Follow-up questionnaires were distributed and addressed a multitude of aspects including demographics, prescriptions, medical history, and family medical history. Survivors reporting pregnancy were supplied an additional questionnaire specific to pregnancy outcomes.

The study identified 21,665 eligible patients and 14,054 agreed to participate in the study. Of the participants, 3,162 patients reported 6,441 pregnancies and 4,157 of them were live births. Controls for the study were selected by including the siblings of the survivors and their offspring who numbered 2,339. This set of survivor offspring and sibling offspring provided an ideal situation to study the possible transgenerational effects of high-dose chemotherapy and ionizing radiation therapy. Any deviation from either the control or expected frequency of genetic disease in the survivor offspring could be evidence for germ cell mutations resulting from the cancer experience.

2.5.2   Results

In a preliminary tabulation not yet completely analyzed, the frequencies of cancer, cytogenetic abnormalities, single-gene disorders, and simple malformations in survivor offspring (3.7%) were not significantly different from the controls (4.1%) or from the expected values.[11] Genetic disease occurred in 157 of the 4,214 survivor offspring versus 95 in the 2,339 control offspring ($P > 0.05$) (Table 2).

The reported outcomes of genetic disease on the questionnaires are still being validated and other confounding factors are to be explored. Radiation dose analyses are being conducted along with calculations to account for confounding factors such as maternal and paternal age, exposures outside therapy, drug and alcohol use, cigarette smoking, and medical conditions other than cancer.

The data are preliminary, but are reassuring that no detectable increase in genetic disease was found in survivors of childhood and adolescent cancers given modern therapies. It is also important to remember that this transgenerational study is evaluating treatments that are now several decades old. Application to current protocols can be made, but only with the understanding that absolute evaluation of new drugs, combinations, and therapies will not be available until this decade's survivors mature into adulthood.

2.6   Future Directions

In September 2004, 89 international scientists met at The Jackson Laboratory in Bar Harbor, Maine, USA, for the purpose of assessing current capabilities and projecting future strategies to (1) detect human germ cell mutagens, (2) identify genetic alterations, and (3) evaluate transgenerational genetic changes in human populations. The workshop was titled, "Assessing Human Germ-Cell Mutagenesis in the Post-Genome Era: A Celebration of the Legacy of William Lawson (Bill) Russell." A pioneer in mammalian germ cell mutagenesis, Bill Russell had passed away in the summer of 2002. Participants called for an international program on the etiology of human genetic variation. The program would be interdisciplinary and multidisciplinary, including germ cell biology, genetic toxicology, bioinformatics, epidemiology, toxicogenomics, and other disciplines. One of the speakers, Professor James F. Crow, a leading

Table 2   *Genetic disease in offspring of survivors and sibling controls (ref 11)*

Type of genetic disease	Survivor offspring (n = 4214)	Sibling offspring (n = 2339)	Frequency ratio
Cancer	10 (0.2%)	4 (0.2%)	1.4
Cytogenetic abnormality	5 (0.1%)	3 (0.1%)	0.9
Single gene disorder	25 (0.6%)	10 (0.4%)	1.4
Simple malformation	117 (2.8%)	78 (3.3%)	0.8
Total	157 (3.7%)	95 (4.1%)	0.9

geneticist from the University of Wisconsin, said, "We now have the tools to probe (human germ cell mutagenesis) deeper; we just need to do it."

The time seems ripe to revisit the possible DNA-based methods for detecting germ cell mutagenesis that were proposed in a pre-genomic time, at a meeting co-sponsored by the US Department of Energy and the International Commission for Protection Against Environmental Agents, Mutagens and Carcinogens.[12] All the proposed strategies called for a collection of DNAs from the cancer survivor, all his or her biologic children, and their other biologic parent. Perhaps some method could be applied, such as subtraction hybridization with artificially constructed oligonucleotides or total genomic comparisons of parents with offspring. Simplistically, could one subtract out all the child's DNA from one parent, and anneal the residual unbound fragments to the DNA of the other parent? In short, all of the child's DNA should be accounted for by the mother's and the father's DNA. Any child's DNA that remains would be *prima facie* evidence of new mutation and would be a small collection of fragments that could be characterized. In the end, the workshop could identify no clear method and wondered if such a study must await cheap, total genomic sequencing, the so-called $1000 genome.

Regardless of the exact laboratory strategy, some protein- or nucleic acid-based search seems necessary to gain sufficient power to rule out, *e.g.*, a relative risk for germ cell mutation, in the order of 1.5. Pilot collection of triad bloods has successfully been done in Denmark.[13] Unless a laboratory method to detect human germ cell mutations throughout the genome is devised, additional cohorts of 10,000 or 25,000 long-term survivors will be needed for assessments of clinical endpoints.

2.7 Counseling

For now, we suggest this outline when clinical advice has to be given to cancer survivors who ask about the genetic risk to their intended offspring.[14] There are enormous theoretical concerns of exposure to ionizing radiation and chemotherapy of the male cancer survivor patient. Cancer therapy is supposed to interfere with DNA. There are limited empirical data on their offspring, but they give room for clinical reassurance about the lack of an excess of birth defects or general diseases by the above definitions.

Germ cell mutation surely does occur in human beings. But, to say, as clinicians do, that mutation is just spontaneous, seems to be antiscientific, akin to allegation of the spontaneous origin of life from another era. Now, there are ever more compelling reasons to press very hard on the issue among cancer survivors, which can be considered to represent a relatively common and, in a sense, worst case exposure. If genetic epidemiology cannot find an environmental germ cell mutagen in cancer survivors by a very sensitive genomics means, strong reassurance may be offered to the larger population about the probable lack of genetic diseases due to potentially harmful environmental exposures.

References

1. M. Hewitt, S. Greenfield and E. Stovall (eds), *From Cancer Patient to Cancer Survivor: Lost in Transition. National Academies Press*, Washington, 2006, 506.
2. J.J. Mulvihill and A. Czeizel, *Mutat. Res.*, 1983, **123**, 345–361.
3. S.A. Rivkees and J.D. Crawford, *JAMA*, 1988, **259**, 2123–2125.
4. F.P. Li, K. Gimbrere, R.D. Gelber, S.E. Sallan, F. Flamant, D.M. Green, R.M. Heyn and A.T. Meadows, *JAMA*, 1987, **257**, 216–219.
5. M.M. Hawkins and R.A. Smith, *Int. J. Cancer*, 1989, **43**, 399–402.
6. J.J. Mulvihill, M.H. Myers, R.R. Connelly, J. Byrne, D.F. Austin, K. Bragg, J.W. Cook, D.D. Hassinger, F.F. Holmes, G.F. Holmes, M.R. Krauss, H.B. Latourette, J.W. Meigs, M.D. Naughton, S.C. Steinhorn, L.C. Strong, M.J. Teta and P.J. Weyer, *Lancet*, 1987, **2**, 813–817.
7. J. Byrne, J.J. Mulvihill, M.H. Myers, R.R. Connelly, M.D. Naughton, M.R. Krauss, S.C. Steinhorn, D.D. Hassinger, D.F. Austin and K. Bragg, *N. Engl. J. Med.*, 1987, **317**, 1315–1321.
8. J. Byrne, J.J. Mulvihill, R.R. Connelly, D.A. Austin, G.E. Holmes, F.F. Holmes, H.B. Latourette, J.W. Meigs, L.C. Strong and M.H. Myers, *Med. Pediatr. Oncol.*, 1988, **16**, 233–240.
9. J. Byrne, S.A. Rasmussen, S.C. Steinhorn, R.R. Connelly, M.H. Myers, C.F. Lynch, J. Flannery, D.F. Austin, F.F. Holmes, G.E. Holmes, L.C. Strong and J.J. Mulvihill, *Am. J. Hum. Genet.*, 1998, **62**, 45–52.
10. L.L. Robison, A.C. Mertens, J.D. Boice, N.E. Breslow, S.S. Donaldson, D.M. Green, F.P. Li, A.T. Meadows, J.J. Mulvihill, J.P. Neglia, M.E. Nesbit, R.J. Packer, J.D. Potter, C.A. Sklar, M.A. Smith, M. Stovall, L.C. Strong, Y. Yasui and L.K. Zeltzer, *Med. Pediatr. Oncol.*, 2002, **38**, 229–239.
11. J.J. Mulvihill, L.C. Strong, L.L. Robison and Investigators of the childhood cancer survivor study, *Am. J. Hum. Genet.*, 2001, **69**, A391.
12. J. Delehanty, R.L. White and M.L. Mendelsohn, *Mutat. Res.*, 1986, **167**, 215–232.
13. J.D. Boice, J.E. Tawn, J.F. Winther, S.S. Donaldson, D.M. Green, A.C. Mertens, J.J. Mulvihill, J.H. Olsen, L.L. Robison and M. Stovall, *Health Phys.*, 2003, **85**, 65–80.
14. J.J. Mulvihill, S. Whitton and P. Horn, In *The Genetic Basis of Common Disease*, R.A. King, J.I. Rotter, A.G. Motulsky (eds), Oxford University Press, New York, 2002, 1023–1032.

CHAPTER 3

Cancer in Siblings of Children with Cancer in the Nordic Countries: A Population-Based Cohort Study
Paediatric Cancer: An Indicator of Familial Cancer Risk?

JEANETTE FALCK WINTHER

Institute of Cancer Epidemiology, Danish Cancer Society, Strandboulevarden 49, Copenhagen, DK-2100, Denmark

3.1   Epidemiological Research in the Nordic Countries

The Nordic countries (Denmark, Finland, Iceland, Norway and Sweden) have unique resources for conducting epidemiological research, as such research is facilitated by the existence of population health registries that can be accessed with unique individual identification numbers assigned to each citizen at birth.[1]

3.1.1   Unique Personal Identification Numbers

The Central Population Register was established in Denmark on 1 April, 1968, when all citizens were assigned a unique 10-digit personal identification number that contains six digits for date of birth and a unique four-digit number ending with an odd digit for males and an even one for females. For citizens born after 1 April, 1968, the identification number was allocated at birth. The system is in universal use and is updated regularly, together with a set of demographic and personal data, such as references to first-degree relatives, information on current and former residential addresses, marital status, vital status and date of death or emigration from Denmark (if applicable). All information is retained in the Register after the death of a citizen. The completeness and accuracy of the identification number system and the associated information are close to 100%. Similar national central population registration systems were started in 1952 in Iceland, 1960 in Norway, 1961 in Sweden and 1967 in Finland.

3.1.2   Danish Health Registries: Valuable Tools in Medical Research

Systematic collection of administrative and health care data is a long-standing tradition in the Nordic countries, and the number of national databases is increasing exponentially. Currently, the Danish Government has nearly 200 databases, 80 of which are medical.[1,2] Some are among the oldest registries in the world; that is, the Danish Cancer Registry (started in 1942; see below), the Multiple Sclerosis Register (1949), the Cerebral Palsy Register (1967) and the Danish Psychiatric Central Register (1969). National systems for registration of health data also cover valuable information such as birth variables and presence of congenital malformations, cytogenetic abnormalities, abortions and cause of death. The personal identification number allows for personal identification of remarkable quality and the possibility of collection of information about the same person in several independent registries.

Danish registries have become valuable tools in medical research, especially because they are population-based, which ensures representativeness. As registry data are collected independently of each research project, there is little room for certain types of bias, such as recall, nonresponse and observation (where the research question might prompt diagnostic tests in an unbalanced manner). Finally, these registries make it possible to examine health conditions with a long lag between exposure and overt disease.[1]

3.1.3   The Danish Cancer Registry

The Danish Cancer Registry is a population-based, nationwide registry containing data on the incidence of cancer throughout Denmark since 1943.[3] The registry was founded in May 1942 by Dr Johannes Clemmensen. In January 1997, cancer registration was integrated with other health registration at the Statistical Office of the Danish National Board of Health. Details of individual cases of cancer are available according to the Danish modified version of the *International Classification of Diseases, Seventh Revision* (ICD-7), which is based mainly on tumour topography,[4] for all years. Since 1 January, 1978, details are also available according to the *International Classification of Diseases for Oncology* (ICD-O-1), which permits coding of neoplasms by topography, histology and behaviour.[5]

The Registry receives notifications of all such diseases from hospital departments and practising specialists, at diagnosis and when changes are made to the initial diagnosis. In addition, reports are received from pathology departments and departments of forensic medicine, giving the results of autopsies of cancer patients. Cases first diagnosed at autopsy, *i.e.* as incidental findings, are also included in the Registry. This information is supplemented by an annual review of all death certificates and, since 1987, by an annual search of the Danish National Hospital Patient Registry for unreported cases of cancer. Cases identified in the latter search are traced by sending a letter to the treating hospital, requesting a standard notification. No case is included on the basis of information from the Patient Registry only.

A core data set is kept on each individual which includes the identification number (or the combination of name and date of birth for persons who died before 1968), date of diagnosis, method of verification and date and cause of death (if applicable). Cancer registration in the four other Nordic countries started in the 1950s; *i.e.* 1953 in Finland and Norway, 1954 in Iceland and 1958 in Sweden.

3.2 Role of Dominant and Recessive Conditions in Cancer Causation

More than 450 single-gene traits associated with specific neoplasms have been identified in clinical and epidemiological studies.[6,7] Most of these family syndromes are autosomal dominant disorders and include neurofibromatosis 1 and 2, von Hippel-Lindau disease, familial breast and ovarian cancer caused by mutations in *BRCA1* and *BRCA2*, Li-Fraumeni syndrome and retinoblastoma (Table 1). Highly penetrant single-gene mutations might be responsible for 5% of all cancers.[8]

The role of dominant conditions in cancer causation has been evaluated in large population-based studies of cancer risk in the parents[9] and offspring of childhood cancer survivors[10] in the Nordic countries. To test the hypothesis that recessive conditions play an important role in cancer development, however, large studies of cancer must be conducted in populations with a high frequency of recessive conditions. Therefore, we designed a Nordic study to assess the relations between childhood cancer and risk in siblings.[11]

Combining the results of this third Nordic study with those of two previously published studies on cancer risk in the parents and offspring of paediatric

Table 1 *Examples of recognizable family cancer syndromes (see Ref. 7)*

Autosomal dominant disorders	*Autosomal recessive disorders*
Familial breast and ovarian cancer caused by mutations in BRCA1 and BRCA2	Ataxia-telangiectasia
Familial melanoma	Bloom syndrome
Familial adenomatous polyposis	Fanconi anaemia
Li-Fraumeni syndrome	Xeroderma pigmentosum
Multiple endocrine neoplasia type 1 and 2	
Neurofibromatosis 1 and 2	
Retinoblastoma	
Von Hippel-Lindau disease	
Wilm's tumour	

cancer patients gave us a powerful means for assessing cancer risk in first-degree relatives of Nordic childhood cancer patients. It also allowed us to evaluate the possible role of both dominant and recessive familial cancer syndromes of importances to the population, including hitherto unknown syndromes (if any). Finally, the proportion of childhood cancer that can be explained by hereditary factors could be estimated. Sections 3.3, 3.4 and 3.5 describes the main findings of these three Nordic settings, focusing on the sibling study.

3.3   Cancer in Siblings of Children with Cancer in the Nordic Countries: A Population-Based Cohort Study

In 2001, Winther *et al.* published the results of a collaborative Nordic cohort study of 42,277 siblings of 25,605 persons who had had cancer in childhood.[11] In this population-based study of about 23 million people in the five Nordic countries, we aimed to assess the relations between childhood cancer and the risk of siblings, providing a powerful opportunity to evaluate the influence of recessive conditions in cancer causation.

3.3.1   Patients, Procedures and Analyses

We identified all 25,605 children and adolescents with a personal identification number in whom cancer was diagnosed before they reached the age of 20, between the start of cancer registration in the five Nordic countries in the 1940s and 1950s and 31 December, 1993 (1987 in Norway). The cases were divided into 12 main diagnostic groups according to the Birch and Marsden classification scheme for childhood cancer.[12] Their parents and their 42,277 siblings were identified from nationwide population registries; adopted siblings and half-siblings were excluded. Cancers in siblings were documented by record linkage with cancer registries and compared with national incidence rates. Follow-up of siblings was begun on the date of diagnosis of the index patient, the date of birth of the sibling or the date of inception of the national population registration, whichever occurred last and ended 31 December, 1994 or the date of death or emigration, whichever occurred first. Cancers diagnosed in siblings under 20 years of age were grouped according to the Birch and Marsden classification, whereas those diagnosed in siblings 20 years and older were grouped according to the Danish modified version of the ICD-7. Cancer incidence in parents was also assessed in order to identify familial cancer syndromes. Sex-, age- and calendar-specific incidence rates were applied to the appropriate person-years of observation to obtain the expected numbers of cancers, and the 95% confidence intervals (CIs) for the standardized incidence ratio (SIR), the ratio of the observed to the expected number, were calculated.

3.3.2 Results

Siblings accrued 694,625 person-years of follow-up (median follow-up for cancer, 16.7 years; range, >0–40). A total of 353 cancers were recorded in siblings, whereas 284.2 were expected, yielding an SIR of 1.24 (95% CI, 1.12–1.38). The risk ratios for siblings were highest in the first decade of life (2.59, 1.89–3.46), decreased with age and approached the expected rates after the age of 30 years (Table 2).

3.3.2.1 Cancer in Childhood and Adolescence

In siblings followed-up over the age range 0–19 years, 82 cancers were observed with 48.2 expected, yielding an SIR of 1.70 (95% CI, 1.35–2.11) (Table 3). The affected siblings tended to have cancers in the same main diagnostic groups as the index patients, which was also true for seven twin sisters and five twin brothers. Several tumours occurred in notable excess among siblings: for example, a 500-fold increase in risk for retinoblastoma, a 34-fold increase in risk for renal tumours and a 17-fold increase for carcinomas and other epithelial neoplasms.

Tumour combinations among sibling pairs that were compatible with well-known hereditary cancer syndromes (framed by boxes in Table 3, $n = 38$) were assessed by review of the original cancer registry registration forms. Ten sibling pairs with retinoblastoma (including two sets of twins), 20 pairs (from 19 families) of the remaining 28 sibling pairs and four further sets of affected twins (not framed by boxes) provided strong evidence of known hereditary cancer syndromes, most of which resulted from autosomal dominant disorders such as neurofibromatosis 1, Li-Fraumeni syndrome, familial Wilms tumour, multiple endocrine neoplasia 2A and von Hippel-Lindau disease, and also recessive disorders such as Bloom syndrome and xeroderma pigmentosum (see original publication[11]). Thus, 34 (33 families) of the 82 cancers diagnosed among

Table 2 *Observed (Obs.) and expected (Exp.) numbers and standardized incidence ratios (SIRs) for all cancers in 42,277 siblings of children with cancer, stratified by age at diagnosis*

	Cancer among siblings		
Age at diagnosis (years)	*Obs.*	*Exp.*	*SIR (95% CI)*
All ages	353	284.2	1.2 (1.1–1.4)
0–4	27	9.3	2.9 (1.9–4.2)
5–9	18	8.1	2.2 (1.3–3.5)
10–19	37	30.8	1.2 (0.8–1.7)
20–29	109	80.9	1.3 (1.1–1.6)
≥ 30	162	155.0	1.0 (0.9–1.2)

Note: CI indicates confidence interval.
Source: from Winther *et al.* (see ref. 11). Copyright © 2001 Elsevier Ltd. All rights reserved.

Table 3 Observed (Obs.) number of sibling pairs with cancer and standardized incidence ratios (SIRs) in siblings of children with cancer, followed up in childhood and adolescence, stratified by type of cancer of index patient

Primary cancer of index patient (main diagnostic group)[a]	Primary cancer of sibling (main diagnostic group)[a]																	
	Leukaemia (I)		Lymphomas (II)		Central nervous system neoplasms (III)		Retino-blastoma (V)		Wilms and other renal tumours (VI)		Soft-tissue sarcomas (IX)		Carcinomas and other malignant epithelial neoplasms (XI)		Other specified and unspecified malignant neoplasms (IV, VII, VIII, X, XII)		All cancers (I–XII)	
	Obs.	SIR	Obs.	SIR	Obs.	SIR	Obs.	SIR	Obs.	SIR	Obs.	SIR	Obs.	SIR	Obs.	SIR	Obs.	SIR
Leukaemia (I)	5	1.5	1	0.5	3	0.9	0	—	1	2.0	0	—	1	0.7	3	1.1	14	1.0
Lymphomas (II)	2	2.1	4[H]	5.9[b]	2	1.9	0	—	0	—	0	—	0	—	0	—	8	1.7
Central nervous system neoplasms (III)	0	—	2	1.2	10[N,L]	3.6[b]	0	—	0	—	1[N,L]	1.3	2	1.5	1	0.4	16	1.3
Sympathetic nervous system tumours (IV)	0	—	1	2.8	2	2.9	0	—	0	—	0	—	0	—	1	2.1	4	1.4
Retinoblastoma (V)	1	3.0	0	—	1	3.0	10[R]	565[b]	0	—	0	—	0	—	0	—	12	8.5[b]
Wilms and other renal tumours (VI)	0	—	0	—	0	—	0	—	4[W]	34.2[b]	1	5.4	1	3.5	1	1.9	7	2.3
Hepatic tumours (VII)	0	—	0	—	0	—	0	—	0	—	0	—	0	—	1	10.4	1	1.9
Malignant bone tumours (VIII)	0	—	0	—	1	2.3	0	—	0	—	0	—	0	—	0	—	1	0.5
Soft-tissue sarcomas (IX)	1	1.7	0	—	1[N,L]	1.6	0	—	1	12.5	2[N,V]	11.7[b]	0	—	1	2.0	6	2.2
Germ-cell, trophoblastic and other gonadal neoplasms (X)	1	2.5	0	—	1	2.3	0	—	0	—	1	8.1	1	4.2	1	2.6	5	2.6
Carcinomas and other malignant epithelial neoplasms (XI)	0	—	0	—	0	—	0	—	0	—	0	-	6[M,E]	16.7[b]	2	4.5	8	3.4[b]
All cancers (I–XII)	10	0.9	8	1.2	21	1.9[b]	10	18.6[b]	6	4.0[b]	5	1.7	11	2.0	11	1.2	82	1.7[b]

Notes: Boxed results show cancer combinations that could indicate known hereditary cancer syndromes: [H]familial lymphoma or histiocytosis (although a viral aetiology cannot be excluded); [N]neurofibromatosis 1; [L]Li-Fraumeni syndrome; [V]Von Hippel-Lindau disease; [M]familial melanoma; [E]multiple endocrine neoplasia 2.

Source: from: Winther et al. (see ref. 11). Copyright © 2001 Elsevier Ltd. All rights reserved.

[a]According to the Birch and Marsden classification scheme for childhood cancer (see ref. 12).

[b]The neutral value 1 is not contained in the 95% CI based on Byar's limits or on the exact Poisson limits (less than ten observed cases) From: Winther et al.[11]

siblings in childhood and adolescence were assumed to be hereditary. With the exclusion of these 33 families, the SIR for cancer occurring before the age of 20 years was reduced from 1.7 to 1.0 (48 observed cases, 48.2 expected; 95% CI, 0.7–1.3).

3.3.2.2 Cancer in Adulthood

In siblings followed-up in adulthood, 271 cases of cancer were observed (SIR, 1.15; 95% CI, 1.02–1.29), resulting mainly from a significant 30% excess risk in the age group of 20–29 years. Most of the combinations of tumour types in index patients and adult siblings provided no evidence of an association (see original publication[11]); however, most of the significant results were for combinations of cancer types indicative of known syndromes predisposing to cancer, such as multiple endocrine neoplasia 2A, familial melanoma and Li-Fraumeni syndrome. After exclusion of 23 families with hereditary cancer syndromes, the SIR for the age interval 20–29 years decreased from 1.3 to 1.0 (0.8–1.3).

3.4 Cancer in Parents of Children with Cancer in Denmark

In 1995, Olsen *et al.* published the results of the first large, population-based cohort study of the cancer risk of the parents of paediatric cancer patients.[9] Cancers in 11,380 parents of 5863 persons in whom cancer was diagnosed when they were under 15 years of age were documented by record linkage with the Danish Cancer Registry and compared with national incidence rates. The 11,380 parents were followed up for a diagnosis of cancer for 320,000 person-years, with follow-up starting on the date of birth of the child; on average, each parent was at risk of cancer for 28 years (range, 3 months to 47 years).

The overall occurrence of cancer in the parents of children with cancer, 1445 observed cases, was remarkably close to that expected from incidence rates in the general adult population: an SIR of 1.0 (1496 expected; 95% CI, 0.9–1.0) was reported for all parents, 1.0 for mothers and 0.9 for fathers (Table 4). The lower rate of cancer among fathers reflected their lower SIR for lung cancer (0.8; 95% CI, 0.6–0.9) and their low ratios for prostatic cancer (0.9) and testicular cancer (0.7).

Parents' risk for cancer was not increased by having a child with any of the 10 major categories of the disease. A noteworthy relation between childhood cancer and specific types of cancer in parents, however, was an increase in the incidence of breast cancer in mothers under the age of 45 whose children had been given a diagnosis of cancer before the age of 3. The association was particularly strong for mothers of children with osteogenic or soft-tissue sarcomas (four observed cases; SIR, 5.4; 1.7–12.9) (Table 5). This finding is in accordance with the familial syndrome of soft-tissue sarcomas, breast cancer and other neoplasms in young adults and children, known as the Li-Fraumeni syndrome. In general, the study revealed few associations between cancer in

Table 4 Observed (Obs.) and expected (Exp.) numbers and standardized incidence ratios (SIRs) for cancer in the parents of 5863 children with cancer

	Mothers (n = 5747)			Fathers (n = 5633)		
	Obs.	Exp.	SIR (95% CI)	Obs.	Exp.	SIR (95% CI)
All sites (140–204)[a]	718	723.1	1.0 (0.9–1.1)	727	772.4	0.9 (0.9–1.0)
Buccal cavity and pharynx (140–148)	10	8.0	1.2 (0.6–2.2)	31	25.3	1.2 (0.8–1.7)
Digestive organs (150–159)	135	128.0	1.1 (0.9–1.3)	198	192.3	1.0 (0.9–1.2)
Respiratory system (160–164)	54	48.9	1.1 (0.8–1.4)	129	171.6	0.8 (0.6–0.9)
Larynx (161)	3	2.4	1.2 (0.2–3.6)	12	13.8	0.8 (0.6–1.5)
Lung (162)	50	43.8	1.1 (0.8–1.5)	114	150.3	0.8 (0.6–0.9)
Breast (170)	168	176.0	1.0 (0.8–1.1)	0	1.2	0.0 (0.0–3.1)
Female genital organs (171–176)	162	168.4	1.0 (0.8–1.1)	–	–	–
Male genital organs (177–179)	–	–	–	71	81.8	0.9 (0.7–1.1)
Prostate (177)	–	–	–	58	65.1	0.9 (0.7–1.2)
Testis (178)	–	–	–	10	13.6	0.7 (0.4–1.3)
Urinary system (180–181)	32	31.2	1.0 (0.7–1.5)	88	90.4	1.0 (0.8–1.2)
Skin (190–191)	79	84.1	0.9 (0.7–1.2)	102	107.2	1.0 (0.8–1.2)
Brain and nervous system (193)	19	21.2	0.9 (0.5–1.4)	22	23.0	1.0 (0.6–1.5)
Bone and connective tissue (196–197)	2	3.4	0.6 (0.1–2.0)	7	4.6	1.5 (0.7–3.0)
Lymphatic and haematopoietic tissues (200–205)	36	33.3	1.1 (0.8–1.5)	57	52.3	1.1 (0.8–1.4)
Other specified organs (192, 194–195)	7	6.8	1.0 (0.5–2.0)	7	5.8	1.2 (0.5–2.4)
Secondary and unspecified sites (198–199)	14	13.8	1.0 (0.6–1.7)	15	16.7	0.9 (0.5–1.4)

Notes: CI indicates confidence interval.

Source: from Olsen et al. (see Ref. 9). Copyright © 1995 Massachusetts Medical Society. All rights reserved.

[a] Classified according to ICD-7.

Table 5 *Standardized incidence ratios (SIRs) for breast cancer in mothers of children with osteogenic or soft-tissue sarcomas, according to the age of the child at diagnosis and the age of the mother*

Age of child at diagnosis (yr)	Age of mother (yr)	Obs.	Exp.	SIR (95% CI)
0–2	All ages	9[a]	3.1	2.9 (1.4–5.3)
3–14	All ages	11	17.2	0.6 (0.4–1.1)
0–14	<45	5[a]	3.7	1.4 (0.5–3.0)
0–14	≥ 45	15	16.6	0.9 (0.5–1.5)
0–2	<45	4[a]	0.7	5.4 (1.7–12.9)

Notes: observed numbers (Obs.); expected numbers (Exp.); confidence interval (CI).
Source: from Olsen *et al.* (see ref. 9). Copyright © 1995 Massachusetts Medical Society. All rights reserved.
[a] One case of bilateral breast cancer is included as a single observation.

children and cancer in their parents. Except for the link between sarcoma in children and early-onset breast cancer in their mothers, which has been reported previously, the associations should be interpreted with caution because of the greater effects of chance in multiple comparisons. More detailed information on the study results and interpretation of findings can be found in the original publication.[9]

3.5 Cancer in Offspring of Cancer Survivors: A Population-Based Cohort Study in the Nordic Countries

The incidence of cancer among 5847 offspring of 14,652 survivors of cancer in childhood or adolescence was assessed on the basis of data from national cancer and population registries in all five Nordic countries and compared with the relevant rates of cancer in the general population. The results of this collaborative study were published in 1998 by Sankila *et al.*[10] Of the survivors, 23% (3369) had children during the follow-up period (Table 6). The proportion of parents was lowest among the survivors of leukaemia (6%) and highest among survivors of carcinomas (43%). The mean number of offspring per survivor ranged from 1.6 to 1.8, depending on the type of cancer, except for survivors of leukaemia, for whom it was 1.3.

The 5847 offspring were followed up for cancer for 86,780 person-years (start of follow-up on date of birth; median age at end of follow-up, 14 years; range, 0–43 years). Forty-four malignant neoplasms were diagnosed among the offspring, yielding an SIR of 2.6 (95% CI, 1.9–3.5); these comprised 17 cases of retinoblastoma (SIR, 37; 95% CI, 22–60) and 27 cases of the other types of neoplasm (SIR, 1.6; 1.1–2.4), representing a small but statistically significant increase (Table 7). The commonest site of primary neoplasms other than retinoblastoma was the brain and nervous system, where eight tumours were

Table 6 *Number of survivors of childhood cancer, survivors who were parents and offspring, according to the primary cancer of the survivors*

Primary cancer of survivor[a]	No. of survivors	No. who were parents	No. of offspring	Ratio of survivor parents to all survivors	Mean no. of offspring/ survivor parent
Leukaemia (I)	2232	136	181	0.06	1.3
Lymphomas (II)	2252	541	924	0.24	1.7
Central nervous system neoplasms (III)	3348	664	1137	0.20	1.7
Sympathetic nervous system tumours (IV)	331	64	101	0.19	1.6
Retinoblastoma (V)	398	134	218	0.34	1.6
Wilms and other renal tumours (VI)	518	117	188	0.23	1.6
Hepatic tumours (VII)	81	6	11	0.07	1.8
Malignant bone tumours (VIII)	1095	229	422	0.21	1.8
Soft-tissue sarcomas (IX)	1090	335	612	0.31	1.8
Germ-cell, trophoblastic and other gonadal tumours (X)	1152	243	418	0.21	1.7
Carcinomas and other malignant epithelial neoplasms (XI)	1927	833	1523	0.43	1.8
Other or unspecified cancers (XII)	228	67	112	0.29	1.7
All cancers (I–XII)	14,652	3369	5847	0.23	1.7

Source: from Sankila *et al.* (see ref. 10). Copyright © 1988 Massachusetts Medical Society. All rights reserved.
[a] according to the Birch and Marsden classification scheme for childhood cancer (see ref. 12).

observed (SIR, 2.0; 95% CI, 0.9–3.9). The SIR for neoplasms other than retinoblastoma was 3.9 (95% CI, 2.1–6.7) for the offspring of survivors whose cancers were diagnosed when they were under 10 years of age and 1.1 (0.6–1.8) for the offspring of survivors whose cancers were diagnosed when they were 10 years or older.

To assess the risk of sporadic cancer among the offspring separately, Sankila *et al.* excluded 16 offspring with hereditary retinoblastoma (one child with sporadic retinoblastoma), two with features suggestive of Li-Fraumeni syndrome, one with von Hippel-Lindau disease and one with neurofibromatosis

Table 7 *Standardized incidence ratios (SIRs) for malignant neoplasms among 5847 offspring of survivors of childhood cancer, according to primary cancer*

Primary cancer of offspring[a]	Obs.	Exp.	SIR (95% CI)
Retinoblastoma	17	0.5	37 (22–60)
Brain and nervous system	8	4.0	2.0 (0.9–3.9)
Connective tissue	4	0.5	8.6 (2.3–22)
Non-Hodgkin lymphoma	3	1.0	3.1 (0.6–9.2)
Leukaemia	3	3.5	0.9 (0.2–2.5)
Kidney	2	0.7	2.8 (0.3–10)
Melanoma of skin	2	0.9	2.4 (0.3–8.5)
Stomach	1	0.0	20 (0.3–110)
Ovary	1	0.3	3.5 (0.1–19)
Testis	1	1.2	0.8 (0–4.6)
Other skin	1	0.4	2.2 (0–12)
Endocrine glands	1	0.3	3.9 (0.1–22)
All other sites	0	3.7	0 (0–1.0)

Notes: observed numbers (Obs.); expected numbers (Exp.); confidence interval (CI).
Source: from Sankila *et al.* (see ref. 10). Copyright © 1988 Massachusetts Medical Society. All rights reserved.
[a] Classified according to ICD-7; two neuroblastomas were included under "brain and nervous system" and one under "endocrine glands" because of different coding practices.

type 1 or 2, as well as two offspring with hereditary retinoblastoma who subsequently had solid tumours that probably arose in external radiotherapy fields. This reduced the observed number to 22 tumours and the SIR to 1.3 (95% CI, 0.8–2.0). More detailed information on the study results and interpretation of findings can be found in the original publication.[10]

3.6 Paediatric Cancer: An Indicator of Familial Cancer Risk?

Known autosomal dominant syndromes predisposing to cancer, such as retinoblastoma, Li-Fraumeni syndrome, von Hippel-Lindau disease, familial Wilms tumour and neurofibromatosis type 1 or 2, were identified in all three studies, as were cancer combinations that might indicate autosomal recessive disorders, such as Bloom syndrome and xeroderma pigmentosum, in the sibling study. The results did not show any features that suggested hitherto unknown genetic cancer syndromes of population importance or evidence that recessive conditions might contribute to cancers not explained by syndromes.

The risk for cancer of the siblings in our Nordic investigation was two to three times higher than that of the general population during the first decade of life. This above-average risk of cancer decreased with age, and approached expected rates after the age of 30 years. Hence, the risk for cancer in close relatives of children and adolescents with a genetically inherited cancer syndrome is generally increased only when they are of a similar age to the index patient at the time of cancer diagnosis. In fact, up to 40% of the cancers that occurred in the siblings of children with cancer before the age of 20 were accounted for by probable inherited germline mutations. An apparent absence of an association between cancers in childhood and cancers in adulthood was also observed in the Danish study of cancer incidence in parents of paediatric cancer patients. After the age of 30, environmental factors are thought to play an increasingly important part in the causation of cancer. Against this background, the absence of associations between cancers in childhood and cancers in adulthood further indicates that shared environmental exposures have little effect on subsequent cancer incidence patterns in parents and siblings in the Nordic countries.

Although large pedigrees were not constructed in any of the three studies, we have no reason to believe that our failure to identify new cancer syndromes was due to low statistical power or bias. As the Nordic cancer registries contain records for the entire population of each country, we were able to identify virtually all persons in whom cancer was diagnosed in childhood or adolescence after the start of cancer registration in the 1940s and 1950s, and who survived until the start of respective population registration in the 1960s or were born thereafter. Furthermore, the identification of first-degree relatives was also nearly complete. Hence, we were unlikely to have been biased in our selection of people. Information bias was also unlikely, as the study groups were established and familial relationship decided before files were searched for evidence of cancer in first-degree relatives, and the study relied on population registers that are kept for administrative purposes. Furthermore, all five Nordic cancer registries have close to 100% coverage.

The overall occurrence of cancer in the parents of children with cancer was remarkably close to that expected from incidence rates in the general adult population. Furthermore, exclusion of only 56 families with well-described familial cancer syndromes from the 25,605 families in the sibling study resulted in observed cancer risks for siblings equivalent to those expected in the general population. In the study of offspring, the risk for nonhereditary cancer among the offspring of survivors of childhood cancer was low and limited to the children of survivors whose cancers were diagnosed when they were under 10 years of age.

As we investigated cancer risk in the offspring of survivors, we were unable to address the potential mixture of familial cancers and cancers that might be due to transgenerational germ-cell mutations related to treatment of surviving parents. Nevertheless, the results of the study in offspring implied that fear that their offspring might develop cancer is no reason for the survivors of sporadic childhood cancer not to have children, and efforts to screen for cancer in the offspring of survivors of nonhereditary cancer in childhood or adolescence are not warranted.

Genetic effects might not be limited to rare, highly penetrant mutations that can be detected as familial cancer, but might extend to polymorphisms that slightly increase risk but are more common in the general population. Such polymorphisms would not be easily detectable in family studies, indicating the need for large molecular epidemiological studies (population-based and in families) to delineate the specific genetic and environmental components of cancer risk.[13]

The aggregated results of these three cohort studies of cancer risk among parents, siblings and offspring of childhood cancer patients, respectively, in the Nordic countries indicate that less than 5% of childhood cancer can be explained by hereditary factors. Apart from rare cancer syndromes, paediatric cancer is not *per se* an indicator of familial cancer risk.

References

1. H.T. Sørensen and S. Schulze, *Dan. Med. Bull.*, 1996, **43**, 463.
2. L. Frank, *Science*, 2000, **287**, 2398–2399.
3. H.H. Storm, E.V. Michelsen, I.H. Clemmensen and J. Pihl, *Dan. Med. Bull.*, 1997, **44**, 549–553.
4. World Health Organization, *Manual of the international statistical classification of diseases, injuries, and causes of death: Based on recommendations of the Seventh Revision conference, 1955, and adopted by the Ninth World Health Assembly under the WHO nomenclature regulations*, Vol. 1, World Health Organization, Geneva, 1957.
5. World Health Organization, *International classification of diseases for oncology*, World Health Organization, Geneva, 1976.
6. J.J. Mulvihill, *Catalog of human cancer genes: McKusick's Mendelian inheritance in man for clinical and research oncologists*, Johns Hopkins University Press, Baltimore, 1999, 1–646.
7. N.M. Lindor and M.H. Greene, *J. Natl. Cancer Inst.*, 1998, **90**, 1039–1071.
8. F.P. Li, *Cancer Epidemiol. Biomarkers Prev.*, 1995, **4**, 579–582.
9. J.H. Olsen, J.D. Boice Jr., N. Seersholm, A. Bautz and J.F. Fraumeni, *N. Engl. J. Med.*, 1995, **333**, 1594–1599.
10. R. Sankila, J.H. Olsen, H. Anderson, S. Garwicz, E. Glattre, H. Hertz, F. Langmark, M. Lanning, T. Møller and H. Tulinius, *N. Engl. J. Med.*, 1998, **338**, 1339–1344.
11. J.F. Winther, R. Sankila, J.D. Boice Jr., H. Tulinius, A. Bautz, L. Barlow, E. Glattre, F. Langmark, T. Møller, J.J. Mulvihill, G.H. Olafsdottir, A. Ritvanen and J.H. Olsen, *Lancet*, 2001, **358**, 711–717.
12. J.M. Birch and H.B. Marsden, *Int. J. Cancer*, 1987, **40**, 620–624.
13. R.N. Hoover, *N. Engl. J. Med.*, 2000, **343**, 135–136.

CHAPTER 4

What Harms the Developing Male Reproductive System?

MICHAEL JOFFE

Department of Epidemiology and Public Health, Imperial College London, London, UK

4.1 Introduction

There has been a great deal of concern in recent years about the possibility that the health of the human male reproductive system has deteriorated over a period of some decades. The publicity has attached to "falling sperm counts",[1,2] but the most robust evidence is for a large and widespread increase in the incidence of testicular cancer,[3,4] a disease of young men that fortunately is rarely fatal nowadays if treated appropriately. Male fertility likely shares causal factors with testicular cancer, and also with two relatively common congenital anomalies of the male reproductive system, cryptorchidism and hypospadias, and it is possible that they form part of a single condition, the "testicular dysgenesis syndrome".[5]

   Most of the research attention has been devoted to the idea that deteriorating trends in these conditions, and also some consistent spatial variations, are due to endocrine factors acting in early life, possibly *in utero*. One possibility is parallel trends and variations in endogenous hormones (maternal oestrogen levels),[6] but this has been little investigated. The major research programme has been concerned with exogenous chemical agents that affect the endocrine system, having either oestrogenic (or anti-oestrogenic) or anti-androgenic activity.[7-10] It is routinely said that there is little concrete evidence that they can explain the trends and spatial variations, but it is less commonly realised just how weak the case is. The purpose of this paper is to briefly review these issues, and then to outline the evidence that all these four conditions have a genetic component; the implications of these findings are then discussed.

4.2   Descriptive Epidemiology of Conditions Affecting the Male Reproductive System

4.2.1   Semen Quality and Fertility

Two types of endpoint can be studied: semen quality, and fertility as measured by the time taken to conceive (Time To Pregnancy, or TTP). TTP reflects the probability of conception for couples having unprotected intercourse. It is a functional measure of biological fertility at the level of the couple, and validity studies have shown that it can be studied retrospectively as well as prospectively.[11] Care in the design and analysis of TTP studies is important as they are prone to numerous sources of bias, including planning bias – arising when one exposure group, such as smokers, has a higher degree of risk taking – and truncation bias, which results from exclusion of part of the TTP distribution with particular patterns of sampling.[12]

Methodological issues also affect the interpretation of studies of semen quality, which is usually taken to include sperm concentration, motility and morphology. All are subject to large degrees of within-person biological variation or measurement error that varies between centres and very likely over time. In addition, representative samples of the general population, which are so important for descriptive epidemiology, are unachievable as participation rates are too low. The best evidence is from candidates for semen donation and for vasectomy; data from men in contact with medical services for a fertility-related problem, or from those accepted for semen donation, are too biased to be reliable for descriptive epidemiology.

The suspicion that semen quality has deteriorated over recent decades was made prominent by a much-cited review of the world literature published in 1992.[1] A more rigorous attempt found the decline in sperm density to be steeper in Europe than in America; studies from elsewhere were too sparse and diverse to draw confident conclusions.[2] However, the 1992 paper stimulated the publication of several single-centre studies, which are more reliable.[13] The main conclusions that emerge are

  (i) declines in semen quality have occurred in some parts of Europe but not in America;
 (ii) where concentration has declined, so usually have sperm motility and morphology;
(iii) at most, the available data go back to the early 1970s;
 (iv) the deterioration lasted for at least two decades, either from the early 1970s as a period effect, or from the mid-1940s as a birth cohort effect; and
  (v) in all affected centres, the observed decline is already visible in the earliest available data so it is impossible to locate the year when the decline started or to estimate what the pre-decline values were.

Three studies on time trends in fertility measured by TTP have not shown a decline in couple fertility, which reflects female as well as male factors. A British study showed that fertility increased in the period 1961–1993.[14] A Swedish study reported a decline in clinical subfertility in 1983–1993,[15] although this may have resulted from truncation bias that had not been allowed for.[16] A Danish study also suggested that fertility had tended to increase with time, for people born in the period 1931–1952.[17]

Substantial spatial variation in sperm concentration has been demonstrated, within both Europe and America, levels being relatively high in New York and Finland, and low in California and north-western Europe including Denmark and Britain.[13,18] Couple fertility assessed by TTP is high in Finland[19] and in parts of southern Europe compared with north-western Europe.[20] The congruence of the findings for Finland suggests that the higher levels of sperm concentration observed there are not due to differences in methodology or to longer abstinence (less frequent intercourse).

4.2.2   Testicular Cancer

Epidemiological information on cancer of the testis is very reliable. As a disease of relatively young men that has unmistakable features, it is likely to be rarely missed or misdiagnosed, so that only an efficient collating system is required to produce high quality ascertainment. Good incidence data have been available from cancer registries in developed countries for some decades. Mortality data are also available for certain countries going back 100 years, and since the disease was invariably fatal if untreated, these are reliable for the early twentieth century, although not more recently as cure rates are now high.

This disease has shown an increasing trend in recent decades throughout the developed world, typically with rates being trebled or more. When analysed according to birth cohort, which is appropriate both for pathological[21] and epidemiological[22] reasons, mortality started rising among men born before 1900 in England and Wales,[23] and incidence in Denmark, Norway and Sweden started rising among men born around 1905.[4] In these latter three countries, rates stabilised or fell for men born during 1935–1945, whereas the rise was rapid and inexorable among men born from 1920 until at least 1960 in East Germany, Finland and Poland.[4]

There is considerable spatial and ethnic variation. Denmark has the highest incidence in the world, the lifetime risk now being almost 1%. However, the Nordic countries do not have a uniformly high risk, as Finnish men have comparatively low rates, with Norway and Sweden in intermediate positions.[3,24] The spatial pattern for testicular cancer in the Nordic countries does not resemble that of other hormone-sensitive carcinomas such as those of the prostate or female breast, but is similar to that of colorectal cancer in both sexes.[24] Other high-risk populations include Switzerland and New Zealand (including Maoris), while the Baltic states and African-Americans have comparatively low rates.[13] The tumour is rare among Chinese and Japanese men.[13]

4.2.3 Anomalies of the Male Genitalia

Both cryptorchidism and hypospadias are likely to be unreliably ascertained at birth, particularly in mild cases, and the study of cryptorchidism is further complicated by the difficulty of distinguishing testes that have not descended from those that readily but reversibly retract back into the abdominal cavity in early infancy. The consequence is that published data from congenital malformation registries cannot be relied on to reflect real variations: reported time trends and differences between registries may both merely reflect differences in ascertainment and reporting.[25] Self-reported data (by mothers) are similarly unreliable.

For hypospadias, the reported apparent increase in many countries may well be due to variations in the registry system rather than a real change,[25] apart from a step increase between 1982 and 1985 in the severe form in Atlanta, Georgia.[26] Studies in Denmark and Finland using strict criteria have shown a higher rate in Denmark.[25]

With cryptorchidism, studies using strict and comparable diagnostic criteria found an almost doubling of the proportion of boys having cryptorchidism between the 1950s and the early 1990s in southern England.[27] A New York study using the same criteria found a similar proportion to the original lower English estimate.[28] A recent high-quality study has shown that the rate in Denmark is higher than that in Finland, and probably higher than that in Denmark in the late 1950s.[29] Unlike for testicular cancer, African-Americans do not appear to have a lower risk.[28,30]

4.2.4 The Question of Linkage

The grouping of male infertility, testicular cancer, cryptorchidism and hypospadias into the "testicular dysgenesis syndrome"[5] appears to be justified, at least for a proportion of cases of each condition, on the basis of shared histological characteristics such as microlithiasis and Sertoli-cell-only tubules.[31] Furthermore, the four attributes occur together in the same individual more often than expected by chance – although it is rare for all four to be present: it has long been recognised that cryptorchidism is associated both with subfertility and with testicular cancer, and more recently it has become clear that men with testicular cancer are subfertile[32] and have poor semen quality,[33] even before the cancer is diagnosed. Cryptorchidism and hypospadias also occur in the same individual more often than would be expected by coincidence.[34] In addition, it is thought that they all originate early in life.[35]

There are some parallels in their descriptive epidemiology, implying that they probably share at least some risk factors, notably the sharp and consistent contrast between Finland and Denmark for the risk of all four endpoints. On the other hand, the trends are not entirely consistent: in northwestern Europe both impaired semen quality and testicular cancer have increased, whereas in America this is only clearly true for testicular cancer. This implies that there must also be some additional harmful or protective factors, which is unsurprising.

4.3   Can Endocrine Disruption Explain these Observations?

4.3.1   Oestrogens

The original version of the endocrine disruption hypothesis was concerned with exposure to endogenous maternal oestrogens such as oestradiol. This was because high maternal weight and excessive vomiting during pregnancy, conditions known to be associated with high oestrogen levels, were observed to increase the risk of testicular cancer[36] and cryptorchidism.[37] It seems unlikely that variation in maternal oestrogen levels (due to exposure to environmental, nutritional or other factors) could explain the observed trends or spatial differences in the incidence of the four components of the testicular dysgenesis syndrome, as the extent of variation is probably insufficient. However, there has been insufficient research in this area.[13]

A more recent version of the oestrogen hypothesis referred to exogenous substances with oestrogenic (or mixed oestrogenic and anti-oestrogenic) activity.[38] However, the exposure levels and the potency of such substances are too low by several orders of magnitude compared with oestradiol.[39] Against this, it was argued that exogenous oestrogens reach the embryo, whereas endogenous maternal oestradiol is protein bound and does not do so, but this is refuted by the evidence just cited for the increased risk associated with high levels of endogenous oestradiol.[13] There are two other conclusive strands of evidence against the hypothesis. First, diethylstilboestrol (DES), which is as potent as oestradiol, was widely used from the late 1940s onwards, especially in the USA, in the (mistaken) belief that it could prevent miscarriage and a range of pregnancy complications. It was given to more than two million women, in pharmacological doses, often at the stage of pregnancy during which the sexual organs develop. Boys exposed *in utero* to DES tended to have cryptorchidism; they possibly also had lower sperm concentration and increased risk of testicular cancer, but of lesser magnitude than the spontaneous historical trends described above. In any case, genital abnormalities were only observed at the highest DES exposure levels.[40] Secondly, apart from DES, by far the highest exposure to exogenous oestrogens comes from dietary phytoestrogens, notably isoflavones in soy, but Chinese and Japanese men who are exposed to high levels *in utero* have a low incidence of testicular cancer. In relation to the "environmental oestrogens" such as bisphenol A, nonylphenol or certain pesticides, which are orders of magnitude less potent again, it is now accepted that their uniformly weak oestrogenicity excludes the possibility that they could induce these disorders.[40]

This is not surprising. While it is superficially plausible that oestrogens "demasculinise" the developing male, this is biologically naïve because mammals are adapted to starting life inside their mothers, whose internal environment is oestrogen-rich (even before the early pregnancy surge). In contrast to other vertebrates the mammalian default sex is female, and masculinisation of the gonads and central nervous system depends on the presence of androgens. In accordance with this, the effects of *in utero* oestrogen exposure are far more marked in female than in male offspring, in toxicological experiments as well as in the human victims of DES.[41]

4.3.2 Anti-androgens

Interference with either the synthesis or the action of androgens could prevent the normal masculinisation of the male fetus, and could also affect male infants postnatally. There is toxicological evidence that p,p'-DDE (1,1-dichloro-2,2-bis (*p* -chlorophenyl) ethylene), the stable breakdown product of DDT (dicophane, or (2,2-bis (*p* -chlorophenyl)-1,1,1-trichloroethane), can block the androgen receptor, as can certain other pesticides, and that some phthalates inhibit testosterone synthesis.[40] More nuanced hypotheses that relate, for example, to the balance between oestrogens and androgens, or to their interconversion *via* aromatase,[40] are interesting, but lack candidate substances that could explain the epidemiological findings.

It is therefore plausible that exposure to anti-androgens can affect male fertility, as well as other related endpoints, and this is supported by toxicological evidence concerning certain phthalates.[40,42] However, it is not clear whether people are exposed to sufficient doses, even though women of reproductive age may have high exposures in hairspray, nail varnish, perfume, *etc.*[42] Duty *et al.* have studied men attending a Boston fertility clinic, with no special source of phthalate exposure.[43-45] They found associations between a specific phthalate and a semen abnormality in three instances with $p < 0.05$ and three instances with $0.05 < p < 0.10$ out of 63 comparisons; no specific pattern emerged. This is exactly what would be expected by chance. A fourth study, on reproductive hormones, commented that the associations found did not fit the expected patterns, and were likely to be the result of multiple comparisons.[46] However, these negative findings were for adult male exposures and do not rule out an effect on the developing male fetus or infant.

4.3.3 The Expected Spectrum of Effects

It is far from clear that endocrine disruption would affect non-quantitative aspects of semen quality, especially morphology. Also, the endocrine disruption hypothesis was based on the idea that Sertoli cell multiplication was inhibited in early life, and this placed a limit on the number of developing spermatogonia that could be supported,[38] yet, as mentioned before, one of the histological criteria of the testicular dysgenesis syndrome is the presence of Sertoli-cell-only tubules.[31]

However, it is necessary to go further. In addition to focusing on the various endpoints and asking "could this be due to endocrine disruption?", it is important also to turn the question around and ask "if an endocrine-disrupting substance were responsible, what spectrum of effects would be predicted?" One plausible expectation is of a coherent pattern in hormone-sensitive cancers, but this is not observed.[13] A second is that endocrine agents would be expected to influence growth and development, secondary sexual characters and the timing of puberty. No such change has been reported among boys, either in Europe or America.[13,47]

4.3.4   Interpretation

It may be true that early exposure to anti-androgens (*e.g.* phthalates), in the doses actually experienced, is toxic to the male reproductive system. There may be other effects of endocrine disrupting agents: for example, semen quality is especially poor in an agriculture area in Missouri where exposure to chemicals such as pesticides is likely,[48] and prostatic and urethral anomalies are observed in mice fed low doses of bisphenol A,[49] a synthetic oestrogen found in plastic products such as babies' feeding bottles.

   Nevertheless, these findings do not necessarily help us answer the separate but related question, could exposure to DDE, phthalates or other anti-androgens explain any of the epidemiological findings presented above? One obvious objection is that the rising trend, at least in testicular cancer, started before any of the known anti-androgens were introduced. Secondly, the striking contrast between Denmark and Finland cannot be explained by exposure to DDE, which has been monitored in human breast milk, and the concentrations were similar in all the Nordic countries.[50] Thirdly, high levels of exposure to DDE in developing countries, in the course of attempts at malaria control, have not resulted in an epidemic of testicular cancer.[13]

   The answer, then, is no: the hypothesis of endocrine disruption, in any of its variants, cannot explain the observations on trends and spatial variation in disorders of the male reproductive system.

4.4   Genetic Epidemiology

4.4.1   Semen Quality and Fertility

Seven studies have been published that examined whether impairment of semen quality tends to aggregate in families, and all have found that it does.[51–57] Czyglik *et al.* studied 581 candidates for semen donation who attended the CECOS (les Centres d'Etudes et de Conservation des Oeufs et du Sperme) clinic in Paris, 36 of whom were brothers of infertile men.[51] They found that the 36 brothers tended to have clearly inferior semen quality compared with the 545 controls, and despite the relatively small group the differences were statistically significant for lower sperm concentration ($p < 0.05$), impairment of motility ($p < 0.01$), and especially of abnormal morphology ($p < 0.00001$). The differences were even greater among the 20 brothers of azoospermic men. The abnormal morphological forms were for all anomalies, rather than a distinct pattern of the familial aggregation of a particular anomaly, which had been observed in previous studies. For sperm concentration, marked heterogeneity was found, with some men having similar levels to the controls but others having extremely low concentrations. The authors suggest the findings could be the result of a shared environmental factor, such as a viral infection or drugs, or of recessive autosomal genes, possibly genes that affect meiosis.

Lilford *et al.* carried out a case control study of men attending a subfertility clinic who had been found to have abnormal semen quality (cases) compared with men who had fathered at least two children (controls), and also performed a segregation analysis.[52] Brothers of cases ($n = 148$) were found to have a higher risk of involuntary childlessness ($p = 0.0005$) than did brothers of controls ($n = 169$). There was also a non-significant ($p = 0.09$) excess of infertility among the uncles of cases (equally in maternal and paternal uncles), but no such excess among sisters – which specificity provided some confidence in the accuracy of recall. In the segregation analysis, the best fitting model was one in which 60% of cases were assumed to be due to a recessive gene, the other 40% being due to random non-genetic factors. In a subset of families in which multiple family members donated semen samples, specific patterns of abnormality were seen, *e.g.* a deficit in motility or in sperm concentration, suggesting that different genetic abnormalities were involved. The lack of subfertility in sisters of cases argued against a genetic impairment affecting both sexes, *e.g.* by disrupting meiosis, and the similarity of rates in maternal and paternal uncles argued against a sex-linked abnormality.

Meschede *et al.* compared 621 infertile couples treated with ICSI (intracytoplasmic sperm injection) and 1302 fertile couples as controls.[53] Involuntary childlessness was reported by close relatives (sibling, half-sib, aunt or uncle) of 11.8% of the infertile couples. A positive family history was more common on the male than the female partner's side, and the family member involved was more likely to be male ($p<0.001$ in both cases), suggesting that familial clustering is largely confined to male-factor infertility. The infertile couples tended to have fewer siblings, especially sisters, than fertile controls, compatible with reduced fertility in the parents due either to a genetic or a non-genetic (environmental or behavioural) cause. They suggest that non-Mendelian multifactorial inheritance is the most likely mode of inheritance, which is supported by the observations among uncles. A specific genetic basis for infertility was identified in 6.4% of cases, *e.g.* CFTR (cystic fibrosis transmembrane conductance regulator) mutations or chromosomal aberrations, with a mixed pattern of inheritance; there was little overlap between this group and the couples with a positive family history. The authors comment that at the time of writing (2000), most fertility-affecting genetic aberrations could not be detected using then-current clinical laboratory methods.

Christensen *et al.*[54] studied functional fertility, measured using TTP, among members of the Danish twins register. They found a clear, highly significant intra-pair correlation among monozygotic twin pairs, ($n = 645$, $r = 0.22$), but not among same-sex dizygotic twins ($n = 826$, $r = 0.00$), suggesting a heritable component in biological fertility. The findings were similar for male and female twins. The correlation for monozygotic twins was mainly due to participants who had taken at least 10 months to conceive. The lack of correlation among dizygotic twins suggests a mode of inheritance that is not simply additive, one that involves gene–gene interaction; this could be due to a recessive allele at a single locus, or could be polygenic.

Storgaard[55] also studied Danish twins, and reported the heritability of sperm concentration, uncorrected for biological variation and measurement error (and therefore an underestimate), as 20%. The heritability of sperm morphology was 41%, and that of chromatin stability was 68%.

Van Golde et al.[56] carried out a case control study of 253 severely subfertile men who were candidates for ICSI, compared with 243 randomly selected controls. The prevalence of male fertility problems was substantially higher among their brothers than among brothers of the control men (10.4% *vs.* 0.5%), and also higher among maternal uncles of the ICSI group (1.7% *vs.* 0.2%). The subfertile men with familial occurrence tended to have lower sperm motility, and also a higher probability of a normal FSH level, than the non-familial cases. Genetic aberrations such as a chromosomal abnormality or a Y microdeletion were present in 13.8% of the severely subfertile men.

Gianotten et al.[57] compared 160 men who had severe impairment of sperma-togenesis of no known cause with 285 men who had normal semen parameters, all of whom attended a fertility clinic. Brothers of the former group were more likely to be subfertile than those of the men with normal semen (OR = 3.2). Among sisters there was a similar tendency, albeit smaller and not statistically significant. Brothers of the men with impaired spermatogenesis were also more likely to have reduced semen parameters. A segregation analysis was compatible with either 47% of the subfertility among the brothers being due to an autosomal recessive gene, or 23% being due to an autosomal dominant gene. The authors concluded that a large number of environmental factors as well as many different genes, including new mutations, may play a role in the aetiology of impaired spermatogenesis.

Thus, there is accumulating evidence for a substantial degree of heritability of fertility in men, as it would be difficult to explain these findings in terms of familial clustering of behavioural characteristics. Although Lilford et al. and Meschede et al. favoured a recessive inheritance pattern, not all authors agree with this. It must be remembered that we are dealing with a highly heterogeneous picture, implying the involvement of numerous different loci for different variants of impairment of semen quality and of fertility, as well as the likelihood of polygenic inheritance in many instances, so that it is more likely that several different modes of inheritance are operating.

Recessivity is one way of explaining how "a gene for infertility" could survive, in evolutionary terms. It would also predict a high correlation of brothers' risks, rather than across the generations (*e.g.* uncle–nephew pairs). One way of looking at this possibility is to study populations containing a high proportion of consanguineous couples. Ober et al.[58] compared the fertility of couples with varying degrees of consanguinity among the Hutterites of South Dakota. They found that women's fertility was associated with their degree of inbreeding, but that the same was not true for male fertility or for the degree of couple relatedness. In the more recent period final family size was not affected, indicating some degree of reproductive compensation, which would slow the loss of deleterious recessive alleles and lead to a situation of balanced polymorphism. Baccetti *et al,*[59] identified 62 men with some degree of consanguinity among 1600 men attending a clinic for fertility problems, and examined their sperm morphology using transmission electron microscopy (TEM). Well-recognised specific genetic sperm defects were observed in 17 out of the 62 men, and in 15 of the 1506 non-consanguineous individuals. These types of defect differ from the pattern of impaired sperm morphology more usually seen in the general population, as 100% of spermatozoa are identically affected in the whole ejaculate for life. The 45 men who did not have one of these morphological syndromes did not differ from the non-consanguineous group in terms of morphology, but they did tend to have impaired sperm motility. Taken together, these two studies suggest that recessive genes play little part in the inheritance of male infertility, apart from specific morphological sperm defects and (possibly) impairment of motility.

A prime candidate for a mode of inheritance of male fertility is through the Y chromosome, because microdeletions of the *AZFa*, *AZFb* and *AZFc* regions of this chromosome are associated with spermatogenic failure. One way of looking at this is to examine haplogroups and haplotypes. A haplogroup is a monophyletic group of Y chromosomes defined by shared allelic states at slowly mutating binary markers, such as single-nucleotide polymorphisms or insertion or deletion events. A Y-chromosome haplotype is defined by a particular combination of allelic states at highly mutating markers, such as microsatellites, within a given haplogroup.[60] A haplogroup has been found in Denmark, hg26, that is associated with having a sperm concentration of less than 20×10^6 sperm ml^{-1} (including azoospermia) in the absence of *AZF* microdeletions.[61] Similar findings have been reported for a different haplogroup from Japan.[62]

4.4.2 Testicular Cancer

While the rapid trends in testicular cancer demonstrate the importance of the environment in the broadest sense, there is also evidence of heritability. There are five main types of study design: migrant studies, investigation of familial clustering, twin studies, segregation analysis and linkage studies. This area has recently been reviewed by Lutke Holzik *et al.*[63]

Tominaga[64] found that the risk of testicular cancer was more than 20 times higher among white Americans than in Japanese men in Japan, with the risk for Americans of Japanese ancestry (mostly living in Hawaii) lying halfway between. It is unclear how long the immigrants had been in the United States, or indeed whether some of them may have been born there. Parkin and Iscovitch[65] compared cancer registrations in Israel in men aged under 30 years between migrants, sons of migrants, sons of one migrant and one Israeli, and sons of two Israelis. Origin in Europe or America was associated with a relatively high risk, the reverse being true of origin in Africa or Asia. These differences in risk were largely retained in the second generation, implying a genetic component, although there was some lessening of risk among sons of

European and American migrants. Hemminki and Li[66] studied immigrants to Sweden, and found that the lower Finnish and higher Danish risks became similar to the Swedish risk in the second generation. In addition, the risk among Americans of African descent is similar to that in African men, and is a quarter of that of white Americans,[63] and the relatively low risk in Italy is paralleled by low risk among people of Italian descent in numerous countries around the world.[67]

Forman *et al.*[68] investigated the familial clustering of testicular cancer, and estimated that 1.5% of cases are familial. The relative risk was found to be ten-fold for brothers and four-fold for father–son pairs. Familial cases were slightly more likely to be bilateral, and to occur at a younger age. Westergaard *et al.*[69] studied all cases in Denmark in the period 1950–1993. Their estimate of the relative risk between brothers was twelve-fold, and for father–son pairs it was two-fold. Among father–son pairs, the relative risk was higher (3.9-fold) if the son was born in 1950–1954, which is compatible with the idea that for later births, the association was diluted by the marked rise in incidence that was occurring during this period. They did not confirm the difference in age at onset in the previous study. The authors comment that the strength of the association between brothers means that it is unlikely to be due to shared environmental or behavioural exposures, and strongly suggests a genetic component in the aetiology. Dieckmann and Pichlmeier[70] carried out a prospective and a retro-spective study, and reviewed the previous literature. They concluded that the proportion of testicular cancer cases that was familial was 1.35%, and that the relative risk for first-degree relatives of patients with the disease was 3–10. In both their studies, the relatives included not only brothers but almost as many father–son pairs, as well as cousins, uncle–nephew pairs and grandfather–grandson pairs. Sonneveld *et al.*[71] estimated that 2.5% of the testicular cancer cases in their clinic population were familial. The familial cases were not markedly more likely to be bilateral, but did appear to have an increased likelihood of cryptorchidism. In the study by Hemminki and Li,[66] the risk was increased nine-fold when a brother and four-fold when the father had the disease; the relative risk for brothers born less than five years apart was 10.8, whereas it was 6.7 if their age difference was larger, suggesting an environ-mental component.

The high relative risk found between related individuals, especially that between brothers, is an unusual finding in cancer epidemiology: for most cancers the relative risk in first degree relatives is usually between 2 and 3, and rarely over 4.[68,72,73] It cannot plausibly be attributed to a shared environ-mental risk factor. Khoury *et al.*[74] calculated that the relative risk of exposure would have to be at least 50 to account for the observations, several times higher than the risk of lung cancer in long-term heavy smokers. Such a risk factor has not been identified.[75]

Hemminki and Li[66] also found that the risk of certain other cancers was raised in the families of men with testicular cancer, especially seminoma. These included colorectal, pancreatic, lung and breast cancer, non-Hodgkin's lymphoma and Hodgkin's disease.

Swerdlow *et al.*[76] report the only study so far to have identified a large enough number of twins with testicular cancer ($n = 194$) to compare the risk in their twin brothers. The relative risks in monozygotic and dizygotic twins were respectively 76.5 and 35.7, which is evidence for a heritable component, but this observation was not statistically significant. The higher relative risk for dizygotic twins than for non-twin brothers, albeit based on small numbers, suggests also the importance of a shared parental environment, which could be a permanent characteristic such as the tendency of their mother to have particular hormonal levels, or it could be a time-specific shared exposure of either parent to some wider environmental factor such as a one-off short-term chemical exposure. In favour of the former idea, dizygotic twins have a significantly higher risk of testicular cancer than do monozygotic twins (OR = 1.5), which fits with the higher maternal oestrogen levels in dizygotic twin pregnancies (Braun *et al.*[77] found the observed to expected ratio to be 1.6). These two ideas are not mutually exclusive.

A segregation analysis has been carried out by Heimdal *et al.*[78], based on 978 patients treated in two hospitals, one in Norway and one in Sweden, of whom 30 had one first-degree relative with testicular cancer. They concluded that a major recessive gene provided the best-fitting model, with a gene frequency of 3.8%. Under the assumption of Hardy–Weinberg equilibrium, this would imply that 7.3% of men carry the mutant allele, 0.1% are homozygous, and 43% of homozygotes will develop the disease. However, the International Testicular Cancer Linkage Consortium (ITCLC), which has the largest collection ($n = 179$) of testicular cancer pedigrees globally, has found that there are at least three susceptibility genes for this disease, and that no single locus can explain as many as 50% of the families.[79] Neither of these studies considered the possibility of sex-linked inheritance.

The first testicular cancer susceptibility gene to be identified is *TGCT1* (TGCT = Testicular Germ Cell Tumour), on chromosome Xq27, based on 99 pedigrees.[80] This gene is particularly associated with bilateral disease ($p = 0.034$). It accounts for about a third of the excess familial risk, and most of the excess risk between brothers rather than father–son pairs. The involvement of the X chromosome accords with the known high risk of testicular cancer in Klinefelter's syndrome, in which the affected men have a second X chromosome. *TGCT1* also appears to predispose to cryptorchidism in families affected by testicular cancer ($p = 0.03$). In contrast, Y haplogroup was not found to be associated with testicular cancer,[60] which accords with the lack of a recognised gene on the Y chromosome for this disease.

4.4.3 Anomalies of the Male Genitalia

The available information on the genetic epidemiology of cryptorchidism and hypospadias is sparse. Weidner *et al.*[81] carried out a case control study of both conditions, identified through the Danish National Patient Register which contains information on all discharge diagnoses and operations performed since 1977, or from the Danish Malformation Register. Among the 6177 boys with cryptorchidism, the risk was increased when an older brother already had the condition (OR = 3.8). Similarly, among the 1345 boys with hypospadias, the risk was increased when an older brother already had the condition (OR = 10.1). These observations accorded with findings in previous studies. There was no increased tendency for the two conditions to occur within the same family (ORs 0.5 and 1.2, both non-significant).

Fredell *et al.*[82] carried out a complex segregation analysis of 2005 pedigrees with at least one member who had hypospadias, excluding cases that were part of a genetic syndrome. They found strong evidence of familial aggregation ($p < 0.001$). A multifactorial model with a heritability of 0.99 gave the best fit. This was compatible with the previous literature, in which smaller studies and studies of individual families had generated a number of hypotheses on the mode of transmission. Some evidence had been found for association with mutations in the 5-alpha reductase and androgen receptor genes, but these tended to be in patients with severe hypospadias in combination with other genital malformations.

Czeizel and Toth[83] compared the number of menstrual cycles taken to conceive among 186 planned pregnancies that had resulted in a boy with hypospadias, and 193 matched controls. Couple fertility was lower in the hypospadias group ($p < 0.01$), and this was not due to medical treatment for their subfertility. This finding was in agreement with previous evidence of testicular anomalies in the fathers of boys with hypospadias. This was confirmed by Fritz and Czeizel, who compared the semen quality of fathers of 25 boys with hypospadias and 50 men who had fathered at least one healthy baby, matched for age and place of residence.[84] The hypospadias group of fathers tended to have lower motility and a higher incidence of abnormal morphology (*e.g.* small sperm head), as well as possibly lower concentration.

4.5 Can Genetic Damage Explain these Observations?

4.5.1 Can Genetic Disorders Increase Rapidly Over Time?

All four endpoints of the testicular dysgenesis syndrome show some degree of heritability. This is compatible with the idea that they are linked as part of the same syndrome. And yet testicular cancer definitely increased rapidly during the twentieth century, and there is strong suspicion that the same is true for at least two of the other three manifestations.

One response is, if a condition has both genetic and environmental determinants, causation can be apportioned between the two [Figure 1(a)] – "environmental or genetic influences" as an opposition. And gene-environment interaction can be considered [Figure 1(b)][85] – here "genetic" and "inherited" are conflated. A third theoretical possibility is an inherited tendency to a particular behaviour, such as excess alcohol consumption [Figure 1(c)]. But the most parsimonious hypothesis in the present context is germ-line genetic damage [Figure 1(d)]: the genetic defect has an environmental cause. This

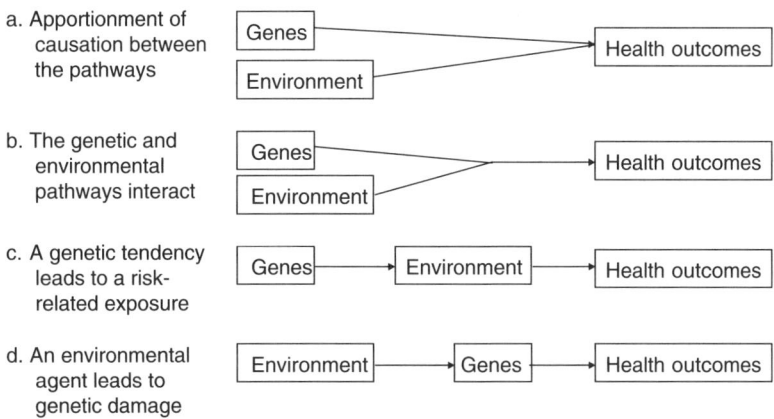

a. Apportionment of causation between the pathways

b. The genetic and environmental pathways interact

c. A genetic tendency leads to a risk-related exposure

d. An environmental agent leads to genetic damage

Figure 1 *Ways in which genes and environment relate to each other.*

can then become heritable if it does not lead to lethality or sterility (obviously the four possibilities are not mutually exclusive).

This suggests a hypothesis that fits with all the observations outlined above, on inheritance of the elements of the testicular dysgenesis syndrome as well as on time trends. Exposure to a toxic agent would lead to some form of mutation, genetic damage or epigenetic process;[86] its survival in subsequent generations would depend on

(i) the degree to which it affects health (including lethality at one extreme), at all stages of life from conception to the end of reproductive life;

(ii) the extent to which it affects biological fertility, the probability of achieving a fertilised ovum, given unprotected intercourse; and

(iii) additional factors that involve volition and control, as well as biology: contraception and achieved family size, and the use of artificial reproductive technology.

The epidemiological implication would be to introduce a degree of inertia into the time trend: while an increase in the health outcome would still directly follow an increase in the causal agent (allowing for latency), there would be a gap of one or more generations between their respective disappearance. At the individual level (*e.g.* in a cohort study), the relevant exposure could be to the individual himself in early life, to his mother for *in utero* exposure, or to either parent for pre-conceptual genetic damage – but it could also well be to a grandparent or earlier ancestors.

4.5.2　Is Heritable Infertility Possible?

In the case of infertility, heritability may at first sight appear to be implausible: a "gene for infertility" is surely impossible on evolutionary grounds? – it would quickly be eliminated, unless it were silent in heterozygotes as with a recessive gene or an X-linked condition, which appears not to be the case here.

An alternative explanation is, however, possible: the genetic basis of infertility is constantly recreated by new mutations. In a steady state, a balance would exist between selection against alleles that impair fertility and their *de novo* creation as a result of genetic damage. As these two processes are not constrained to be equal, new damage could occur at a rate greater than elimination, leading to an increase in incidence. The same argument applies, but with less strong elimination, to testicular cancer, hypospadias and cryptorchidism, which are associated with subfertility, and which show evidence of heritability.

The selection process can also vary in intensity. Czeizel and Rothman[87] have pointed out that as family size decreased markedly during the twentieth century, births to the biologically most fertile couples became a less dominant proportion of births at the population level. Secondly, towards the end of the century, assisted reproduction meant that the proportion of births to clinically subfertile couples increased. These two tendencies would have the effect of decreasing the rate of removal from the population of alleles that reduce fertility and that might also predispose to testicular cancer or one of the other endpoints.

The first of these tendencies at least may well be important to consider in relation to historical trends. However, it is unlikely that on its own it would be strong enough to bring about, for example, a three-fold rise in testicular cancer incidence.[88] Furthermore, this hypothesis depends on the existence of alleles that decrease fertility, raising the question, why had they not already been eliminated from the population, even before 1900? New generation of such alleles is required to complete the picture, and once this is accepted, it becomes important to consider what agent(s) could increase the mutation rate, and could account for the trends and spatial variations already described.

An alternative variant relating to selection for "a gene for infertility" is that such a gene is selected for because it confers higher inclusive fitness. For example, less frequent conception could lead to a higher chance of infant survival into adulthood – although it might be objected that other methods of child spacing are more likely to have evolved. A more likely possibility is that some pathogenic mutations have a selective advantage for the spermatogonial cells in which they arise.[89] There is some evidence for premeiotic selection in humans.[90]

4.5.3   So What is Going on?

It is clear that there are multiple genes involved, even for each endpoint, with various modes of inheritance. The Y and X chromosomes may have some importance for disorders of male fertility and for testicular cancer, respectively. The observed stronger association in brothers is compatible with (a) autosomal recessive inheritance; (b) X-linked inheritance; (c) lower fertility or earlier mortality among potential fathers; (d) parental mutation affecting stem cells, or at least a sizeable proportion of them (if paternal, without affecting his phenotype). In the case of testicular cancer, the X-linked gene explains a proportion of the excess rate in brothers compared with father–son pairs. The evidence appears to be that autosomal recessive inheritance plays little or no part in either of these two conditions.

A possible causal agent would need to have exposure characteristics that correspond to the epidemiological observations. It would need to be genotoxic and to be absorbed. If female (*in utero*) exposure were responsible, it would have to cross the placenta. And it would need to localise in the testis; this could happen if the agent binds to oestrogen or androgen receptors that are expressed in the testis while retaining its mutagenic activity,[91] just as radioactive iodine causes thyroid cancer by combining localising and genotoxic attributes.

The sequence of events may be something like this. In the embryonic testis the germ cells undergo many mitotic divisions, providing opportunities for error. The induced abnormality is not a simple mutation at a specific site, but something that interferes with one or more fundamental aspects of the cellular mechanism such as the spindle, DNA repair or apoptosis. This affects a large proportion of the cell population, especially if it occurs early in development. The resulting cellular defect is characteristic of testicular dysgenesis syndrome; as the primary defect is in the germ cells and reduces their number, this includes Sertoli-cell-only tubules. It includes carcinoma-*in-situ* (CIS) in the more severe cases. Once meiosis starts at puberty, the malfunctioning cellular apparatus leads to structural and numerical abnormalities in subsequent stages of spermatogenesis, and this is manifest as reduced semen quality, i.e. number and quality (motility, morphology, *etc.*) of spermatozoa, or even of azoospermia. If present, CIS may develop into carcinoma. In a small proportion of cases, the cellular abnormalities, which may include a tendency to carcinoma, are passed onto the man's male offspring. A proportion of sons also develop cryptorchidism or hypospadias. Figure 2 is a simplified diagram of this process, starting with *in utero* exposure.

This would account for the diversity of genetic abnormalities. It would also explain several observations: genetic damage,[92] and structural and numerical abnormalities,[93,94] in the sperm of oligospermic men, and extended asynapsis in their testes,[95] as well as sperm chromosome aneuploidy in the husbands of women with unexplained recurrent pregnancy loss.[96] It also fits with the deregulated cell cycle[97] and the pattern of structural chromosomal abnormalities and hyperploidy[98] seen in testicular cancer, together with the absence of known specific gene defects, such as that for the androgen receptor,[99] and AZF.[100] Furthermore, the occurrence of hypospadias in the following generation recalls that found in the sons of women who had been exposed *in utero* to DES[101] (although that was female- not male-mediated), with DES here acting as a genotoxic agent, possibly by generating reactive oxygen species.[102]

There are also unanswered questions. What is the mechanism of the abnormality? How is it transmitted to the next generation, and why in only a small proportion of cases? Why does it sometimes lead to cryptorchidism and hypospadias? What unifying principle ensures that it leads only to the conditions discussed in this paper? – or could the health implications extend beyond these to include a broader range of abnormalities in the offspring and even in future generations (Figure 3)?[103] With such heterogeneous outcomes, each

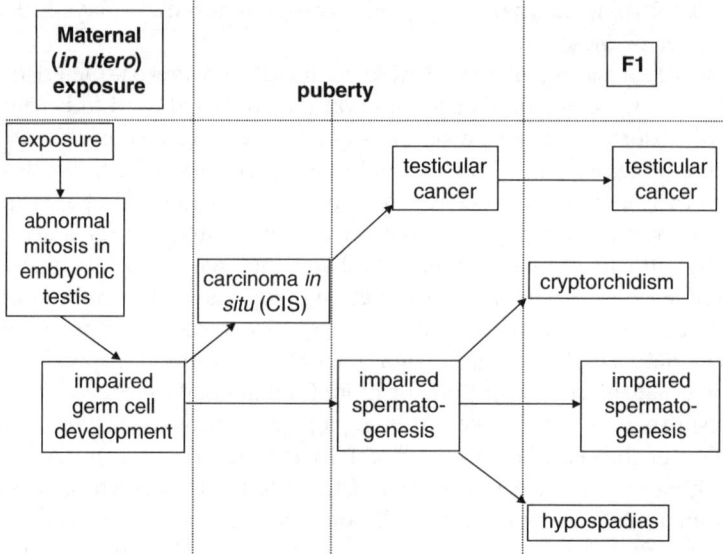

Figure 2 *Harming the developing male reproductive system: A suggested sequence of events.*

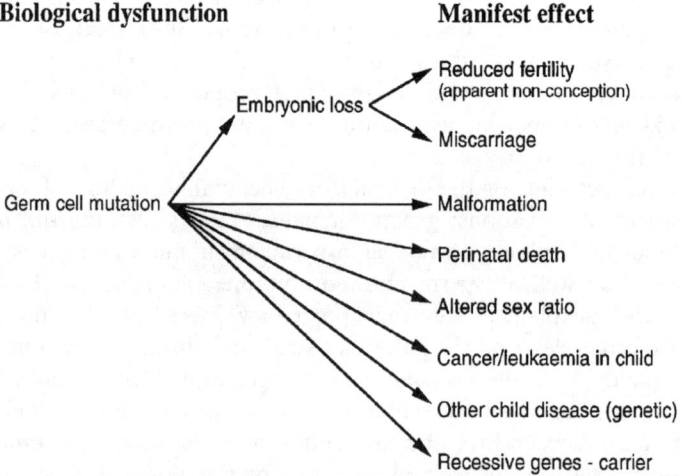

Figure 3 *Possible consequences of germ line genetic damage*
(Reproduced with permission from M. Joffe, Asclepios, Time to pregnancy: A measure of reproductive function in either sex, Occup. Environ. Med., 1997, **54**, 289–295).[11]

being uncommon, it is possible that increasing trends have escaped detection. And are all cases of these four conditions linked to testicular dysgenesis syndrome, or are some isolated cases? Is testicular cancer of X-linked origin a completely independent entity?

4.6 Conclusions

While the observed deterioration in semen quality, and in other possibly linked conditions affecting the male reproductive tract, have been widely discussed in relation to the "oestrogen hypothesis", pollution with weak environmental oestrogens cannot plausibly be responsible. The anti-androgen variant of the endocrine disruption hypothesis, or androgen to oestrogen balance, may be important in relation to male reproduction, but cannot explain the existing epidemiological findings.

Several lines of evidence point towards genetic damage as an explanation of various types of impairment of the male reproductive system, including a high degree of clustering in families, with intergenerational transmission; a strong degree of heritability in twin studies; and slow adaptation of testicular cancer risk among migrants. New mutations would be balanced by their elimination due to reduced fertility, the speed of elimination being affected by reproductive behaviour. Multiple genes are involved, with many possible modes of inheritance. The causal agent would localise in the testis by binding to steroidal hormone receptors. It would cause disruption of the cell mechanism, at least in a large proportion of tubules, leading to impaired spermatogenesis and sometimes to CIS. However, the mechanism of damage is unclear, and several unanswered questions remain.

References

1. E. Carlsen, A. Giwercman, N. Keiding and N.E. Skakkebaek, *BMJ*, 1992, **305**, 609–613.
2. S.H. Swan, E.P. Elkin and L. Fenster, *Environ. Health. Perspect.*, 2000, **108**, 961–966.
3. H.-O. Adami, R. Bergstrom, M. Mohner, W. Zatonski, H. Storm, A. Ekborn, S. Tretli, L. Teppo, H. Ziegler and M. Rahu, *Int. J. Cancer*, 1994, **59**, 33–38.
4. R. Bergström, H.-O. Adami, M. Möhner, W. Zatonski, H. Storm, A. Ekborn, S. Tretli, L. Teppo, O. Akre and T. Hakulinen, *J. Natl. Cancer. Inst.*, 1996, **88**, 727–733.
5. N.E. Skakkebaek, E. Rajperts-de Meyts and K.M. Main, *Hum Reprod.*, 2001, **16**, 972–978.
6. H. Adlercreutz, *Ann. NY Acad. Sci.*, 1990, **595**, 281–290.
7. J. Toppari, J.C. Larsen, P. Christiansen, A. Giwercman, P. Grandjean, L.J. Guillette, B. Jégon, T.K. Jensen, P. Jonannet, N. Keiding, H. Leffers, J.A. McLachlan, O. Meyer, J. Müller, E. Rajpert-De Meyts, T. Scheike, R. Sharpe, J. Sumpter and N.E. Skakkeback, *Male Reproductive Health and Environmental Chemicals with Estrogenic Effects*, Danish Environmental Protection Agency, Copenhagen, 1995.
8. Institute for Environment and Health (IEH). *Assessment on Environmental Oestrogens: Consequences to Human Health and Wildlife*, Institute for Environment and Health, Leicester, 1995.

9. R.J. Kavlock, G.P. Daston, C. DeRosa, P. Fenner-Crisp, L.E. Gray, S. Kaattari, G. Lucier, M. Luster, M.J. Mac, C. Maczka, R. Miller, J. Moore, R. Rolland, G. Scott, D.M. Sheeham, T. Sinks and H.A. Tilson, *Environ. Health. Perspect.*, 1996, **104**(suppl 4), 715–740.

10. P. Holmes, P. Harrison and C. Humfrey, *European Workshop on the Impact of Endocrine Disrupters on Human Health and Wildlife*, Commission of the European Community, Brussels, 1997.

11. M. Joffe, Asclepios, *Occup Environ. Med.*, 1997, **54**, 289–295.

12. M. Joffe, J. Key, N. Best, N. Keiding, T. Scheike and T. K. Jensen, *Am. J. Epidemiol.*, 2005, **162**, 115–124.

13. M. Joffe, *Occup. Environ. Med.*, 2001, **58**, 281–288.

14. M. Joffe, *Lancet*, 2000, **355**, 1961–1965.

15. O. Akre, S. Cnattingius, R. Bergström, U. Kvist, D. Trichopoulous and A. Ekbom, *Fertil. Steril.*, 1999, **71**, 1066–1069.

16. T.K. Jensen, N. Keiding, T. Scheike, R. Slama and A. Spira, *Fertil Steril.*, 2000, **73**, 421–422.

17. T.K. Jensen, M. Joffe, T. Scheike, A. Skytthe, D. Gaist and K. Christensen, *Hum. Reprod.*, 2005, **20**, 955–964.

18. N. Jørgensen, E. Carlsen, I. Nermoen, M. Punab, J. Suominen, A.G. Andersen, A.M. Andersson, T.B. Hangen, A. Horte, T.K. Jensen, O. Magnus, J.H. Peterson, M. Vierula, J. Toppari and N.E. Skakkebaek, *Hum. Reprod.*, 2002, **17**, 2199–2208.

19. M. Joffe, *Lancet*, 1996, **347**, 1519–1520.

20. W. Karmaus and S. Juul, *Eur. J. Public Health*, 1999, **9**, 229–235.

21. N.E. Skakkebaek, J.G. Berthelsen, A. Giwercman and J. Muller, *Int. J. Androl.*, 1987, **10**, 19–28.

22. R. Bergström, H.-O. Adami, M. Möhner, W. Zatonski, H. Storm, A. Ekbom, S. Tretli, L. Teppo, O. Akre and T. Hakulinen, *J. Natl. Cancer Inst.*, 1996, **88**, 727–733.

23. J.M. Davies, *Lancet*, 1981, **i**, 928–932.

24. O.M. Jensen, B. Carstensen, E. Glattre, B. Malker, E. Pukkala and H. Tulinius, *Atlas of Cancer Incidence in the Nordic Countries*, Nordic Cancer Union, Helsinki, 1988.

25. J. Toppari, M. Kaleva and H.E. Virtanen, *Hum. Reprod. Update*, 2001, **7**, 282–286.

26. L.J. Paulozzi, *Environ. Health Perspect.*, 1999, **107**, 297–302.

27. John Radcliffe Hospital Cryptorchidism Study Group, 1984–1988, *Arch. Dis. Child,* 1992, **67**, 892–899.

28. G.S. Berkowitz, R.H. Lapinski, S.E. Dolgin, J.G. Gazella, C.A. Bodian and I.R. Holzman, *Pediatrics*, 1993, **92**, 44–49.

29. K.A. Boisen, M. Kaleva, K.M. Main, H.E. Virtanen, A.-M. Haavisto, I.M. Schmidt, M. Chellakooty, I.N. Damgaard, C. Mau, M. Reunanen, N.E. Skakkebaek and J. Toppari, *Lancet*, 2004, **363**, 1264–1269.

30. R.H. Depue, *Int. J. Epidemiol.*, 1984, **13**, 311–318.

31. N.E. Skakkebaek, M. Holm, C. Hoel-Hansen, N. Jørgensen and E. Rajpert-De Meyts, *APMIS*, 2003, **111**, 1–11.

32. H. Møller and N.E. Skakkebaek, *BMJ*, 1999, **318**, 559–562.
33. R. Jacobsen, E. Bostofte, G. Engholm, J. Hansen, J.H. Olsen, N.E. Skakkebaek and H. Møller, *BMJ*, 2000, **321**, 789–792.
34. I.S. Weidner, H. Møller, T.K. Jensen and N.E. Skakkebaek, Risk factors for cryptorchidism and hypospadias, *J. Urol.*, 1999, **161**, 1606–1609.
35. C. Asklund, N. Jørgensen, T.K. Jensen and N.E. Skakkebaek, *BJU Int.*, 2004, **93**(suppl 3), 6–11.
36. B.E. Henderson, B. Benton, J. Jing, M.C. Yu and M.C. Pike, *Int. J. Cancer*, 1979, **23**, 598–602.
37. R.H. Depue, *Int. J. Epidemiol.*, 1984, **13**, 311–318.
38. R.M. Sharpe and N.E. Skakkebaek, *Lancet*, 1993, **341**, 1392–1395.
39. S.H. Safe, *Environ. Health Perspect.*, 1995, **103**, 346–351.
40. R.M. Sharpe, *Int. J. Androl.*, 2003, **26**, 2–15.
41. Committee on Toxicity of Chemicals in Food, Consumer Products and the Environment, *Phytoestrogens and health*, Food Standards Agency, London, 2003.
42. J.S. Fisher, *Reproduction*, 2004, **127**, 305–315.
43. S.M. Duty, M.J. Silva, D.B. Barr, J.W. Brock, L. Ryan, Z. Chen, R.F. Herrick, D.C. Christiani and R. Hauser, *Epidemiology*, 2003, **14**, 269–277.
44. S.M. Duty, N.P. Singh, M.J. Silva, D.B. Barr, J.W. Brock, L. Ryan, R.F. Herrick, D.C. Christiani and R. Hauser, *Environ. Health. Perspect.*, 2003, **111**, 1164–1169.
45. S.M. Duty, A.M. Calafat, M.J. Silva, J.W. Brock, L. Ryan, Z. Chen, J. Overstreet and R. Hauser, *J. Androl.*, 2004, **25**, 293–302.
46. S.M. Duty, A.M. Calafat, M.J. Silva, L. Ryan and R. Hauser, *Hum. Reprod.*, 2005, **20**, 604–610.
47. A.-S. Parent, G. Teilmann, A. Juul, N.E. Skakkebaek, J. Toppari and J-P. Bourguignon, *Endocrine Rev.*, 2003, **24**, 668–693.
48. S.H. Swan, C. Brazil, E.Z. Drobnis, F. Liu, R.L. Kruse, M. Hatch, J.B. Redmon, C. Wang and J.W. Overstreet, *Environ. Health Perspect.*, 2003, **111**, 414–420.
49. B.G. Timms, K.L. Howdeshell, L. Barton, S. Bradley, C.A. Richter and F.S. vom Saal, *PNAS*, 2005, **102**, 7014–7019.
50. A. Ekbom, A. Wicklund-Glynn and H.-O. Adami, *Lancet*, 1996, **347**, 553–554.
51. F. Czyglik, M.-J. Mayaux, M.-L. Guihard-Moscato, G. David and D. Schwartz, *Fertil. Steril.*, 1986, **45**, 255–258.
52. R. Lilford, A.M. Jones, D.T. Bishop, J. Thornton and R. Mueller, *BMJ*, 1994, **309**, 570–573.
53. D. Meschede, B. Lemke, H.M. Behre, C. De Geyter, E. Nieschlag and J. Horst, *Hum. Reprod.*, 2000, **15**, 1604–1608.
54. K. Christensen, H.-P. Kohler, O. Basso, J. Olsen, J.W. Vaupel and J.L. Rodgers, *Epidemiology*, 2003, **14**, 60–64.
55. L. Storgaard, J.P. Bonde, E. Ernst, C.Y. Andersen, K.O. Kivik and J. Olsen, in L. Storgaard (ed), *Genetical and prenatal determinants for*

semen quality: An epidemiological twin study, (Ph.D Thesis), University of Aarhus, Aarhus, 2003.

56. R.J. Van Golde, I.A. van der Avoort, J.H. Tuerlings, L.A. Kiemeney, E.J. Meuleman, D.D. Braat and J.A. Kremer, *J. Androl.*, 2004, **25**, 819–823.

57. J. Gianotten, G.H. Westerveld, N.J. Leschot, M.W.T. Tanck, R.J. Lilford, M.P. Lombardi and F. van der Veen, *Hum. Reprod.*, 2004, **19**, 71–76.

58. C. Ober, T. Hyslop and W.W. Hauck, *Am. J. Hum. Genet.*, 1999, **64** 225–231.

59. B. Baccetti, S. Capitani, G. Collodel, G. Di. Cairano, L. Gambera, E. Moretti and P. Piomboni, *Hum. Reprod.*, 2001, **16**, 1365–1371.

60. K. McElreavey and L. Quintana-Murci, *APMIS*, 2003, **111**, 106–114.

61. C. Krause, L. Quintana-Murci, E. Rajpert-De-Meyts, N. Jorgensen, M.A. Jobling, Z.H. Rosser, N.E. Skakkeback and K. McElreavy, *Hum. Mol. Genet.*, 2001, **10**, 1873–1877.

62. Y. Kuroki, T. Iwamoto and J. Lee *et al.*, *J. Hum. Genet.*, 1999, **44**, 289–292.

63. M.F. Lutke Holzik, E.A. Rapley, H.J. Hoekstra, D.T. Sleijfer, I.M. Nolte and R.H. Sijmons, *Lancet Oncol.*, 2004, **5**, 363–371.

64. S. Tominaga, *Natl. Cancer Inst. Monogr.*, 1985, **69**, 83–92.

65. D.M. Parkin and J. Iscovich, *Int. J. Cancer*, 1997, **70**, 654–660.

66. K. Hemminki and X. Li, *Br. J. Cancer*, 2004, **90**, 1765–1770.

67. M. Geddes, D.M. Parkin, M. Khlat, D. Balzi and E. Buatti (eds), *Cancer in Italian Migrant Populations*, No. 123, IARC, Lyon, 1993.

68. D. Forman, R.T. Oliver, A.R. Brett, S.G. Marsh, J.H. Moses, J.G. Bodmer, C.E. Chilvers and M.C. Pike, *Brit. J. Cancer*, 1992, **65**, 255–262.

69. T. Westergaard, J.H. Olsen, M. Frisch, N. Kroman, J.W. Nielsen and M. Melbye, *Int. J. Cancer*, 1996, **66**, 627–631.

70. K.P. Dieckmann and U. Pichlmeier, *Cancer*, 1997, **80**, 1954–1960.

71. D.J.A. Sonneveld, D.Th. Sleijfer, H. Schraffordt Koops, R.H. Sijmons, W.T. van der Graaf, W.J. Sluiter and H.J. Hoekstra, *Eur. J. Cancer*, 1999, **35**, 1368–1373.

72. F.P. Li, *Br. J. Cancer*, 1993, **68**, 217–219.

73. D.E. Goldgar, D.F. Easton, L.A. Cannon-Albright and M.H. Skolnick, *J. Natl. Cancer Inst.*, 1994, **86**, 1600–1608.

74. M.J. Khoury, T.H. Beaty and K.Y. Liang, *Am. J. Epidemiol.*, 1988, **127**, 674–683.

75. K. Heimdaland and S.D. Fossa, *World J. Urol.*, 1994, **12**, 178–181.

76. A.J. Swerdlow, B.L. De Stavola, M.A. Swanwick and N.E.S. Maconochie, *Lancet*, 1997, **350**, 1723–1728.

77. M.M. Braun, A. Alhbom, B. Floderus, L.A. Brinton and R.N. Hoover, *Cancer Causes Control.*, 1995, **6**, 519–524.

78. K. Heimdal, H. Olsson, S. Tretli, S.D. Fossa, A.L. Borresen and D.T. Bishop, *Br. J. Cancer*, 1997, **75**, 1084–1087.

79. E.A. Rapley, G.P. Crockford, D.F. Easton, M.R. Stratton and D.T. Bishop, *APMIS*, 2003, **111**, 128–135.

80. A.E. Rapley, G.P. Crockford, G. Teare, P. Biggs, S. Seal, R. Barfoot, S. Edwards, R. Hamoudi, K. Heimdal, S.D. Fossa, K. Tucker, J. Donald, F. Collins, M. Friedlandler, D. Hogg, P. Goss, A. Heidenreich, W. Ormiston, P.A. Daly, D. Forman, T.D. Oliver, M. Leahy, R. Huddart, C.S. Cooper, J.G. Bodmer, D.F. Easton, M.R. Stratton and D.T. Bishop, *Nat. Genet.*, 2000, **24**, 197–200.

81. I.S. Weidner, H. Møller, T.K. Jensen and N.E. Skakkebaek, *J. Urol.*, 1999, **161**, 1606–1609.

82. L. Fredell, L. Iselius, A. Collins, E. Hansson, S. Holmner, L. Lundquist, G. Lackgren, J. Pedersen, A. Sternberg, G. Westbacke and A. Nordenskjold, *Hum. Genet.*, 2002, **111**, 231–234.

83. A. Czeizel and J. Toth, *Teratology*, 1990, **41**, 167–172.

84. G. Fritz and A.E. Czeizel, *J. Reprod. Fertil.*, 1996, **106**, 63–66.

85. S.J. Harland, *Lancet*, 2000, **356**, 1455–1456.

86. M.D. Anway, A.S. Cupp, M. Uzumcu and M.K. Skinner, *Science*, 2005, **308**, 1466–1469.

87. A.E. Czeizel and K.J. Rothman, *Epidemiology*, 2002, **13**, 113–114.

88. R. Slama and H. Leridon, *Epidemiology*, 2002, **13**, 613–615.

89. A. Goriely, G.A.T. McVean, M. Röjmyr, B. Ingemarsson and A.O.M. Wilkie, *Science*, 2003, **301**, 643–646.

90. S. Zöllner, X. Wen, N.A. Hanchard, M.A. Herbert, C. Ober and J.K. Pritchard, *Am. J. Hum. Genet.*, 2004, **74**, 62–72.

91. M. Joffe, *Brit. Med. Bull.*, 2003, **68**, 47–70.

92. T.E. Schmid, A. Kamischke, H. Bollwein, E. Nieschlag and M.H. Brinkworth, *Hum. Reprod.*, 2003, **18**, 1474–1480.

93. T.E. Schmid, M.H. Brinkworth, F. Hill, E. Sloter, A. Kamischke, F. Marchetti, E. Nieschlag and A.J. Wyrobek, *Hum. Reprod.*, 2004, **19**, 1395–1400.

94. D.T. Carrell, B.R. Emery, A.L. Wilcox, B. Campbell, L. Erickson, H.H. Hatasaka, K.P. Jones and C.M. Peterson, *Arch. Androl.*, 2004, **50**, 181–185.

95. M.R. Guichaoua, J. Perrin, C. Metzler-Guillemain, J. Saias-Magnan, R. Giorgi and J.M. Grillo, *Hum. Reprod.*, 2005, **20**, 1897–1902.

96. D.T. Carrell, A.L. Wilcox, L. Lowy, C.M. Peterson, K.P. Jones, L. Erickson, B. Campbell, D.W. Branch and H.H. Hatasaka, *Obstet. Gynecol.*, 2003, **101**, 1229–1235.

97. J. Bartkova, E. Rajpert-de-Meyts, N.E. Skakkebaek, J. Lukas and J. Bartek, *APMIS*, 2003, **111**, 252–266.

98. A.M. Ottesen, J. Larsen, T. Gerdes, J.K. Larsen, C. Lundsteen, N.E. Skakkebaek and E. Rajpert-de-Meyts, *Cancer. Genet. Cytogenet.*, 2004, **149**, 89–97.

99. E. Rajpert-de-Meyts, H. Leffers, G. Daugaard, C.B. Andersen, P.M. Petersen, J. Hinrichsen, L.G. Pedersen and N.E. Skakkebaek, *Int. J. Cancer*, 2002, **102**, 201–204.

100. L. Frydelund-Larsen, P.H. Vogt, H. Leffers, A. Schadwinkel, G. Daugaard, N.E. Skakkebaek and E. Rajpert-de-Meyts, *Molecular Hum. Reprod.*, 2003, **9**, 517–521.
101. H. Klip, J. Verloop, J.D. van Gool, M.E. Koster, C.W. Burger and F.E. van Leeuwen, OMEGA Project Group, *Lancet*, 2002, **359**, 1102–1107.
102. M.M. Dobrzynska, A. Baumgartner and D. Anderson, *Mutagenesis*, 2004, **19**, 325–330.
103. M.H. Brinkworth, *Int. J. Androl.*, 2000, **23**, 123–135.

CHAPTER 5

Links between Paternal Smoking and Childhood Cancer

TOM SORAHAN

Institute of Occupational and Environmental Medicine, University of Birmingham, Edgbaston, B15 2TT, Birmingham, UK

5.1 Introduction

In 1992, Pershagen and colleagues reported the results of a large prospective study ($n = 497,051$ births) concerning maternal smoking during pregnancy and risks of childhood cancer.[1] A total of 327 cancers were identified and the overall risk for mothers reporting smoking during pregnancy was 0.99 (95% confidence interval (CI) 0.78–1.27). The maximum follow-up age in this study, however, was only 5 years and the authors noted that further studies could be carried out to examine any risks relating to later ages at presentation of disease. The report referenced 16 epidemiological studies (including 14 case-control studies) concerning cancer risks in children of mothers who smoked during pregnancy and summarised the evidence from these studies as inconclusive.

A further review of literature on childhood cancer risks and parental use of tobacco was published in 1994 and concluded that "the associations between maternal smoking during pregnancy and childhood cancer have been studied intensively, but there is no clear association overall, or for specific sites".[2] The review also summarised information on parental smoking from 13 case-control studies. Many of these studies were small in size and a total of only 1953 childhood cancers (various diagnostic groups) formed the combined case series. The review concluded that "no clear associations have been identified". In the 1990s, the current author embarked on a series of analyses of this topic making use of the historical case-control study known as the Oxford Survey of Childhood Cancers (OSCC).

5.2 The Oxford Survey of Childhood Cancers

The OSCC is an historical nationwide case-control study into the aetiology of childhood cancers.[3,4] The survey began in Oxford in 1956, but has been located

51

at the University of Birmingham since 1975. The OSCC is one of the largest case-control studies in the history of medicine and in 1989 Bithell calculated that 73% of the available analytical epidemiological data on the topic of prenatal X-rays and childhood cancer is to be found in this single survey.[5]

The survey has sought to interview the parents (usually the mother) of all children dying of cancer (including leukaemia) before their sixteenth birthday in England and Wales, and Scotland in the period 1953–1984. If the parents of a case child were willing to cooperate with the survey, a 'control list' of six healthy control children, matched for sex and date of birth, was selected from the birth register of the local authority area in which the case parents were living at the time of the interview (for 98% of interviewed case children, this was the local authority area in which the case child died). Control parents were contacted in turn until one family agreed to be interviewed. In later years, the case and control parents within each matched pair were always interviewed by the same person, often a physician or nurse from the local health authority. A number of standard questionnaires, covering a wide range of social and medical topics, have been used during the course of this prolonged study. Parental smoking data were not collected for all years of the survey but were available for deaths and matching controls in three periods: 1953–1955, 1971–1976 and 1977–1981.

Findings for these three periods are summarised in the order that they were first reported.

5.2.1   OSCC Data: 1977–1981

There were 3364 childhood cancer deaths in England and Wales for the period from 1977 to 1981.[6] Interview data were obtained from the parents of 1816 (54%) of these children, and for 1641 case children interview data were also available from the parents of healthy control children (1641 matched pairs, or 49% of all cases). Parents of 519 case children had refused to participate with the survey; current address had not been found for a further group of 360 case parents. Of the remaining case parents, 669 had not replied to survey requests, their general practitioner (GP) had advised the survey against approaching them, or arrangements to carry out interviews had fallen through. The response range from case parents approached was thus 61%.

The interview folders of all 1641 matched pairs were reviewed both to abstract full information on smoking habits and parental alcohol consumption. Two standard questionnaires were used for these matched pairs. The first questionnaire sought information from both parents on 'smoking habits' before the relevant pregnancy. Information was to be supplied in terms of 'daily quantity' and 'date of starting'. The second questionnaire used the same wording except that 'before relevant pregnancy' was implied rather than stated and information on any maternal changes in smoking habits (and alcohol consumption) during the relevant pregnancy was sought in terms of same, less or more. In addition, a pre-interview form was sent to the parents who agreed to participate in the survey which asked, "Do you smoke? If the answer is yes, please say how much each day." This question was directed at current rather than past smoking habits.

Information on parental smoking habits before the pregnancy of the survey child was re-abstracted in terms of daily consumption of cigarettes, use of pipe or cigars and duration of cigarette smoking. Ex-smokers were defined as parents who stopped smoking at least 2 years before the survey child was born. For this analysis (and all OSCC analyses), when the 'daily quantity' was reported with upper and lower values, the upper value was selected. Birth weight data were re-abstracted for each child; weights previously coded to the nearest pound were recoded as weights in ounces. Also, birth weights obtained from clinic records were allowed to take precedence over the weights given by mothers.

In all of the OSCC reports, matched pairs in which the case child was adopted were excluded, and case and control data relating to tobacco consumption were compared (with and without adjustment for other variables) by means of (multiple) conditional logistic regression using the EGRET programme. Smoking habits of mothers and fathers were analysed separately, simultaneously, and simultaneously with additional adjustment for other variables. The purpose of the simultaneous analyses was to allow for the effects of other variables, so that the independent effects of each smoking habit could be examined. The odds ratio was used to obtain estimates of relative risk (RR). Risks are shown relative to a baseline risk of unity for the non-smokers.

A summary of the main findings for paternal cigarette smoking is shown in Table 1. A highly significant trend ($p < 0.001$) is shown for childhood cancer risk in relation to daily consumption of cigarettes. Similar findings were not obtained for maternal smoking and the paternal findings were little influenced by simultaneous adjustment for maternal smoking, social class and maternal age at the birth of the child.

5.2.2 OSCC Data: 1953–1955

Preliminary analyses of these data (the first 3 years of data collection) were published many years ago, but the amount of smoking was not considered.[3] There were 1952 childhood cancer deaths in England, Wales and Scotland for the period 1953–1955.[7] Interview data had been obtained from the parents of 1631 (84%) of these children. Parents of 112 case children had refused to participate with the survey, a further group of 94 case parents had moved abroad or to an unknown address, and the remaining 115 case parents had not replied to survey request, their GP had advised the survey not to approach them, or arrangements to carry out interviews that had fallen through. The response rate from case parents approached was thus 88%. Some 16% of the interviewed case parents had moved local authority area between the birth and the death of the survey child.

Interview data were obtained for 1622 control children (907 first choices, 342 second choices and 373 later choices). Only 56% of first choices may seem a low percentage but the birth registers from which the controls were selected had

Table 1 *Childhood cancer risks in relation to paternal smoking: selected findings from the OSCC and IRESCC surveys[a]*

Paternal smoking[b]	Cases	Controls	RR	95% CI	p value for trend
OSCC data: 1977–1981 deaths					
Non-smoker	632	732	1.0		
< 10 cpd	53	52	1.20	0.81–7.78	
10–19 cpd	190	181	1.24	0.98–1.56	
20–29 cpd	402	375	1.26*	1.05–1.50	$p < 0.01$
30–39 cpd	133	114	1.35*	1.03–1.78	
≥ 40 cpd	102	81	1.47*	1.07–2.01	
OSCC data[c] 1953–1955 deaths					
< 1 cpd	263	302	1.0		
1–9 cpd	356	409	0.99	0.80–1.24	
10–20 cpd	677	623	1.26*	1.03–1.55	$p < 0.001$
> 20 cpd	203	170	1.38*	1.08–1.79	
OSCC data: 1971–1976 deaths					
Non-smoker	1008	1179	1.0		
1–9 cpd	118	139	0.99	0.77–1.29	
10–19 cpd	326	289	1.33**	1.11–1.60	
20–29 cpd	579	533	1.30***	1.12–1.51	$p < 0.001$
30–39 cpd	157	133	1.43**	1.12–1.84	
≥ 40 cpd	144	105	1.62***	1.24–2.11	
IRESCC data[d] 1980–1983 diagnoses					
Non-smoker	184	218	1.0		
< 10 cpd	26	34	0.94	0.53–1.66	
10–19 cpd	79	60	1.63*	1.10–2.41	
20–29 cpd	143	122	1.46*	1.05–2.03	$p = 0.02$
30–39 cpd	23	32	0.95	0.52–1.73	
≥ 40 cpd	28	21	1.77	0.94–3.34	

Notes: *$p<0.05$, **$p<0.01$, ***$p<0.001$.
[a] See acknowledgments for details for permissions from original publishers.
[b] See text for detail on paternal smoking variables.
[c] Also includes 'corresponding' use of pipe or cigars.
[d] Findings only shown for GP controls, not for hospital controls.

been compiled, on average, 6 or 7 years before the survey began and about 25% of the required control families were found to have definitely left the district; only 6% of the control mothers approached refused to cooperate with the survey (Stewart *et al.*)[3]. For some 94% of the 1622 matched pairs, the case and control parents within each pair were interviewed by the same person, usually a physician or nurse from the local health authority. For the remaining matched pairs, the case parents had moved locality between the death of the child and the time of the interview, and different interviewers were used for case and control parents.

The interview folders of all matched pairs were reviewed and information on parental use of tobacco was abstracted and amalgamated with existing study computer files. The interview questionnaire requested responses in terms of 'nil', 'slight', 'moderate' or 'heavy'; definitions of these terms had been supplied to the interviewers. The question was directed at current rather than past smoking habits, and for fathers, responses could refer to use of cigarettes or pipe tobacco.

A summary of the main findings in this second study for paternal smoking is also shown in Table 1. A highly significant trend ($p < 0.001$) is shown for childhood cancer risk in relation to daily tobacco consumption. Similar findings were also obtained for maternal smoking, although some of the maternal effect could be explained as confounding from the paternal habit (not shown in Table). The point estimates of risk for levels of paternal smoking habit were increased slightly following simultaneous adjustment for maternal smoking, social class, sibship position and maternal age at the birth of the child.

5.2.3 OSCC Data: 1971–1976

There were 5111 childhood cancer deaths in England, Wales and Scotland for the period 1971–1976.[8] Interview data had been obtained from the parents of 2933 (57%) of these children. Parents of 819 case children had refused to participate with the survey, a further group of 428 case parents had moved abroad or to an unknown address, and the remaining 931 case parents had not replied to survey requests, their GP had advised the survey not to approach them or arrangements to carry out interviews had fallen through. The overwhelming majority of the last group of case parents had not replied to survey requests; the response rate from case parents approached was thus at least 63%. Some 25% of the interviewed case parents had moved local authority area between the birth and death of the survey child. Some 97% of the interviews with case parents took place before the fourth anniversary of the death of the child (median interval, 21 months). The median interval between the birth of the case child and the parental interview was 8 years.

Interview data were obtained for 2628 control children (1371 first choices, 472 second choices and 785 later choices). (Control interviews were not obtained for 305 case children with interview data; these cases did not feature in the analysis.) Only 52% of first choices is a relatively low percentage but the birth registers from which the controls were selected had been compiled, on average, some 8 or 9 years before the interviews were arranged.

The interview folders of all matched pairs were reviewed and information on parental use of tobacco was re-abstracted and amalgamated with existing study computer files. A pre-interview form (postal questionnaire) had been sent to those parents (cases and controls) who agreed to participate in the survey which asked, "Do you smoke? If yes, please say about how much each day". The main interview questionnaire requested information on 'smoking' in terms of 'daily quantity'; the question was also directed at current rather than past smoking habits. Information was abstracted in terms of daily consumption of cigarettes, use of pipe and use of cigars. Given that all the smoking questions were directed at current habits, there was no requirement for ex-smokers to identify themselves. A small number did so, and for these analyses, ex-smokers were defined as parents who stopped smoking at least 2 years before the survey child was born (23 mothers and 37 fathers). Other ex-smokers were included with the smokers (*i.e.* smokers in the 2 year period before birth of the survey child). A response limited to ounces of tobacco was assumed to relate to a pipe smoker. A total of 79 mothers and 208 fathers were reported to be smokers but no information on daily consumption was supplied. The smoking questions were left unanswered for a further group of 24 mothers and 219 fathers; most of these fathers were not living with their children. Birth weight data were re-abstracted for each child as described previously.

A summary of the main findings for paternal cigarette smoking is also shown in Table 1. A highly significant trend ($p < 0.001$) is shown for childhood cancer risk in relation to daily consumption of cigarettes. Similar findings were not obtained for maternal smoking and the paternal findings were little influenced by simultaneous adjustment for maternal smoking, social class and maternal age at the birth of the child.

5.3   Inter Regional Epidemiological Survey of Childhood Cancers

Following these three OSCC reports, attempts were made to identify other existing studies that could be used to cast further light on the topic of parental smoking and childhood cancer risks. The Inter Regional Epidemiological Survey of Childhood Cancers (IRESCC) had been established to investigate the role of possible aetiological factors in childhood cancer with particular emphasis on environmental exposures to the foetus and family history of diseases.[9] Study design, control selection and data collection procedures have been published in some considerable detail.[10] The survey sought to interview the parents of all 761 children resident in the Yorkshire, West Midlands and North Western Regional Health Authority areas who were first diagnosed with malignant disease before their fifteen birthday; diagnoses relate to the period January 1980 to January 1983.[11] Children who were not living with their natural mother were excluded and a random sample of certain types of cancer was excluded to reduce the workload. Of the 615 cases eligible for interview, parents of 19 cases were not approached on the advice of their GP or consultant and parents of 41 cases declined to take part; interview data were obtained for 555 cases. It thus proved possible to approach most case parents soon after their children had been diagnosed with cancer.

For each child with interview data, interview data were sought for two control children matched for sex and date of birth. One set of potential controls was selected from the practice lists of the case GPs, a second set of potential controls was selected from lists of acute surgical and accident cases from six large hospitals; hospital controls were drawn from hospitals in the same region as their respective cases. Control parents from each list were contacted in turn until one control family agreed to be interviewed. Interview data were obtained for 555 GP controls (400 first choices (72%), 111 second choices (20%) and 44 later choices (8%)). Interview data were obtained for 555 hospital controls

(355 first choices (64%), 122 second choices (22%) and 78 later choices (14%)). Participation rates for approached parents were about 97% for cases, 74% for GP controls and 64% for hospital controls. Both parents were present at the interview for 59% of the cases, 50% of the GP controls and 50% of the hospital controls. Interviews were carried out by a small number of trained interviewers and all parents in any given case-control set were always interviewed by the same person.

The micro-filmed interview records of all study subjects were reviewed and information on parental cigarette smoking habits was re-abstracted; the IRESCC computer files developed in the 1980s were in a machine-specific format not compatible with computers currently in use. The interview sought information for mothers on the question "Did you smoke before and/or during your pregnancy?", and information for fathers on the question "Do you smoke or have you ever smoked?" Positive responses for both parents were to be given in terms of 'type of product', 'quantity and frequency' and 'dates'. Some interviewers collected detailed smoking histories with 'dates' given in terms of ages (*e.g.* 10 cigarettes per day (cpd) at age 17–19, 10–20 cpd at ages 19–24, gave up when pregnancy was confirmed). Other interviewers collected summary information (*e.g.* before pregnancy 20–25 cpd, during pregnancy 10 cpd). All available information on consumption of cigarettes was computerised in text form. The coding system applied to this analysis was that used in the earlier OSCC reports so that when the daily consumption of cigarettes was reported with upper and lower values, the upper value was selected. Parental age at the time of conception was calculated and the relevant daily smoking habits at this age were evaluated (smoking before the pregnancy). The microfilm for one hospital control was not found. The smoking questions were left unanswered for one hospital control mother, 29 case fathers, 17 GP control fathers and 30 hospital control fathers; most of these fathers were not living with their children. Birth weight data were also re-abstracted for each child as described previously.

A summary of the main findings for paternal cigarette smoking is shown in Table 1, from comparisons of cases and GP controls. A significant trend ($p = 0.02$) is shown for childhood cancer risk in relation to daily consumption of cigarettes. Similar findings were not obtained for maternal smoking and the paternal findings were little influenced by simultaneous adjustment for maternal smoking, social class and maternal age at the birth of the child. There were in fact significant differences in the responses of GP controls and hospital controls and there was no support for the paternal smoking hypothesis when cases were compared to hospital controls.

5.4 Other Studies

Many other studies are available on the topic of childhood cancer risks and parental smoking, although some of these studies are small in comparison with the OSCC data. Some other large studies are available including two

US case-control studies of childhood leukaemia.[12,13] Neither study provides any support for the paternal smoking hypothesis. Similarly negative overall findings were also obtained from the recent UK Childhood Cancer Study (UKCCS),[14] although positive findings were obtained from a smaller Chinese case-control study.[15] The UKCCS study did report positive findings for risks of hepatoblastoma in relation to smoking by both parents; a similar result has also been obtained from the OSCC.[16]

5.5   Discussion

The OSCC and IRESCC studies provide evidence of an association between the smoking of cigarettes by fathers and cancer in their offspring; the smoking of cigarettes by mothers can, with some confidence, be excluded as an important risk factor for the generality of childhood cancers.

If the paternal smoking association is causal in nature, this might be due either to pre-conception effects or to the effects of passive smoking on young infants or both. A passive smoking effect seems unlikely because of the weight of evidence against maternal smoking being a risk factor for childhood cancers; it might be imagined that, in general, the infant has more contact with passive smoke from the mother than from the father. A pre-conception effect may not be biologically implausible and evidence for potential mechanisms has been reviewed.[17-19]

There is no reason to believe that any risk presented by paternal smoking before conception would only affect one type of childhood cancer. The combined OSCC and IRESCC reports suggest that the risk factor may be operating across the spectrum of childhood cancers. Risks may be more pronounced for lymphomas and neuroblastomas, although much of the variation in the ranking of site-specific risks from study to study may represent no more than chance fluctuations. It does not follow, of course, that each and every subtype of childhood cancer is necessarily affected by paternal smoking.

The paternal results are most unlikely to be due to chance because in each of the three relevant OSCC studies, trends with smoking habit have been highly significant. Confounding also presents an unlikely sole explanation. Those potential confounders that have been considered (social class, age of father, 'family mobility', sibship position) had little effect on the paternal smoking findings and the use of alcohol can be excluded on the basis of previous work.[6,15] If an unknown variable was confounding the paternal smoking effect, it would need, by definition, to be associated with higher risks than paternal smoking, both for point estimates of RR and for attributable risk. The confounder would, therefore, need to be responsible for some 15% (or more) of all childhood cancers; an unusual occupational exposure would not, therefore, provide a likely candidate.

One key issue in evaluating the importance of these findings is the reliability of OSCC data. For the data relating to mothers' smoking habits there was one successful test of their reliability, namely a demonstration of the inverse relation of smoking habit with birth weight. For the father and smoking habits there was no similar test. Other issues need to be considered. Some of the OSCC studies are limited by modest response rates, and the effects of having to ignore the non-responders are not known. The method of selecting OSCC controls means that 'mobile' families tend to be under-represented in the control series although analyses restricted to 'non-mobile' families suggested that this feature of control selection was not an important issue for this analysis. The case series in this study comprised childhood cancer deaths rather than all incident cases, and some improvement in survival rates past the age of 16 years did take place in the later survey years. The inclusion of childhood cancer survivors could have led to materially different results if paternal smoking only increased mortality rates in children diagnosed with cancer. It would be difficult to maintain such a hypothesis given that the paternal smoking findings were reasonably consistent across calendar periods. Before these analyses had been carried out, it was predicted that a paternal smoking effect would be more pronounced for younger ages at presentation of childhood cancer, and that bias would offer an unlikely explanation for such a finding. No evidence of such an effect was found. It could be argued that the paternal findings in the new series merely reflect changes in paternal smoking brought on by the death of a child. However, a change in alcohol consumption would seem even more likely, and as mentioned above, there is no evidence for a paternal alcohol effect. Caution is still required, however, in interpreting these findings because it is not possible to exclude all potential biases from the findings, and there is a lack of consistency in the available literature.

More information on the subject is required. The paternal smoking data available to many case-control studies of childhood cancer have not yet been fully analysed and reported. Even more useful would be the results of investigations of cancer in the offspring of subjects whose smoking habits were collected in contexts other than case-control studies.

Acknowledgments

I would like to thank the many researchers who have worked with me on the OSCC and IRESCC studies. Material from ref 6 is reprinted with permission of Elsevier Ltd (the publishers of the Annals of Epidemiology) and material from refs 7, 8, 11 and 16 are reprinted with permission of Churchill Livingstone (the publishers of the British Journal of Cancer). I also thank Margaret Williams for word processing.

References

1. G. Pershagen, *Int. J. Epidemiol.*, 1992, **21**, 1.
2. J. Tredaniel, *Paediatr. Perinat. Epidemiol.*, 1994, **8**, 233.
3. A.M. Stewart, *Br. Med. J.*, 1958, **1**, 1495.
4. E.A. Gilman, *J. Soc. Radiol. Prot.*, 1988, **8**, 9.

5. J.F. Bithell, in *Low dose Radiation: Biological Bases of Risk Assessment*, K.F. Baverstock and J.W. Stather (eds), Taylor and Frances, London, 1989, 77.

6. T. Sorahan, *Ann. Epidemiol.*, 1995, **5**, 354.

7. T. Sorahan, *Br. J. Cancer*, 1997, **75**, 134.

8. T. Sorahan, *Br. J. Cancer*, 1997, **76**, 1525.

9. R.A. Cartwright, *Lancet*, 1984, **ii**, 999.

10. J.M. Birch, *Br. J. Cancer*, 1985, **52**, 915.

11. T. Sorahan, *Br. J. Cancer*, 2001, **84**, 141.

12. R.K. Severson, *Cancer Epidemiol. Biomarkers Prev.*, 1993, **2**, 433.

13. J. Brondum, *Cancer*, 1999, **85**, 1380.

14. D. Pang, *Br. J. Cancer*, 2003, **88**, 373.

15. B.T. Ji, *Natl. Cancer Inst.*, 1997, **89**, 238.

16. T. Sorahan, *Br. J. Cancer*, 2004, **90**, 1016.

17. A.J. Wyrobek, *Reprod. Toxicol.*, 1993, 7(Suppl 1), 3.

18. A.J. Wyrobek, *Mutat. Res.*, 1996, **352**, 173.

19. A.A. Woodall, in *Preventive Nutrition: The Comprehensive Guide for Health Professionals*, A. Bendich and R.J. Deckelbaum (eds), Humana Press, Totowa, NJ, USA, 1997.

CHAPTER 6

Feasibility Study of Metal Effects on the X:Y Ratio in Human Sperm

WENDIE A. ROBBINS,[a] KAREN E. YOUNG,[a] FUSHENG WEI[b] AND THE BORON EPIDEMIOLOGY RESEARCH GROUP[a,b]

[a] University of California, Los Angeles, California, USA
[b] China National Environmental Monitoring Station, Beijing, China

6.1 Introduction

Direct effects of metals on human male reproductive health have been investigated and reported for years, primarily through correlation of occupation with conventional semen parameters or fertility. Additionally, a number of male reproductive health studies have evaluated levels of metals in human blood, whole semen, seminal plasma, or sperm cells (Table 1). Fewer studies have been published on the effects of metal exposures on human sperm DNA.[1–8] Given extensive use of metals in modern society, it is reasonable to investigate not only their potential to affect male fertility but also their potential to induce changes in sperm DNA that might be transmissible to offspring and to explore potential modifiers of these relationships.

One way to address the important question of the contribution of metal exposures to male mediated effects on human offspring is to measure metal content in blood or semen and look for correlations with adverse changes in sperm DNA. Multiple DNA/chromatin assays are available to measure specific toxic effects on sperm,[9] some of which effects are capable of being transmitted to offspring. In this chapter we describe our research in NE China that explored levels of seven different metals in blood and semen across five different occupational categories and looked for associations with the X:Y ratio in sperm.

6.2 Research Methods

In 2002, a study group of 1185 male workers living in NE China gave human-subjects' consent to participate in a research study investigating occupational

Table 1 Selected male reproductive health studies reporting metals in human blood and semen

Metal	Method of analysis	Concentration in semen (µg mL⁻¹)	Concentration in seminal plasma (µg mL⁻¹)	Concentration in spermatozoa (µg kg⁻¹)	Concentration in blood (µg mL⁻¹)	Reference
Ca	ICP-OES	245 ± 79 (fertile) 264 ± 125 (infertile)[a]				Umeyama et al.[13]
	ICP–MS		167 ± 44 (normo.)[b] 202 ± 58 (oligo.) 179 ± 50 (severe oligo.)			Abou-Shakra et al.[15]
	ICP–MS		195 ± 63 (azo.) 140 ± 59 (Pb-exposed) 155 ± 51 (unexposed)	285,000 ± 127,000 (Pb-exposed) 432,000 ± 109,000[a] (unexposed)		Apostoli et al.[16]
	Flame AAS		533 (450; 672) (healthy men, S-TTP)[c] 470 (391; 541) (healthy men, L-TTP)			Sorensen et al.[21]
Cr	ICP–MS ICP–OES	352.13 ± 124 0.115 ± 0.198 (fertile) 0.120 ± 0.207 (infertile)[a]			98.6 ± 4.7	Current study Umeyama et al.[13]
	ICP–MS		0.32 ± 0.17 (normo.)[b] 0.55 ± 0.29 (oligo.) 0.51 ± 0.32 (severe oligo.) 0.41 ± 0.20 (azo.)			Abou-Shakra et al.[15]

Metal	Method	Concentration				Reference
	ICP–MS		0.0012 ± 0.0008 (Pb-exposed) 0.0011 ± 0.0009 (unexposed)	10 ± 8.3 (Pb-exposed) 9 ± 7.5 (unexposed)		Apostoli *et al.*[16]
	ICP–MS				0.131 ± 0.0526 (welders) 0.0174 ± 0.0089 (controls)	Danadevi *et al.*[22]
Cu	ICP–MS AAS	0.085 ± 0.047	{0.030–0.200} (suspected infertile)	{1430–11,000} (suspected infertile)	0.121 ± 0.016	Current study Pleban and Mei[23]
	ICP–OES	0.034 ± 0.070 (fertile) 0.074 ± 0.137 (infertile)[a]				Umeyama *et al.*[13]
	ICP–MS		0.17 ± 0.06 (normo.)[b] 0.16 ± 0.07 (oligos.) 0.18 ± 0.07 (severe oligo.) 0.17 ± 0.04 (azo.)			Abou-Shakra *et al.*[15]
	ICP–MS		0.085 ± 0.039 (Pb-exposed) 0.077 ± 0.033 (unexposed)	906 ± 631 (Pb-exposed) 690 ± 372 µg kg⁻¹ (unexposed)		Apostoli *et al.*[16]
Mg	ICP–MS ICP–OES	0.222 ± 0.247			0.912 ± 0.121	Current study
		78.9 ± 33.3 (fertile) 77.5 ± 49.8 (infertile)[a]				Umeyama *et al.*[13]
	AAS ICP–MS		103.5 ± 49.2 (donors) 54.4 ± 33.6 (normo.)[b] 59.5 ± 38.2 (oligo.) 60.0 ± 43.1 (severe oligo.) 59.2 ± 32.0 (azo.)			Saaranen *et al.*[24] Abou-Shakra *et al.*[15]

Table 1 *(continued)*

Metal	Method of analysis	Concentration in semen (μg mL^{-1})	Concentration in seminal plasma (μg mL^{-1})	Concentration in spermatozoa (μg kg^{-1})	Concentration in blood (μg mL^{-1})	Reference
	ICP-MS		80 ± 89 (Pb-exposed) 72 ± 27 (unexposed)	170,000 ± 46,000 (Pb-exposed) 150,000 ± 42,000 (unexposed)		Apostoli et al.[16]
	ICP-MS		86 (57; 134) (healthy men, S-TTP)[c] 100 (56; 118) (healthy men, L-TTP)			Sorensen et al.[21]
Se	ICP-MS AAS GFAAS	125.2 ± 5.5	0.0288 ± 0.0095 (donors) 0.0715 ± 0.0198 (donors)		21.88 ± 2.0	Current study Pleban and Mei[23] Saaranen et al.[24]
	AAS				0.1636 ± 0.0281 (donors)	Xu et al.[12]
	ICP-MS		0.056 ± 0.021 (Pb-exposed) 0.079 ± 0.025[a] (unexposed)	171 ± 0.1 (Pb-exposed) 313 ± 107[a] (unexposed)		Apostoli et al.[16]
	AAS		0.0513 (non-smokers) [95% CI: 0.0261–0.1007]			Xu et al.[7]
Sr	ICP-MS ICP-OES	0.08 ± 0.12 0.063 ± 0.020 (fertile) 0.078 ± 0.116 (infertile)[a]			0.108 ± 0.019	Current study Umeyama et al.[13]
	ICP-MS	0.091 ± 0.060			0.048 ± 0.013	Current study

Zn			
Pleban and Mei[23]	AAS	{8–270) (suspected infertile)	{911,000–4,205,000} (suspected infertile)
Madding et al.[14]	AAS	125 ± 71 (fertile) 116 ± 53 (infertile) 124 ± 54 (fertile) 129 ± 72 (infertile)^a	
Umeyama et al.[13]	ICP–OES		
Saaranen et al.[24] Abou-Shakra et al.[15]	AAS ICP–MS	141.0 ± 71.7 (donors) 105 ± 53 (normo.)^b 93 ± 51 (oligo.) 117 ± 63 (severe oligo.) 129 ± 60 (azo.)	
Xu et al.[12]	Ion chromatography	6.2 ± 1.3 (donors)	202.1 ± 96.7 (donors)
Apostoli et al.[16]	ICP–MS	88 ± 41 (Pb-exposed) 96 ± 48 (unexposed)	215,000 ± 124,000 (Pb-exposed) 320,000 ± 136,000^a (unexposed)
Sorensen et al.[21]	ICP–MS	106 (72; 183) (healthy men, S-TTP)^c 113 (68; 212) (healthy men, L-TTP)	
Benoff et al.[25]	GFAAS	50.800 ± 24.485 (donors)	
Current study	ICP–MS	192.04 ± 92.07	0.981 ± 0.195

Notes: $^a p \leq 0.05$; methods are abbreviated as ICP-OES, inductively coupled plasma–optical emission spectrometry; ICP–MS, inductively coupled plasma–mass spectrometry; AAS, atomic absorption spectrometry; GFAAS, graphite furnace atomic absorption spectrometry; and flame AAS, flame atomic absorption spectrometry.
Values are listed as mean ± SD; median (25th percentile; 75th percentile); [range]; or geometric mean [95% CI].
^{a} Pooled infertile groups (#2–5).
^{b} Normo.= $n > 40 \times 10^6$ sperm mL^{-1}; Oligo. =$10 < n < 40 \times 10^6$ sperm mL^{-1}; Severe Oligo. = $n < 10 \times 10^6$ sperm mL^{-1}; and Azo. ($n = 0$).
^{c} S-TTP = short time-to-pregnancy; L-TTP = long time-to-pregnancy.

exposures and male reproductive health. Questionnaires were administered to collect demographic information, work history, diet and lifestyle exposures, general health history, and reproductive history. Based on findings from the questionnaires, in the summer of 2003, samples of blood, semen, and urine were collected and analyzed from a targeted subset of workers ($n = 69$ men). Samples were also collected in 2004 ($n = 203$ men). Questionnaire data were updated for each man who had been enrolled in 2002 if he participated in the biological sampling conducted during 2003 and 2004. As some new men consented and were enrolled for biological sampling in 2003 and 2004 that were not in the original interviewed cohort of 1185 subjects, the final study group total was 1376 men.

Biological samples were coded in the field when collected so that laboratories conducting the metal and semen analyses were blinded as to occupational or other exposures to the participants. Blood serum and whole semen were analyzed for metals in a subset of 203 participants. Metals assayed included B, Ca, Mg, P, Cr, Cu, Sr, Zn, and Se by inductively coupled plasma–mass spectrometry (ICP–MS) or inductively coupled plasma–atomic emission spectrometry (ICP–AES). Results for boron in blood and semen of men working in the boron industry are reported elsewhere leaving 129 non-boron industry workers for the analyses reported here. Selected characteristics of the 129 men are shown in Table 2. In addition to metals, an immuno-chemoluminescent assay for blood nicotine level (Immunlite 1000, DPC, New Jersey, USA) was conducted to determine exposure to cigarette smoke because greater than 92% of the total 1376 participants reported exposure to cigarette smoke through personal smoking or environmentally at home, work, or both. Semen samples from a subset of men matched on smoking, alcohol, and age were evaluated for proportion of X *vs.* Y bearing sperm cells. Hybridization methods were as described in Robbins *et al.*[10] and 1000 sperm cells were scored per subject for X *vs.* Y bearing sperm. One person scored all the slides. The scorer was blinded to occupational group or other information about the study subjects.

The study population was divided into five different occupational categories clustered around potential toxic exposures. The potential exposures were assessed during walk-through observations by the research team in collaboration with industrial hygienists either employed at the worksite or at the local government office in the area of the worksite. The five groups were agricultural workers including farmers and forestry workers; mechanics and truck drivers; production workers inside a factory working on production lines; professional and office workers including teachers, policemen, post-office workers, and government officials; and raw ore workers from marble and coal mines.

Summary statistics, data transformation, and statistical testing were done using SAS software version 8.2.[11]

6.3 Findings

Levels of metals in blood and semen of the participants in the current study are reported in Table 1. Blood values were within the range reported in

Table 2 *Characteristics of the subset of 129 workers evaluated for metals in blood and semen*

Characteristics	Work category	n	Mean	Std. dev	Min.	Max.	p-value
	Mechanics/drivers	41	201.01	206.89	10	500	
	Professional/office	16	95.13	150.71	10	500	
Blood nicotine (ng mL^{-1})	Raw ore workers	10	287.90	223.42	10	500	0.0067
	Production	16	159.16	188.11	10	500	
	Agriculture	44	281.51	199.74	10	500	
	Mechanics/drivers	41	31.80	6.51	20	41	
	Professional/office	16	32.81	5.13	24	40	
Age (years)	Raw ore workers	10	31.10	6.74	23	40	0.0720
	Production	16	29.50	5.15	18	39	
	Agriculture	46	33.91	4.58	23	41	
	Mechanics/drivers	41	24 (58.54%)				0.0637
	Professional/office	16	7 (43.75%)				
Smoker=yes	Raw ore workers	10	7 (70.00%)				
	Production	16	8 (50.00%)				
	Agriculture	46	36 (78.26%)				
	Mechanics/drivers	41	23 (56.10%)				0.2932
	Professional/office	16	5 (31.25%)				
Alcohol=yes	Raw ore workers	10	7 (70.00%)				
	Production	16	10 (62.50%)				
	Agriculture	46	24 (52.17%)				

literature from studies of male reproductive health except for blood zinc, which was lower in the current study group compared with the single other study available for comparison.[12] Values for semen zinc, however, were similar to those reported by two previous studies.[13,14] Metals in semen vary according to the specific fraction measured: seminal plasma, sperm cells, or whole semen. In the present study, whole semen was evaluated and in all cases where whole semen values were available in literature for studies of male reproductive health, the current study group fell within the ranges reported for healthy men except for copper. Men in the present study had higher levels of semen copper compared with levels reported by Umeyama *et al.*[13] using ICP–OES, but similar levels compared with seminal plasma concentrations reported by Abou-Shakra *et al.*[15] using ICP–MS. The relationship of metals in semen and blood to work category are depicted in Figures 1–7. (Selenium in

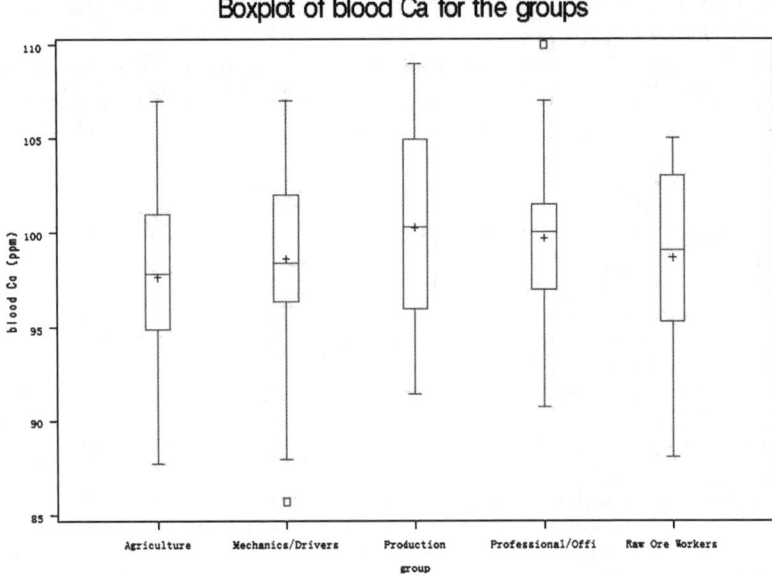

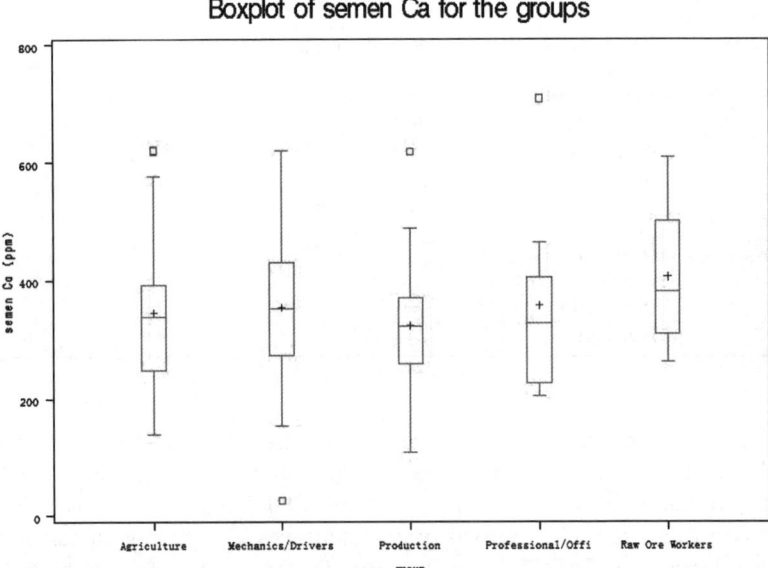

Figure 1 *Calcium in blood and semen (ppm) across five work categories.*

semen is not shown because 67% of the values were at or below the level of detection.)

Related to potential male mediated effects, questionnaire data indicated no statistically significant difference in ratio of male to female offspring across the five work categories although the number of births was small ($p = 0.77$

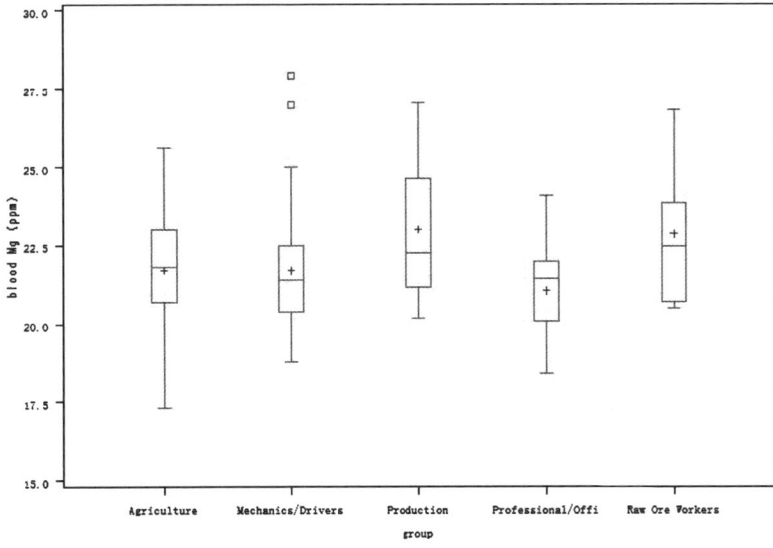

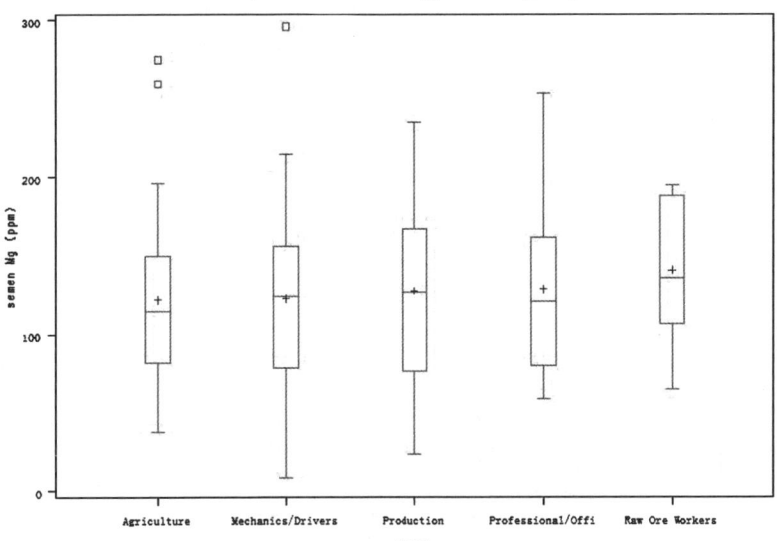

Figure 2 *Magnesium in blood and semen (ppm) across five work categories.*

based on 125 children). None of the five categories of work were found to have significant deviations from the expected ratio of X *vs.* Y bearing sperm cells in ejaculated semen specimens. Additionally, levels of metals in neither semen nor blood were significantly correlated with X to Y ratio in ejaculated sperm cells.

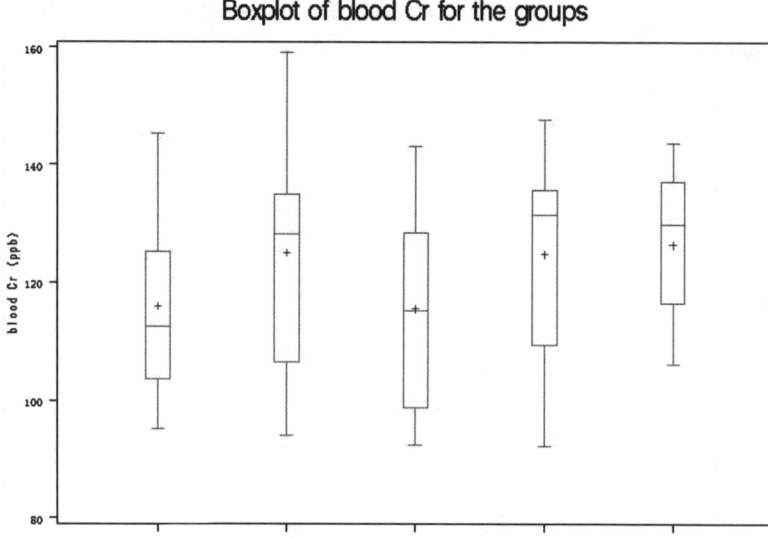

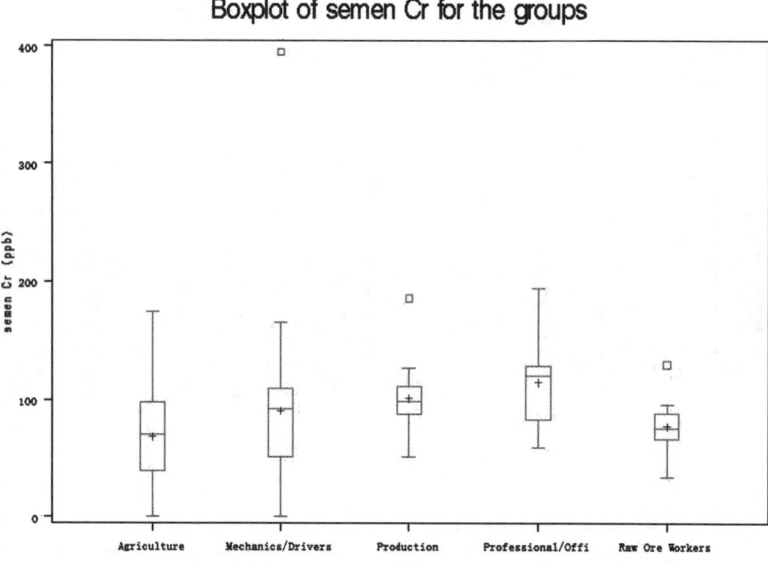

Figure 3 *Chromium in blood and semen (ppb) across five work categories.*

6.4 Discussion

Elemental determinations in blood, semen, or fractions thereof are useful in the study of reproductive health effects in men exposed to metals. Identifying the most appropriate matrix (seminal plasma, sperm cells, whole semen, or blood) is the first step to investigating the relationship between essential and toxic

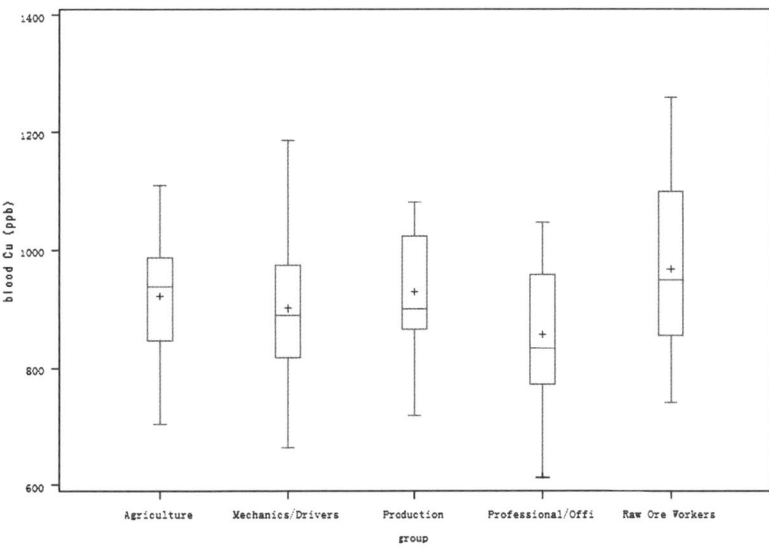

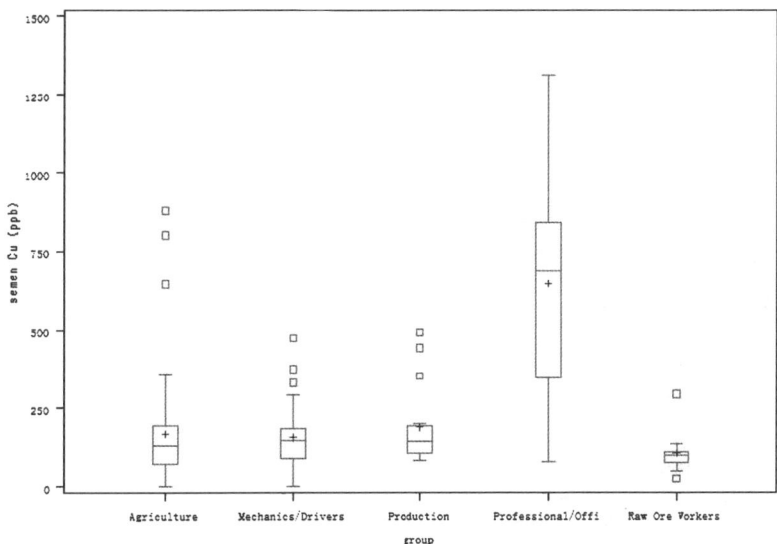

Figure 4 *Copper in blood and semen (ppb) across five work categories.*

metal concentrations since the source and magnitude of the elements may vary between matrices.[16] The most widely reported methods of analysis for studying metals in human blood and semen are atomic absorption spectrometry (AAS) and inductively coupled argon plasma (ICP) with optical emission spectrometry (ICP–OES), or mass spectrometry (ICP–MS). While considering semen and blood analysis, ICP has several advantages over AAS including

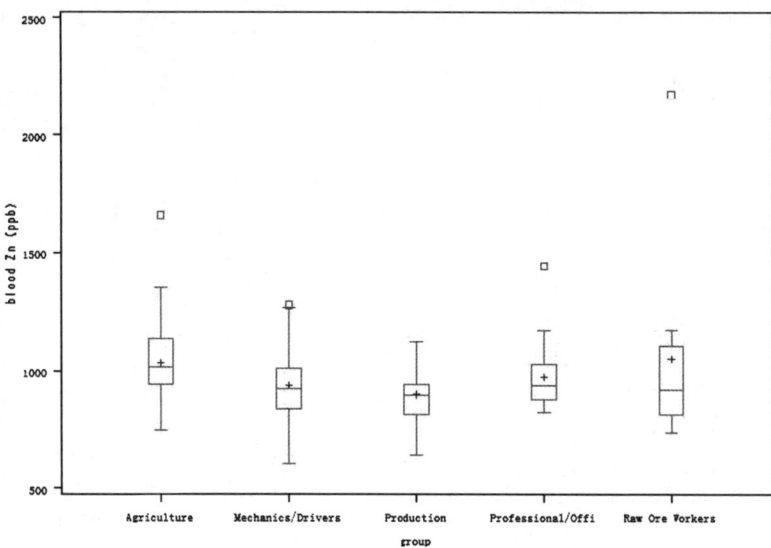

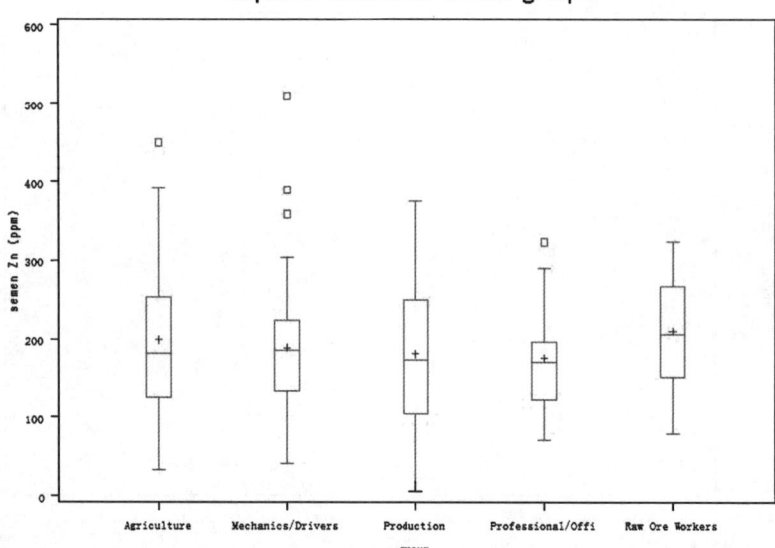

Figure 5   *Zinc in blood (ppb) and semen (ppm) across five work categories.*

simultaneous excitation and analysis of multiple elements[17] as would be expected to be present in human biological fluids, more complete breakdown of chemical compounds due to the increased temperatures resulting in decreased matrix effects,[18] and detection over a linear dynamic range of six orders of magnitude.[19]

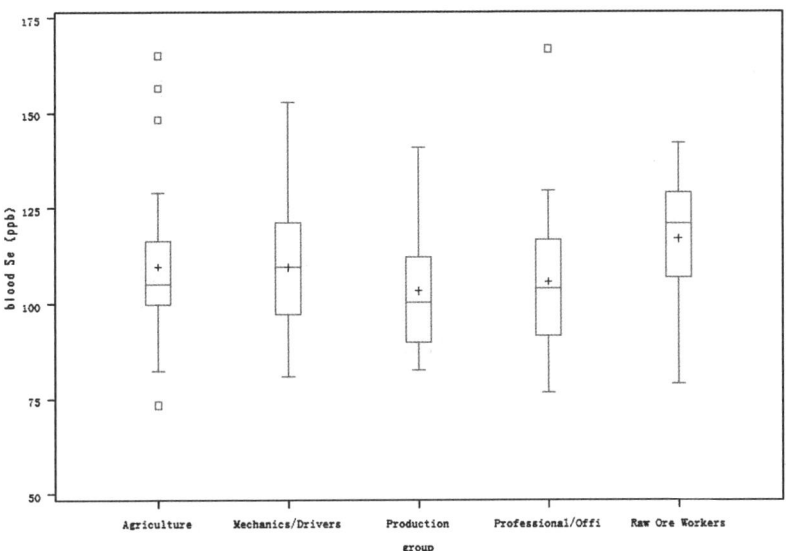

Figure 6 *Selenium in blood (ppb) across five work categories.*

Values for metals in blood and semen for men included in the current study fell within ranges reported in the male reproductive health literature (Table 1). Only copper appeared elevated in semen when compared with the single other study available.[13] Blood zinc was lower than the single other study for comparison although zinc in semen was similar to previous reports.[13,14]

None of the metals or occupational categories evaluated in this study of 129 workers was associated with changes in ratio of male to female offspring. Although it is relatively easy to gather data in human studies on differences in gender ratio at birth, it requires a sample size of 17,000 births to exposed men to have enough power to detect statistically significant differences from the expected sex ratio. This kind of study would be logistically difficult and subject to the influences of social, economic, and political forces on birthing practices in the populations studied. For example, currently in China, selection pressure for boy babies is great. Traditional preference for male children, especially in rural areas, is compounded by the 'one child' rule that has been in effect in China for 25 years.[20] Currently, the male to female sex ratio at birth in China is 1.18 according to the 2000 Census data. In most other nations of the world, 103–105 males are born to each 100 females.

To try and get around non-biological influences on sex ratio at birth, semen was collected from workers and sperm cells in the ejaculate evaluated for X *vs.* Y chromosomes by fluorescence *in situ* hybridization (FISH). It was determined that none of the work categories evaluated differed from each other in terms of the ratio of X bearing to Y bearing sperm with all groups ranging from 1.01 to 1.02. Additionally, there was no statistically significant correlation between any

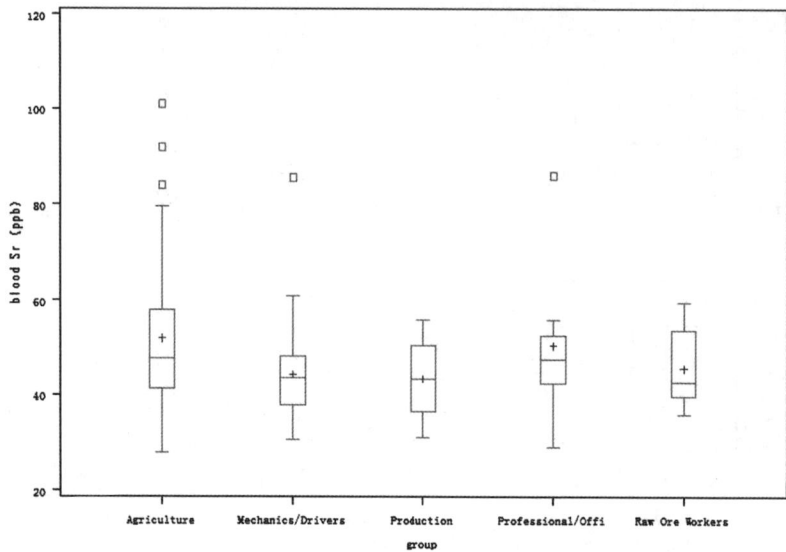

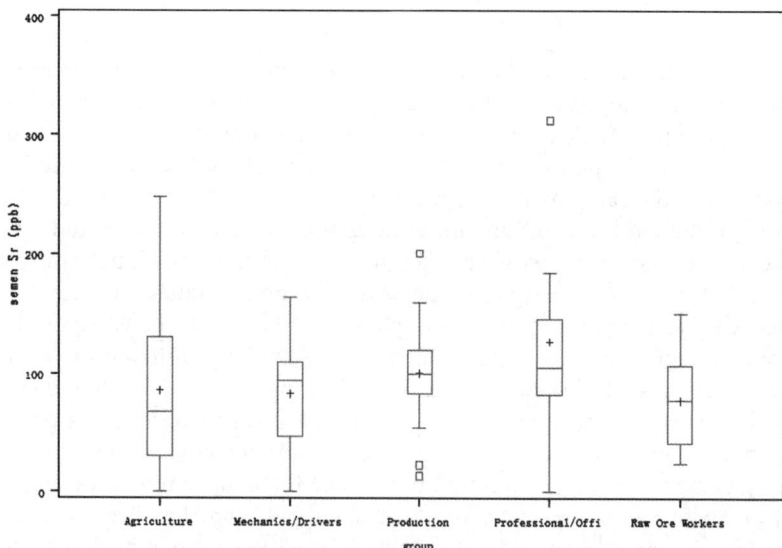

Figure 7 *Strontium in blood and semen (ppb) across five work categories.*

of the metals evaluated and X to Y ratio in sperm. However, this study is limited in evaluating effects of high exposure to metals because only copper was found to be elevated in semen above expected background levels. Other metals remained essentially within ranges reported in the male reproductive health literature for healthy men. It remains to be seen whether the metals evaluated in the current study might result in changes in X:Y ratio in men with high workplace or environmental exposures.

6.5 Conclusion

Given extensive use of metals in modern society, it is reasonable to investigate their potential to induce changes in sperm DNA that could affect offspring and to explore potential modifiers of this relationship. One approach is to correlate levels of metals in blood and semen with specific DNA changes in sperm cells that are of a nature transmissible to offspring. In the present work, questionnaire data on sex ratios at birth in offspring was collected and found not to differ across five different categories of work. Levels of metals in blood and semen were analyzed and found not to correlate with sex ratios of offspring at birth or with ratio of X to Y bearing sperm in ejaculates. However, the study does not address high metal exposures. It does demonstrate feasibility of investigation of X and Y bearing sperm in ejaculates as a viable alternative to the traditional method of assessing sex ratios at birth when trying to assess male mediated effects.

Acknowledgments

This work was supported by grant RO1 OH007575 from DHHS/CDC/NIOSH and the UCLA Center for Occupational and Environmental Health.

References

1. J.B. Brodsky, E.N. Cohen, C. Whitcher, B.W. Brown Jr. and M.L. Wu, *J. Am. Dent. Assoc.*, 1985, **111**, 779.
2. K.H. Alcser, K.A. Brix, L.J. Fine, L.R. Kallenbach and R.A. Wolfe, *Am. J. Ind. Med.*, 1989, **15**, 517.
3. M.L. Lindbohm, M. Sallmen, A. Anttila, H. Taskinen and K. Hemminki, *Scand. J. Work Environ. Health*, 1991, **17**, 95.
4. S. Cordier, F. Deplan, L. Mandereau and D. Hemon, *Br. J. Ind. Med.*, 1991, **48**, 375.
5. N.H. Hjollund, J.P. Bonde, T.K. Jensen, E. Ernst, T.B. Henriksen, H.A. Kolstad, A. Giwercman, N.E. Skakkebaek and J. Olsen, *Reprod. Toxicol.*, 1998, **12**, 91.
6. I. Figa-Talamanca and G. Petrelli, *Int. J. Epidemiol.*, 2000, **29**, 381.
7. D.X. Xu, H.M. Shen, Q.X. Zhu, L. Chua, Q.N. Wang, S.E. Chia and C.N. Ong, *Mutat. Res.*, 2003, **534**, 155.
8. J.P. Bonde, M. Joffe, P. Apostoli, A. Dale, P. Kiss, M. Spano, F. Caruso, A. Giwercman, L. Bisanti, S. Porru, M. Vanhoorne, F. Comhaire and W. Zschiesche, *Occup. Environ. Med.*, 2002, **59**, 234.
9. S.D. Perreault, J. Rubes, W.A. Robbins, D.P. Evenson and S.G. Selevan, *Andrologia*, 2000, **32**, 247.

10. W.A. Robbins, M.F. Vine, K.Y. Truong and R.B. Everson, *Environ. Mol. Mutagen.*, 1997, **30**, 175.
11. SAS (r) Proprietary Software Release 8.2 (TS2M0), copyright (c) 1999–2001 by SAS Institute Inc., Cary, NC, USA.
12. B. Xu, S.E. Chia and C.N. Ong, *Biol. Trace Elem. Res.*, 1994, **40**, 49.
13. T. Umeyama, H. Ishikawa, H. Takeshima, S. Yoshii and K. Koiso, *Fertil. Steril.*, 1986, **46**, 494.
14. C.I. Madding, M. Jacob, V.P. Ramsay and R.Z. Sokol, *Ann. Nutr. Metab.*, 1986, **30**, 213.
15. F.R. Abou-Shakra, N.I. Ward and D.M. Everard, *Fertil. Steril.*, 1989, **52**, 307.
16. P. Apostoli, S. Porru and L. Bisanti, *Scand. J. Work Environ. Health*, 1999, **25**(Suppl 1), 40.
17. A. Paudyn, D.M. Templeton and A.D. Baines, *Symposium on Recent Advances in Biological Trace Element Analysis,* Ottawa, Ontario, Canada, 1988, **89**, 343.
18. G.J. Shugar and J.T. Ballinger, *Chemical Technicians' Ready Reference Handbook*, McGraw-Hill, New York, NY, 1996.
19. D.M. Templeton, in *Handbook on Metals in Clinical and Analytical Chemistry*, H.G. Seiler, A. Sigel and H. Sigel (eds), Marcel Dekker, Inc., New York, NY, 1994, vol. 12, 167.
20. T. Plafker, *Br. Med. J.*, 2002, **324**, 1233.
21. M.B. Sorensen, I.A. Bergdahl, N.H. Hjollund, J.P. Bonde, M. Stoltenberg and E. Ernst, *Mol. Hum. Reprod.*, 1999, **5**, 331.
22. K. Danadevi, R. Rozati, P.P. Reddy and P. Grover, *Reprod. Toxicol.*, 2003, **17**, 451.
23. P.A. Pleban and D.S. Mei, *Clin. Chim. Acta*, 1983, **133**, 43.
24. M. Saaranen, U. Suistomaa, M. Kantola, S. Saarikoski and T. Vanha-Perttula, *Hum. Reprod.*, 1987, **2**, 475.
25. S. Benoff, G.M. Centola, C. Millan, B. Napolitano, J.L. Marmar and I.R. Hurley, *Hum. Reprod.*, 2003, **18**, 374.

CHAPTER 7

Use of the Sperm Chromatin Structure Assay (SCSA®) as a Diagnostic Tool in the Human Infertility Clinic

DONALD P. EVENSON AND REGINA L. WIXON

Department of Biology, Box 2140D Northern Plains Biostress, South Dakota State University, Brookings SD 57007, USA

7.1 Introduction

Elevated sperm DNA fragmentation can be attributed to various pathological conditions including cryptorchidism, cancer, varicocele, fever, age, infection, and leukocytospermia among others.[1–6] Many environmental conditions can also affect sperm DNA fragmentation such as chemotherapy, radiation, prescription drugs, air pollution, smoking, pesticides, chemicals, heat, and ART preparation protocols.[7–18] Reactive oxygen species (ROS) activity may be a major factor in DNA strand breakage.[19] It is now recognized that elevated sperm DNA fragmentation has a significant effect on reproductive outcome.

The pioneering manuscript published by Evenson and colleagues[18] showed a significant relationship between human and bull sperm DNA fragmentation and loss of fertility potential. This was followed by a series of papers showing that sperm retrieved from mice exposed to reproductive toxicants had elevated sperm chromatin structure assay (SCSA) defined DNA fragmentation values. Exposure of mice to methyl methanesulfonate led to a dramatic increase (100% DNA fragmentation index (DFI)) in SCSA defined DNA fragmentation 3 days post exposure.[15] Exposure to thiotepa, hydroxyurea, triethylenemelamine, and ethylnitrosourea in mice all showed alterations in testicular cell kinetics and an increase in sperm DNA fragmentation.[14–17] The SCSA appears to be the most sensitive assay for the detection of DNA fragmentation due to X-ray damage. Forty-days after testicular exposure from 5 to 400 rads of radiation, mouse epididymal sperm were removed and analyzed by SCSA. The lowest level of detection was at 12.5 rads with a dose response increase.[8]

There are now hundreds of manuscripts on multiple species regarding time of toxicant-induced DNA damage to developing sperm cells and mechanisms of damage as related to toxicant exposure and pregnancy outcomes. Polychlorinated biphenyls (PCBs) and the insecticide dichlorodiphenyltrichloroethane (DDT) and its major metabolite, dichlorodiphenyldichloroethylene (p,p'-DDE) are a concern in the environment due to their resistance to degradation and their ability to bioaccumulate with negative effects on male reproduction. Rignell-Hydborn and colleagues found a significantly lower %DFI in the lowest CB-153 quintile compared with the other quintiles.[13] Men exposed to insecticides such as carbaryl and pesticides such as Fenvalerate, showed significantly increased levels of sperm DNA fragmentation.[20–21] A dramatic effect of organophosphorous was reported by Sanchez-Pena and colleagues,[22] where nearly 3/4 of these operators had DFI values $>30\%$. Significantly, higher levels of sperm DNA fragmentation were found in factory workers exposed to styrene which is used to make plastics, rubber, and resins. Residents of Teplice, Czech Republic, a town with heavy winter time air pollution generated by burning soft brown coal, experienced a higher than normal rate of infertility and spontaneous miscarriages.[11] Czech army conscripts, 18 years of age, provided semen samples for a cross-sectional study and then a 2 year longitudinal study that went through periods of clean and polluted air. Sperm DNA fragmentation as measured by the SCSA was the only measure to detect a correlation between air pollution and semen quality in 18-year old army conscripts. One-fourth of these young men had %DFI >30, placing them in a statistical group known to be at an increased risk for infertility.

Carrell and colleagues[23] measured DNA fragmentation by the Tunel assay in men whose partners had repeated pregnancy losses (RPL). The observation that sperm donors, the general population, and RPL patients had about 12, 21, and 39%, respectively, of sperm with fragmented DNA supports the hypothesis that elevated sperm DNA fragmentation can cause miscarriage as well as our observations that couples having a DFI $>30\%$ have a near doubling of spontaneous miscarriages.[24] Patients with $>30\%$ DFI had about twice the level of miscarriage as those with $<30\%$ DFI.[25] A $\geq 30\%$ DFI score was associated with a higher rate of spontaneous abortions at 12 weeks of gestation $(p<0.01)$.[26]

Preliminary data indicate that cholesterol-lowering medications $(p<0.05)$ and anti-ulcer agents $(p<0.0001)$ significantly increased sperm DNA fragmentation in comparison to men taking no medications. Men taking 5α reductase inhibitors showed significantly higher DNA stainability (HDS) than those not on medications. HDS has been associated with a longer time to pregnancy. The cholesterol lowering medications, anti-ulcer agents, and 5α reductase inhibitors had no effect on sperm count or morphology. The ingestion of SSRIs (Prozac) in smoking men significantly increased sperm DNA fragmentation, which did not return to baseline values $(p<0.001)$, possibly indicating a genetic mutation.[9] We have reported a case study where a man had a semen analysis as well as an SCSA evaluation for eight consecutive months. The SCSA analysis for 7 of the 8 months showed sperm DNA fragmentation consistent with excellent fertility. After a back injury, the man was given cortisone injections and the following month the SCSA analysis showed poor sperm DNA integrity.[10]

For the human infertility clinic, odds ratios on data from thousands of semen samples show that for samples containing less than 27–30% sperm with fragmented DNA there was a 6.5–10×, 7.0–8.7×, ~2×, and ~1.5× greater probability of a successful pregnancy by *in vivo*, IUI, routine IVF, and ICSI fertilizations, respectively, in comparison to semen samples >27–30% DFI. The above-mentioned studies show that many environmental toxicants can elevate sperm DNA fragmentation and as a consequence, significantly affect pregnancy outcome.

7.2   Materials and Methods

The primary principle of the SCSA is that sperm in buffer-diluted raw semen are exposed for 30 s to a low pH (1.2) that denatures DNA at sites of DNA strand breaks.[17,27] The sperm sample is then stained with acridine orange (AO) which is a metachromatic DNA dye that fluoresces green when intercalated into native DNA and shifts to a red fluorescence when associated with collapsed single-stranded DNA. These stained samples are then measured by flow cytometry that collects green light (515–530 nm) and red light (>630 nm). The amounts of green and red fluorescence of 5000 individual sperm are quantitated by a multichannel analyzer with 1024 channels of intensity.

The resulting scattergram (cytogram) data are processed with SCSAsoft® software (SCSA® Diagnostics) (see Figures 1 and 2). The DFI is processed as red fluorescence/(red) + (green) fluorescence with a total of 1024 DFI units.

Clinical Results of the SCSA®

Pregnancy Outcome = Pregnant

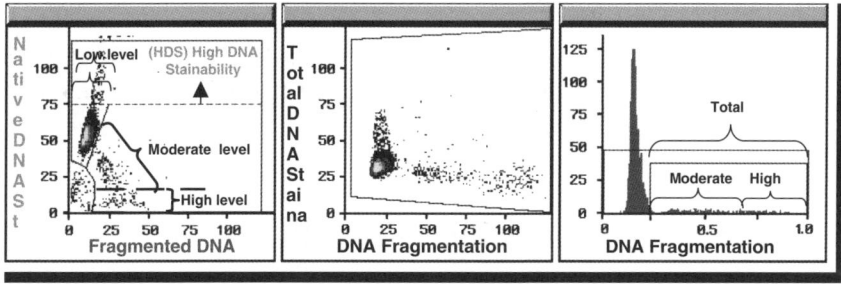

Patient	Date	Measurement	X DFI	SD DFI	DFI (%)	HDS (%)
7272-87	####	1	213.5	111.6	6.8	5.0
		2	221.2	118.1	8.3	5.4
		mean	217.4	114.8	7.5	5.2
		sd	5.4	4.6	1.1	0.2

Figure 1 *Cytogram of a patient who initiated a pregnancy*

Clinical Results of the SCSA®

Pregnancy Outcome = NOT Pregnant

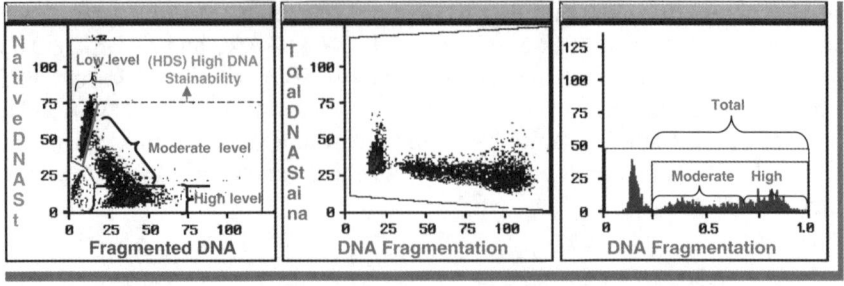

Patient	Date	Measurement	X DFI	SD DFI	DFI (%)	HDS (%)
7272-113	####	1	563.7	307.0	64.9	6.4
		2	561.4	304.8	64.9	7.2
		mean	562.6	305.9	64.9	6.8
		sd	1.2	1.1	0.0	0.4

Figure 2   *Cytogram of a patient who did not initiate a pregnancy*

Note that this is not the same as red/total (515 long pass filter). The mean (0–1024 channels or units) of DFI units and standard deviation of DFI (SD DFI) are calculated by SCSAsoft®. The percent of sperm with fragmented DNA is %DFI. The percent of sperm with high DNA stainability are listed as %HDS. These sperm have an increased DNA fluorescence due to lack of complete sperm nuclear condensation thus making more DNA accessible to AO staining. Five thousand sperm are measured per sample with two independent measurements of the same thawed sample. The mean of the two samples provides the clinical reported data of mean DFI, SD of DFI, %DFI, and %HDS. Previous and current research continue to support the view that $\sim$ >30% DFI and >15% sperm with HDS, (immature sperm) in raw semen are statistically significant with regards to a reduced pregnancy outcome for *in vivo*, IUI, and routine IVF fertilizations as well as an increased risk for early spontaneous miscarriage.

7.3   SCSA Data and Clinical Results

Previous and current research continues to support the utility of the SCSA in the infertility and urology clinics. Odds ratios from *in vivo* and IUI studies ranged from 6.5× to 10× increased pregnancy rate if %DFI <30 (significance range 0.01–0.002), whereas, odds ratios from several IVF studies ranged from 1.5× to 9.5× increased pregnancy rate if %DFI <30 (significance range: N.S.–0.003). ICSI studies ranged from 1.5× to 2.0× increased pregnancy/delivery rate if %DFI was $\approx$ <30 (significance range: N.S.) see Table 1. The

Table 1 *For ease of reading the table, an example for the first row is provided, e.g., a patient is 6.5 times more likely to become pregnant and/or deliver if <30% of the sperm have fragmented DNA using an in vivo procedure*

Odds Ratio	%DFI	ART Procedure	Patient No.	Chi Square[a]	Analysis	Author
6.5x	<30	In vivo	N = 147	0.002	SCSA	Evenson et al. 1999
10x	<40	In vivo	N = 215	N/A	SCSA	Spano et al. 2000
8.7x	≤27	IUI	N = 131	0.01	SCSA	Bungum et al. 2004
~2x	≤36.5	IVF	N = 167	0.03	TUNEL	Henkel et al. 2004
2x	<30%	IVF	N = 249	0.01	SCSA	Virro et al. 2004[b]
1.5x	<27%	IVF	N = 109	NS	SCSA	Bungum et al. 2004
1.5x	Low DNA damage	IVF or ICSI	N = 60	NS	Comet	Morris et al. 2002
2x	<30%	IVF or ICSI	N = 52	NS	SCSA, TUNEL, SCD	Chohan ASRM 2004 presentation
3.1x	>27%	ICSI vs IVF	N = 66	NS	SCSA	Bungum et al. 2004
9.5x	<30%	ICSI/IVF	50 donor egg cycles	0.003	SCSA	Adams et al. ASRM 2004 presentation[c]

[a] Results were significant at $P < 0.05$

[b] Patients were 2x as likely to carry a pregnancy >12 weeks if %DFI was <30%

[c] Patients were 9.5x more likely to carry a pregnancy with high quality donor eggs when sperm DFI was <30%

following meta-analyses were conducted to clarify the utility of the SCSA®
sperm DNA fragmentation test for the human infertility clinic.

A meta-analysis of three studies ($n = 1575$) was conducted to investigate the
relationship of %DFI on pregnancy outcome using *in vivo* and IUI procedures.[24,28–29] The meta-analysis indicated that patients were 7.3 times more
likely to achieve a pregnancy/delivery if the %DFI <30 ($p = 0.0001$). Results
from the Breslow Day Test (BDT) showed that the odds ratio for all studies
tested was not significantly different ($p = 0.96$) and showed similar trends.

When routine IVF fertilization as well as *in vivo* and IUI procedures were
considered, couples were 3.9 times more likely to become pregnant if their DFI
was $<30\%$ ($n = 1990$, $p = 0.0001$).[24,28–32] Results from the BDT showed that
the odds ratio for all studies tested was not significantly different ($p = 0.20$) and
showed similar trends.

When routine IVF alone was considered, couples were 2.2 times more likely
to become pregnant if their DFI was $<30\%$ ($n = 521$, $p = 0.0008$).[29–32] Results
from the BDT showed that the odds ratio for all studies tested was not
significantly different ($p = 0.26$) and showed similar trends.

A meta-analysis of five studies using ICSI and IVF ($n = 216$) showed a
non-significant trend where patients were 1.7 times more likely to achieve
a pregnancy/delivery if the %DFI was $<30\%$ ($p = 0.11$).[29,31,33–35] With
the BDT, the odds ratio for all studies tested was significantly different
($p = 0.04$).

7.4   Conclusions

The SCSA test has been proven over the past quarter century to be a rapid,
machine objective, and statistically robust procedure to measure fragmented
DNA in multiple species as related to toxicant-induced damage and infertility
potential. The above meta-analysis shows that the SCSA data are significantly
predictive for reduced pregnancy success using *in vivo*, IUI, and to a lesser
extent routine IVF and ICSI. A $>30\%$ DFI places a man into a statistical
category of a longer time to natural pregnancy, more ART cycles, an increased
level of spontaneous abortions, or no pregnancy. Some clinics use the SCSA
test as part of the initial infertility workup while other clinics use it for those
patients with failed IUI and routine IVF cycles potentially to direct that patient
to ICSI.

References

1. O. Stahl, J. Eberhard, K. Jepson, M. Spano, M. Cwikiel, E. Cavallin-Stahl
   and A. Giwercman, *Cancer*, 2004, **6**, 1137.
2. A. Zini, A. Blumenfeld, J. Libman and J. Willis, *Hum. Reprod.*, 2005, **4**,
   1018.
3. D.P. Evenson, L.K. Jost, M. Corzett and R. Balhorn, *J. Androl.*, 2000, **5**,
   739.

4. R.J. Aitken, M.A. Baker and D. Sawyer, *Reprod. Biomed. Online*, 2003, **1**, 65.
5. N. Burrello, A.E. Calogero, A. Perdichizzi, M. Salmeri, R. D'Agata and E. Vicari, *Reprod. Biomed. Online*, 2004, **5**, 569.
6. J.G. Alvarez, R.K. Sharma, M. Ollero, R.A. Saleh, M.C. Lopez, A.J. Thomas Jr., D.P. Evenson and A. Agarwal, *Fertil. Steril.*, 2002, **2**, 319.
7. S.D. Fossa, P. De Angelis, S.M. Kraggerud, D. Evenson, L. Theodorsen and O.P. Clausen, *Cytometry*, 1997, **4**, 192.
8. B.L. Sailer, L.K. Jost, K.R. Erickson, M.A. Tajiran and D.P. Evenson, *Environ. Mol. Mutagen.*, 1995, **1**, 23.
9. C.A. Adams, J.C. Anderson, S.H. Juanengo and S.H. Wood, Abstract, Reproductive Sciences Center, La Jolla, CA, 2004.
10. D.P. Evenson, L.K. Jost, R.K. Baer, T.W. Turner and S.M. Schrader, *Reprod. Toxicol.*, 1991, **5**, 115.
11. J. Rubes, S.G. Selevan, D.P. Evenson, D. Zudova, M. Vozdova, Z. Zudova, W.A. Robbins and S.D. Perreault, *Hum. Reprod.*, 2005, Epub ahead of print.
12. R.J. Potts, C.J. Newbury, G. Smith, L.J. Notarianni and T.M. Jefferies, *Mutat. Res.*, 1999, **1–2**, 103.
13. A. Rignell-Hydbom, L. Rylander, A. Giwercman, B.A. Jonsson, C. Lindh, P. Eleuteri, M. Rescia, G. Leter, E. Cordelli, M. Spano and L. Hagmar, *Environ. Health Perspect.*, 2005, **113**, 175.
14. D.P. Evenson, R.K. Baer and L.K. Jost, *Toxicol. Appl. Pharmacol.*, 1986, **82**, 151.
15. D.P. Evenson and L.K. Jost, *Cell Prolif.*, 1993, **26**, 147.
16. D.P. Evenson, R.K. Baer and L.K. Jost, *Environ. Mol. Mutagen.*, 1989 **14**, 79.
17. D.P. Evenson, P.J. Higgins, D. Grueneberg and B.E. Ballachey, *Cytometry*, 1985, **6**, 238.
18. D.P. Evenson, Z. Darzynkiewicz and M.R. Melamed, *Science*, 1980, **240**, 1131.
19. R.J. Aitken, *Adv. Exp. Med. Biol.*, 2003, **518**, 85.
20. Y. Xia, S. Cheng, Q. Bian, L. Xu, M.D. Collins, H.C. Chang, L. Song, J. Liu, S. Wang and X. Wang, *Toxicol. Sci.*, 2005, **85**, 615.
21. Q. Bian, L.C. Xu, S.L. Wang, Y.K. Xia, L.F. Tan, J.F. Chen, L. Song, H.C. Chang and X.R. Wang, *Occup. Environ. Med.*, 2004, **61**, 999.
22. C. Sanchez-Pena, B.E. Reyes, L. Lopez-Carrillo, R. Recio, J. Moran-Martinez, M.E. Cebrian and B. Quintanilla-Vega, *Toxicol. Appl. Pharmacol.*, 2004, **196**, 108.
23. D.T. Carrell, L. Liu, C.M. Peterson, K.P. Jones, H.H. Hatasaka, L. Erickson and B. Campbell, *Arch. Androl.*, 2003, **49**, 49.
24. D.P. Evenson, L.K. Jost, M.J. Zinaman, E. Clegg, K. Purvis, P. de Angelis and O.P. Clausen, *Hum. Reprod.*, 1999, **14**, 1039.
25. J.H. Check, V. Graziano, R. Cohen, J. Krotec and M.L. Check, *Arch. Androl.*, 2005, **51**, 121.

26. M.R. Virro, K.L. Larson-Cook and D.P. Evenson, *Fertil. Steril.*, 2004, **8**, 1289.
27. D.P. Evenson, K.L. Larson and J.K. Jost, *J. Androl.*, 2002, **23**, 25.
28. M. Spano, J. Bonde, H.I. Hjollund, H.A. Kolstatd, E. Cordelli and G. Leter, *Fertil. Steril.*, 2000, **73**, 43.
29. M. Bungum, P. Humaidan, M. Spano, K. Jepson, L. Bungum and A. Giwercman, *Hum. Reprod.*, 2004, **19**, 1401.
30. R. Henkel, M. Hajimohammad, T. Stalf, C. Hoogendijk, C. Mehnert, R. Menkveld, H. Gips, W.B. Schill and T.F. Kruger, *Fertil. Steril.*, 2004, **81**, 965.
31. K.L. Larson-Cook, J.D. Brannian, K.A. Hansen, K. Kasperson, E.T. Aamold and D.P. Evenson, *Fertil. Steril.*, 2003, **80**, 895.
32. C. Adams, L. Anderson and S. Wood, *Abstracts of the Scientific Oral & Poster Sessions*. Philadelphia, PA, *Journal of the American Society for Reproductive Medicine* (with permission), 2004, S44.
33. K.R. Chohan, J.T. Fiffin, M. Lafromboise, C.J. DeJonge and D.T. Carrell, *Abstracts of the Scientific Oral & Poster Sessions*. Philadelphia, PA, *Journal of the American Society for Reproductive Medicine* (with permission), 2004, S55.
34. K.L. Larson, C.J. DeJonge, A.M. Barnes, L.K. Jost and D.P. Evenson, *Hum. Reprod.*, 2000, **15**, 1717.
35. I.D. Morris, S. Ilott, L. Dixon and D.R. Brison, *Hum. Reprod.*, 2002, **17**, 990.

CHAPTER 8

Safety of Sperm for Use in Intra-Cytoplasmic Sperm Injection

D. SAKKAS,[a] E. SELI,[a] D. BIZZARO,[b] G.C. MANICARDI,[c] A. JAKAB[d] AND G. HUSZAR[a]

[a] Department of Obstetrics, Gynecology and Reproductive Sciences, Yale University School of Medicine, PO Box 208063, New Haven, CT 06520-8063, USA

[b] Institute of Biology and Genetics, University of Ancona, Ancona, Italy

[c] Department of Animal Biology, University of Modena and Reggio Emilia, Reggio Emilia, Italy

[d] Department of Obstetrics and Gynaecology, University of Debrecen, Hungary

8.1 Introduction

The impact an abnormal paternal genome may have on reproductive outcome is unquestionably less when compared to its female counterpart's role. The egg's importance has been well established, as shown by the success of donor oocyte programs. It can be estimated that in about 80% of cycles, egg quality plays the major driving force in respect to the chances of a patient achieving a pregnancy. In contrast, the influence of the human sperm on reproductive outcome has been less well characterized. A number of studies using an egg-share model have now shown that a paternal factor exists and that the influence is far less than that of the egg.[1,2] The paternal importance however rises significantly with the increased use of intra-cytoplasmic sperm injection (ICSI), where the quality of spermatozoa is generally accepted to be poorer. In light of this, the safety of ICSI is being increasingly examined as concerns have arisen that aberrant paternal inheritance can be derived at the chromosomal, epigenetic and nuclear DNA level (reviewed in ref 3). Many of the influential factors in the paternal genome that impact on poor reproductive outcome are still theoretical, however one area that has been more rigorously examined in the last decade is the quality of the sperm nuclear DNA. In this chapter, we will examine the experimental evidence linking abnormal sperm to poor reproductive outcome in relation to the safety of ICSI and how we can improve selection methods to avoid harmful outcomes.

8.2   Incidence of Sperm Chromosome Aneuploidy and Implications on Reproduction

Based on FISH studies, paternal errors account for 5–10% of autosomal trisomies, while maternal MI errors are the predominant aetiology.[4] Paternal effect on sex chromosome trisomies is higher because 100% of 47,XYY, and nearly 50% of 47,XXY are paternal in origin.[4,5]

Hansen *et al.*[6] reported compiled data from the registries in Western Australia, involving 301 infants conceived with ICSI, 837 infants conceived with *in vitro* fertilization (IVF), and 4000 naturally conceived controls between 1993 and 1997. They found the incidence of major birth defects to be more than 2-fold higher for ICSI and IVF groups (8.6% and 9.0%, respectively) compared to normal controls (4.2%). Their data show an increased incidence of chromosomal abnormalities in the ICSI group (1.0% for all infants and 1.6% for singletons only) compared to IVF (0.7% for all infants and 0.6% for singletons only; the difference is not statistically significant) and normal controls (0.2% for all infants and 0.2% for singletons only; $p < 0.05$). The potentially increased risk of birth defects after IVF and ICSI was also addressed in a recent review based on 26 studies from various countries. They concluded that although differences in birth-defect classification warrant further interpretation, six of the reports suggested that there might be an elevated risk for birth defects in children treated with reproductive technologies; however, the risks seem to be comparable when considering ICSI or IVF.[7] Van Steirteghem and colleagues summarized data from seven studies reporting karyotype analyses performed for prenatal diagnosis in a total of 2139 pregnancies conceived with ICSI.[8] In comparison with the general population, they calculated a slight but significant increase in *de novo* sex chromosomal aneuploidy (0.6% *vs.* 0.2%), structural autosomal abnormalities (0.7% *vs.* 0.04%), and an increased number of inherited (mostly from the father) structural aberrations. More recently, the same group presented a review of 10 years' experience with ICSI and concluded that there is a slight increase in *de novo* chromosomal abnormalities, the major congenital malformation rate is similar for IVF and ICSI (between 3% and 4%), and at approximately 2 years of age the developmental outcome as assessed by the Bayley scale is similar for IVF and ICSI.[9]

8.3   Epigenetic Effects and Their Relation to Intra-Cytoplasmic Sperm Injection

Epigenetics refers to the covalent modifications of DNA or nucleoproteins that regulate gene activity without altering DNA sequence. These disorders manifest themselves in what are known as imprinting disorders. Some recent publications have associated assisted reproductive treatments with a number of children who were affected by diseases caused by imprinting disorders.[10–12] The most concerning study was published by Halliday *et al.*,[13] who reported the first case-control study in an Australian population. Among 1,316,500 live births in Victoria between 1983 and 2003, they identified 37 cases of Beckwith–Wiedemann syndrome. For each Beckwith–Wiedemann syndrome case, they randomly selected four live-born controls. IVF was the method of conception in four Beckwith–Wiedemann syndrome cases and in one control. Their results indicated that if a child has Beckwith–Wiedemann syndrome, the odds that the child was conceived using IVF was 18 times greater than that for a child without Beckwith–Wiedemann syndrome. The calculated risk of Beckwith–Wiedemann syndrome in the IVF population was 1/4000, or nine times greater than the general population. Although this study has shortcomings including a large confidence interval, its results are concerning. Future studies are needed to assess the association between specific assisted reproductive technologies and imprinting disorders. Of course, the use of abnormal male gametes and a proposed association with imprinting disorders is one of major concern; however, an association with the use of ICSI, specifically, is yet to be shown as the cases reported have arisen from an array of ART-associated treatments.

Imprinting in spermatozoa of men with abnormal semen parameters has yet to be shown conclusively: an initial study using PCR-based techniques to analyze DNA extracted from spermatozoa of men with normal semen analysis and those undergoing ICSI, failed to detect a difference in methylation status.[14] More recently, Marques *et al.*[15] reported that maternal imprinting was correctly erased in all patients, however, methylation of the *H19* gene changed in 30% of the severe oligozoospermic patients tested. Their findings suggested an association between abnormal genomic imprinting and hypospermatogenesis. They concluded that spermatozoa from oligozoospermic patients carry a raised risk of transmitting imprinting errors. In respect to epigenetic alterations, some animal studies have raised the greatest concern. In particular, a recent study by Anway *et al.*[16] showed that altered DNA methylation patterns in the germ line were transferred through the male germ line to nearly all males of the next four subsequent generations examined.

8.4 The Impact of Sperm Nuclear DNA Strand Breaks on Reproductive Outcome

An area of sperm integrity that has been examined more closely in the past decade has been the presence of nuclear DNA strand breaks in ejaculated spermatozoa. Their presence was initially reported in the early 1990s,[17–19] while their impact is still not completely understood. Reproductive parameters that could be affected by an increased presence of DNA strand breaks in ejaculated spermatozoa include fertilization, blastocyst development, and pregnancy rates. Investigation of the possible association between DNA strand breaks in spermatozoa and fertilization rates in patients undergoing ART found no correlation between DNA integrity of ejaculated spermatozoa and IVF and ICSI fertilization rates.[20–24] In contrast to these reports, a negative correlation between sperm DNA strand breaks and IVF[25] and ICSI[26] fertilization rates has been reported, using the TUNEL assay. Activation of embryonic genome expression occurs at the four- to eight-cell stage in human embryos,[27] suggesting that the paternal genome may not be effective until that stage. Therefore, the lack of correlation between elevated DNA strand breaks in spermatozoa and fertilization rates is not surprising as they seem to be much more important later on.[2,28]

As expected, a negative correlation between the extent of nuclear DNA damage in ejaculated spermatozoa and blastocyst development after IVF and ICSI was observed using both the TUNEL assay to evaluate spermatozoa processed for IVF[29] and the SCSA to evaluate unprocessed spermatozoa.[30] In addition, pregnancy rates after IVF are reduced in couples who have higher percentages of spermatozoa with DNA strand breaks detected by *in situ* nick translation,[20] and as discussed above there is a tendency toward lower pregnancy rates in patients exhibiting high DFI values as assessed by the SCSA.[30] Finally, Carrel *et al.*[31] found that the percentage of sperm staining positive for DNA fragmentation using TUNEL, was significantly increased in men whose wives suffered recurrent pregnancy loss (38 +/− 4.2) compared with donor sperm (11.9 +/− 1.0) or general population (22 +/− 2.0) control groups. In the recurrent pregnancy-loss group, no correlation was observed between semen quality parameters and the TUNEL data. They concluded that their data indicated some recurrent pregnancy-loss couples have a significant increase of sperm DNA fragmentation, which may be causative of pregnancy loss. Very recently, ICSI outcomes were compared in two sequential attempts performed with ejaculated and testicular spermatozoa, in a group of 18 men. Sperm DNA fragmentation was assessed by the TUNEL assay and the incidence of abnormal sperm was found to be markedly lower in testicular spermatozoa as compared with ejaculated spermatozoa. Even though no differences in fertilization and cleavage rates and in embryo morphological grade were found between the ICSI attempts performed with ejaculated and testicular spermatozoa, a 44.4% pregnancy rate was achieved with testicular spermatozoa, whereas ICSI with ejaculated spermatozoa led to only one pregnancy, which was spontaneously aborted.[32]

The increasing number of publications in this field indicates that the relevance of sperm nuclear DNA is not completely black and white. We have made a number of conclusions from the ever-increasing wealth of collected data about the various sperm nuclear DNA integrity tests and their predictive power in ART.[33] Briefly, the conclusions can be summarized by the following points:

- an increased fraction of sperm showing DNA damage is a negative trait that reduces the chances to father a child;
- an absolute number or percentage of DNA strand breaks not compatible with pregnancy is far from being established;
- the predictive power of the current sperm DNA integrity tests seem to lose their strength from natural conception to ICSI, passing through intra-uterine insemination (IUI) and IVF.

8.5 Improving the Safety of Intra-Cytoplasmic Sperm Injection

Human semen is heterogeneous in quality, not only between males but also within a single ejaculate. Differences in quality are evident, both when examining the classical parameters of sperm number, motility, and morphology and in the integrity of the sperm nucleus. The ability to (a) improve the efficiency of preparation techniques to eliminate spermatozoa with nuclear anomalies, (b) improve the selection of the best spermatozoa prior to ICSI, and/or (c) improve the selection of embryos that may have an abnormal paternal complement, may all assist in making ICSI a safer technique. A number of studies, including our own, have shown that spermatozoa prepared using a density gradient centrifugation technique significantly improves the quality of spermatozoa in the preparation.[20,34] In our own study, we showed that there was a significant ($p < 0.001$) decrease in both chromomycin A3 (CMA3) positivity and DNA strand breakage in sperm samples from different men after preparation by density gradient centrifugation. An increase in CMA3 positivity indirectly demonstrates a decreased presence of protamines. Our results indicated that density gradient centrifugation could enrich the sperm population by separating out those with nicked DNA and poorly condensed chromatin. Another technique that we proposed in 1999 was to culture ICSI embryos postembryonic genome activation to the blastocyst stage.[35] A modification of this initial model is shown in Figure 1.

A final methodology is to improve the selection of spermatozoa prior to ICSI. One technique that has been reported is the selection of spermatozoa under high magnification. Bartoov *et al.*[36] reported that they were able to achieve a pregnancy rate of 58% in 24 patients who had previously failed at least five consecutive routine cycles of IVF and ICSI. They have also reported a follow-up study showing improved pregnancy rates with ICSI and morphologically selected sperm compared with conventional ICSI.[37] The group of Aitken [38] has also reported a novel electrophoretic sperm isolation technique for the isolation of functional human spermatozoa free from significant DNA damage. Briefly, the separation system consists of a cassette comprising two chambers. Semen is introduced into one chamber and a current applied that within seconds leads to a purified suspension of spermatozoa collecting on one side of the chamber. Suspensions generated by the electrophoretic separation technique contain motile, viable, morphologically normal spermatozoa, which exhibit lower levels of DNA damage.

A more promising sperm selection technique has recently been reported by Huszar and collaborators. They had previously reported that sperm that are able to bind to hyaluronic acid (HA) are mature and have completed the spermiogenetic process of sperm plasma membrane remodelling, cytoplasmic extrusion, and nuclear histone−protamine replacement.[39,40] Testing of the newly invented, ICSI sperm selection method based on the binding ability of spermatozoa to HA found that in the HA-bound sperm *vs.* the unselected

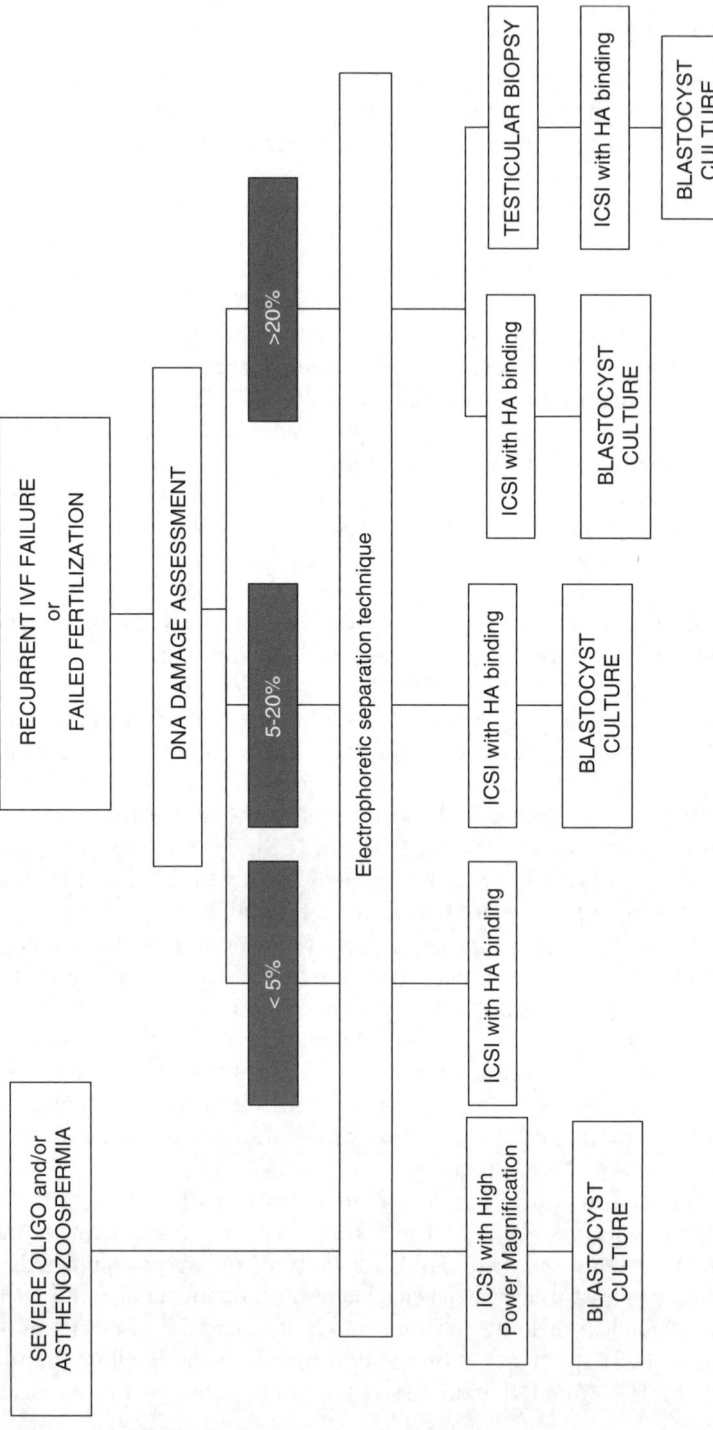

Figure 1   *A model for attributing ICSI treatment protocols to male infertility patients, using assessment of sperm nuclear DNA damage as the defining parameter.*

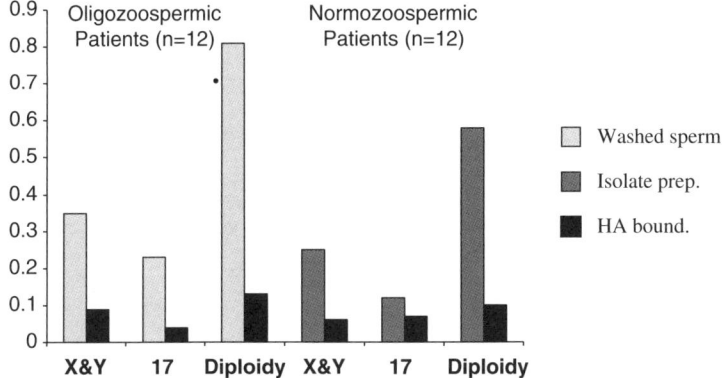

Figure 2 *Aneuploidy (X, Y, and 17) and diploidy rates in ejaculated human spermatozoa after a simple wash with centifugation, density gradient preparation using isolate and after selection using hyaluorinc acid binding. The experiment was performed on 12 men with Oligozoospemia and 12 normozoospermic men.*

sperm, the chromosomal disomy frequencies were reduced to 0.16 from 0.52%, diploidy to 0.09 from 0.51%, and sex chromosome disomy to 0.05 from 0.27% (a 5.4-fold reduction) (Figure 2). They concluded that the ICSI sperm selection method is likely to reduce the potential genetic complications and adverse public health effects of ICSI.[41]

8.6 Conclusion

There is accumulating evidence linking sperm nuclear anomalies to poor reproductive outcome in relation to ICSI. The tests currently available only provide an inkling of the impact of sperm nuclear DNA abnormalities on outcomes. More research is needed to improve our current knowledge in relation to the DNA anomalies in spermatozoa, how to detect them more accurately, and how they may relate to failed or abnormal reproductive outcomes. Finally, improvement in the detection and selection techniques of abnormal spermatozoa prior to choosing them for ICSI should alleviate the growing concerns over the safety of ICSI.

References

1. D. Sakkas, Y. D'Arcy, G. Percival, L. Sinclair, M. Afnan and K. Sharif, *Fertil. Steril.*, 2004, **82**, 74.
2. J. Tesarik, E. Greco and C. Mendoza, *Hum. Reprod.*, 2004, **19**, 611.
3. E. Seli and D. Sakkas, *Hum. Reprod. Update*, 2005, **11**, 337.
4. T. Hassold, M. Abruzzo, K. Adkins, D. Griffin, M. Merrill, E. Millie, D. Saker, J. Shen and M. Zaragoza, *Environ. Mol. Mutagen.*, 1996, **28**, 167.
5. M. MacDonald, T. Hassold, J. Harvey, L.H. Wang, N.E. Morton and P. Jacobs, *Hum. Mol. Genet.*, 1994, **3**, 1365.

6. M. Hansen, J.J. Kurinczuk, C. Bower and S. Webb, *N. Engl. J. Med.*, 2002, **346**, 725.

7. J.J. Kurinczuk, M. Hansen and C. Bower, *Curr. Opin. Obstet. Gynecol.*, 2004, **16**, 201.

8. A. Van Steirteghem, M. Bonduelle, P. Devroey and I. Liebaers, *Hum. Reprod. Update*, 2002, **8**, 111.

9. P. Devroey and A. Van Steirteghem, *Hum. Reprod. Update*, 2004, **10**, 19.

10. G.F. Cox, J. Burger, V. Lip, U.A. Mau, K. Sperling, B.L. Wu and B. Horsthemke, *Am. J. Hum. Genet.*, 2002, **71**, 162.

11. M.R. DeBaun, E.L. Niemitz and A.P. Feinberg, *Am. J. Hum. Genet.*, 2003, **72**, 156.

12. E.R. Maher, L.A. Brueton, S.C. Bowdin, A. Luharia, W. Cooper, T.R. Cole, F. Macdonald, J.R. Sampson, C.L. Barratt, W. Reik and M.M. Hawkins, *J. Med. Genet.*, 2003, **40**, 62.

13. J. Halliday, K. Oke, S. Breheny, E. Algar and D.J. Amor, *Am. J. Hum. Genet.*, 2004, **75**, 526.

14. M. Manning, W. Lissens, I. Liebaers, A. Van Steirteghem and W. Weidner, *Int. J. Androl.*, 2001, **24**, 87.

15. C.J. Marques, F. Carvalho, M. Sousa and A. Barros, *Lancet*, 2004, **363**, 1700.

16. M.D. Anway, A.S. Cupp, M. Uzumcu and M.K. Skinner, *Science*, 2005, **308**, 1466.

17. P.G. Bianchi, G.C. Manicardi, D. Bizzaro, U. Bianchi and D. Sakkas, *Biol. Reprod.*, 1993, **49**, 1083.

18. W. Gorczyca, F. Traganos, H. Jesionowska and Z. Darzynkiewicz, *Exp. Cell Res.*, 1993, **207**, 202.

19. G.C. Manicardi, P.G. Bianchi, S. Pantano, P. Azzoni, D. Bizzaro, U. Bianchi and D. Sakkas, *Biol. Reprod.*, 1995, **52**, 864.

20. M.J. Tomlinson, O. Moffatt, G.C. Manicardi, D. Bizzaro, M. Afnan and D. Sakkas, *Hum. Reprod.*, 2001, **16**, 2160.

21. I.D. Morris, S. Ilott, L. Dixon and D.R. Brison, *Hum. Reprod.*, 2002, **17**, 990.

22. A. Ahmadi and S.-C. Ng, *J. Exp. Zoolog.*, 1999, **284**, 696.

23. K.L. Larson-Cook, J.D. Brannian, K.A. Hansen, K.M. Kasperson, E.T. Aamold and D.P. Evenson, *Fertil. Steril.*, 2003, **80**, 895.

24. K.L. Larson, C.J. DeJonge, A.M. Barnes, L.K. Jost and D.P. Evenson, *Hum. Reprod.*, 2000, **15**, 1717.

25. J.G. Sun, A. Jurisicova and R.F. Casper, *Biol. Reprod.*, 1997, **56**, 602.

26. S. Lopes, J.G. Sun, A. Jurisicova, J. Meriano and R.F. Casper, *Fertil. Steril.*, 1998, **69**, 528.

27. P. Braude, V. Bolton and S. Moore, *Nature*, 1988, **332**, 459.

28. J.P. Twigg, D.S. Irvine and R.J. Aitken, *Hum. Reprod.*, 1998, **13**, 1864.

29. E. Seli, D.K. Gardner, W.B. Schoolcraft, O. Moffatt and D. Sakkas, *Fertil. Steril.*, 2004, **82**, 378.

30. M.R. Virro, K.L. Larson-Cook and D.P. Evenson, *Fertil. Steril.*, 2004, **81**, 1289.

31. D.T. Carrell, A.L. Wilcox, L. Lowy, C.M. Peterson, K.P. Jones, L. Erickson, B. Campbell, W. Branch and H.H. Hatasaka, *Obstet. Gynecol.*, 2003, **101**, 1229.
32. E. Greco, F. Scarselli, M. Iacobelli, L. Rienzi, F. Ubaldi, S. Ferrero, G. Franco, N. Anniballo, C. Mendoza and J. Tesarik, *Hum. Reprod.*, 2005, **20**, 226.
33. M. Spano, E. Seli, D. Bizzaro, G.C. Manicardi and D. Sakkas, *Curr. Opin. Obstet. Gynecol.*, 2005, **17**, 255.
34. D. Sakkas, G.C. Manicardi, M. Tomlinson, M. Mandrioli, D. Bizzaro, P.G. Bianchi and U. Bianchi, *Hum. Reprod.*, 2000, **15**, 1112.
35. D. Sakkas, *The Male Gamete: From Basic Science to Clinical Applications*, C. Gagnon (ed), Cache River Press, Vienna, 1999, Chapter 34.
36. B. Bartoov, A. Berkovitz and F. Eltes, *N. Engl. J. Med.*, 2001, **345**, 1067.
37. B. Bartoov, A. Berkovitz, F. Eltes, A. Kogosovsky, A. Yagoda, H. Lederman, S. Artzi, M. Gross and Y. Barak, *Fertil. Steril.*, 2003, **80**, 1413.
38. C. Ainsworth, B. Nixon and R.J. Aitken, *Hum. Reprod.*, 2005, **20**, 2261.
39. G. Huszar, C.C. Ozenci, S. Cayli, Z. Zavaczki, E. Hansch and L. Vigue, *Fertil. Steril.*, 2003, **79**(3), 1616.
40. G. Huszar, M. Sbracia, L. Vigue, D.J. Miller and B.D. Shur, *Biol. Reprod.*, 1997, **56**, 1020.
41. A. Jakab, D. Sakkas, E. Delpiano, S. Cayli, E. Kovanci, D. Ward and G. Huszar, *Fertil. Steril.*, 2005, **84**(6), 1665.

Animal Models

CHAPTER 9

Male-Mediated F_1 Effects in Mice Exposed to Di(2-ethylhexyl)phthalate (DEHP)

MAŁGORZATA M. DOBRZYŃSKA,[a] URSZULA CZAJKA[a] AND EWA J. TYRKIEL[b]

[a] Department of Radiation Protection and Radiobiology, National Institute of Hygiene, 00-789 Warsaw, 24 Chocimska StreetPoland
[b] Department of Environmental Toxicology, National Institute of Hygiene, Warsaw, Poland

9.1 Introduction

Phthalates (phthalate diesters) are the group of multifunctional industrial chemicals widely used in consumer products (*e.g.* soaps, perfumes, lotions, shampoos and cosmetics) and as solvents (*e.g.* in paints, glue, insect repellents and lubricants). They are used to soften a wide range of plastics, including medical products such as polyvinyl chloride (PVC) blood products and intravenous bags as well as dialysate bags and tubing.[1-3] Phthalates are also used in production of plastic goods, toys for children and food wrappings.[4] The highest concentration of phthalates with levels of up to 30% is found in baby teething rings.[5]

People are potentially exposed to many products containing phthalates. Due to constant occupational exposure, workers in plastic and rubber industries and those involved in automobile and aircraft manufacturing may have the greatest health risk.[5,6] The primary exposure of the general population to phthalates comes from ingestion of food, especially fatty foods, such as milk, butter and meat. Human exposure to phthalates can also occur *via* inhalation and dermal routes and intravenous and parenteral absorption, for example, in the case of patients undergoing medical procedures that involve the use of medical devices containing phthalates.[1,3,7] Phthalates are also found in drinking water.[4]

Di(2-ethylhexyl)phthalate (DEHP) is used in consumer products for instance in teething rings, pacifiers and toys for young children, as well as in food containers and a variety of buildings, household and automotive products.[13,15]

Some phthalates are reproductive and developmental toxicants. They are known to have adverse effects on the reproductive system and its function and on developing foetuses. Some can cause damage to the testis and decrease sperm production.[8–12]

DEHP is a chemical that has become widely distributed in the environment and our bodies. DEHP constitutes approximately half the total production volume of the approximately 20 phthalate esters in common use. More than 2 million tons of this phthalate are produced each year wordwide.[13,14] DEHP is the main plasticiser for PVC products.[14]

Typical exposure of the general human population is estimated to be 3–30 μg DEHP kg^{-1} day^{-1}.[15–17] Some people receive greater exposure coming from DEHP plasticised medical devices such as blood bags, haemodialysis tubing and membranes, autophoresis equipment and nasogastric feeding tubes.[15] For instance, dialysis patients may be exposed to approximately 12 g of DEHP over the course of a year.[18]

DEHP has antiandrogenic properties and is suspected to be responsible for endocrine disruptor-like effects.[11] DEHP is a known reproductive and developmental toxicant in animals.[16] Effects on young and adult rodents include reduction of testosterone and sperm production, reduction in testis and epididymis weights and pathological effects on the testis.[13,16,19–21] Utero and lactational exposure reduces testis weight and sperm count in rats.[22,23] The aim of the present study was to investigate the possibility of transmission of mutations induced by subchronic exposure, covering the full spermatogenesis cycle, to the F1 generation.

9.2   Methods

Animals were housed in plastic cages in a room designed for control of temperature, humidity and light cycle. Tap water and rodent LSM diet were available *ad libidum*. 42–45 days old Pzh:Sfis outbred male mice were exposed by gavage to olive oil or DEHP solution in olive oil for 8 weeks at the rate of 3 days per week. The doses of DEHP were 2000 mg kg^{-1} bw ($1/16$ LD_{50}) and 8000 mg kg^{-1} bw ($1/4$ LD_{50}) daily.

9.2.1   Effects in Exposed Generation

For this study, 5 male mice in each group were weighed and killed at 24 h and at 4 weeks after the last treatment. Both testes and epididymides were removed and weighed from each male. One epididymis was macerated in 0.2 ml of 1% solution of trisodium citrate for 5–8 min and minced. Then the solution was made up to 2 ml and mixed for about 1 min. The sperm suspension was diluted 1:1 in 10% buffered formalin. The spermatozoa were counted using an improved Neubauer haemocytometer.[24,25]

The contents of the second epididymides were placed into 0.2 ml of warm (37°C) physiological saline. An aliquot was placed on warm (37°C) microscope slide and covered with a coverslip. Two hundred cells per animal were evaluated for motility within 5 min after the removal of the epididymis.[26]

The remaining sperm was dispersed evenly in saline. The study of frequency of morphologically abnormal spermatozoa was performed according to the procedure described by Wyrobek and Bruce.[27] Smears were prepared on microscope slides, air-dried overnight and stained with eosin Y. Then 500 spermatozoa per mouse were analysed using a light microscope, and abnormal sperm heads (*e.g.* lacking hook, amorphous, banana-shaped head) were recorded.

For comet assay analysis, one testis from each animal was decapsulated and placed in RMPI 1640 medium and minced with scissors. Before using the cells, tubes were swirled so that single cells remained in suspension. The basic technique of Singh *et al.*[28] and further described by Anderson *et al.*[29] was used. 5 µl of cell suspension was mixed in an Eppendorf tube with 75 µl low melting point agarose (LMA) for embedding on slides. The slides were immersed in lysing solution overnight at 4°C. Then they were drained and placed in a gel electrophoresis tank, and left in the solution for 20 min. The electrophoresis was conducted at 4°C for 20 min using 19 V and 300 mA. After neutralisation, slides were stained with EtBr and examined using a fluorescence microscope. Images of 200 randomly selected cells from each animal were analysed. According to the method of Anderson *et al.*,[29] cells were graded by eye into 5 categories corresponding to the following amounts of deoxyribonucleic acid (DNA) in the tail: (A) no damage, $<5\%$; (B) low-level damage, 5–20%; (C) medium-level damage, 20–40%; (D) high-level damage, 40–95%; (E) total damage, $>95\%$.

To obtain a semi-quantitative analysis of the data, the score of DNA damage (the migration) of DNA was calculated as follows: percentage of cells with category B × 2, plus percentage of cells with category C × 3, plus percentage of cells with category D × 4, plus percentage of cells with category E × 5.

9.2.2   Effects in the Offspring of Exposed Males

Immediately after the end of an 8-week exposure, each male from the control as well as experimental groups was caged for one week with two unexposed, virgin females. They were checked daily for the presence of a vaginal plug. This determined day 0 of pregnancy. Three-quarters of the mated females from each group were humanely killed a day before parturition. The other quarter were allowed to deliver and rear their litters.

9.2.2.1   *Dominant Lethal and Congenital Malformation Study*

The standard protocol for dominant lethal assay[30] with modifications proposed by Anderson *et al.*[31] was used.

A male mating with at least one female was defined as fertile. A female with at least one live or dead implantation was defined as pregnant.

Females were examined for the number of implantations, the number of live foetuses and the number of early and late postimplantation deaths. Postimplantation deaths were classified as early, if the embryo had died and had been resorbed, or late, if the dead embryo was at a stage beyond the onest of organogenesis.

The dominant lethal mutations (DLM) were calculated according to the formula:

$$\%\text{DLM} = \left[1 - \frac{\text{living embryos/pregnant treated female}}{\text{living embryos/pregnant control female}}\right] \times 100$$

Live embryos were weighed and analysed for the presence and type of gross malformations (*e.g.* exencephaly). Runts were defined as live foetuses having a body weight less than 75% of the mean of their litter mates.[32] Malformed foetuses and half that number of randomly selected normal foetuses, from each of the exposed and control groups, were processed for skeletal malformations after alcian blue and alizarin red staining.

9.2.2.2 Effect on the Postnatal Development of F1 Generation

Pups were counted and weighed at birth, then weighed weekly up to 8 weeks of age. They were observed for physiological markers and growth parameters.

Mortality was recorded from birth to the age of 8 weeks, and the percent of mortality was calculated as follows:

$$\% \text{Mortality} = \frac{\text{Total number of deaths}}{\text{Number of live births}} \times 100$$

Mean body weight (g) of the individual litters and of each group was calculated weekly. Pups weighing less than 2 standard deviations of the mean body weight of the control group were considered growth retarded.[33] The percent of growth-retarded pups was calculated.

$$\% \text{Growth-retarded pups} = \frac{\text{Number of growth retarded pups}}{\text{Total number of live pups}} \times 100$$

Animals of the F1 generation were observed for physiological markers such as fur development, pinna detachment, eye opening, vaginal opening and testes descent.

The appearance of pinna detachment unfolding was recorded as the age (in days) when pinnae of both ears unfolded to a fully erect position. Eye opening is defined as any visible break in the membrane covering the eye. Vaginal opening was defined as any visible break in the membrane when the vaginal lips were gently pulled laterally. Testes descent was recorded when the testes descended to lie in the scrotal sac (Hossain *et al.* 1999).

9.2.3 Statistical Analysis

Statistical analysis was performed by using Student *t*-test, Fisher test and χ^2 test.

9.3 Results

The results of the mean testes and epididymis weights are shown in Table 1. Exposure to DEHP for 8 weeks reduced testes weight and relative testes weight. Results were statistically significant at 24 h after the end of exposure to 2000 mg kg^{-1} daily DEHP, only. At 4 weeks after the end of exposure the mean testes weight after exposure to $1/16$ LD_{50} was slightly decreased, but relative testes weight was similar to control value. Epididymides weights were slightly reduced in some cases, but results were not statistically significant. The relative epididymides weight was slightly reduced at 4 weeks after the last treatment to 8000 mg kg^{-1} bw daily.

The results of sperm count, motility, morphology and for the score of DNA damage are presented in Table 2. Sperm counts were the lowest after an 8-week exposure to 2000 mg kg^{-1} bw of DEHP, but the result was not statistically significant. Sperm motility was not changed at 24 h after the end of exposure and was dose dependent and decreased at 4 weeks after the last treatment. After exposure to 8000 mg kg^{-1} there was about 20% less motile spermatozoa as in control group. Exposure to DEHP for 8 weeks induced malformation of sperm. At 24 h after the last treatment, an increased frequency of abnormal spermatozoa was noted in both experimental groups. The results were dose dependent 4 weeks after the end of exposure. The highest frequency of abnormal spermatozoa was observed after exposure to 8000 mg kg^{-1} bw per day DEHP. There were no effects on the induction of DNA damage measured by Comet assay analysis at 24 h after the end of exposure. At 4 weeks after the last treatment, DNA migration was slightly increased in both the exposed groups, but results were not statistically significant.

Effects on male fertility and on the frequency of pregnant females were not observed (Table 3). A decrease in total implantation as well as in the mean of live foetuses after exposure to the higher dose was noted, but only for live foetuses was the result statistically significant. A slight effect was found on the frequency of dead implants. The highest frequency of dead implants was observed after the exposure to the low dose of DEHP. The percent of early deaths increased with dose, and the percent of late deaths decreased with dose. Exposure to the lower dose for 8 weeks induced 6% DLM, whereas exposure to the higher dose induced 21% DLM.

The results of the bodyweight, gross and skeletal malformations of surviving foetuses are shown in Table 4. There were no effects on the mean body weight of surviving foetuses. Some of the live foetuses in the control and in experimental groups showed congenital defects, but we did not find any statistical differences in their incidence between the groups. The skeletal examination of

Table 1 *Mean testes and epididymides weights after 8 weeks exposure of male mice to DEHP*

Dose	Time	Mean bodyweight (g)	Mean testes weight (mg) ±SD	[a] Relative testes weight [%]	[b] Mean epididymides weight (mg) ±SD	Relative epididymides weight [%]
Control	8 weeks	38.65	249 ± 36.7	0.64	50.0 ± 10.7	0.13
1/16 LD50 DEHP	8 weeks	34.48	168 ± 12.9**	0.49	45.5 ± 5.5 ns	0.13
1/4 LD50 DEHP	8 weeks	39.20	235 ± 55.5 ns	0.60	52.8 ± 4.9 ns	0.13
Control	8+4 weeks	41.62	248 ± 37.7	0.60	58.2 ± 10.1	0.14
1/16 LD50 DEHP	8+4 weeks	38.30	216 ± 59.9 ns	0.56	54.2 ± 6.3 ns	0.14
1/4 LD50 DEHP	8+4 weeks	41.14	236 ± 29.7 ns	0.57	49.6 ± 8.6 ns	0.12

Notes: Student *t*-test: ns stands for not significant, **$p < 0.01$.

[a] Relative testes weight $= \dfrac{\text{Testes weight}}{\text{Bodyweight}} \times 100$

[b] Relative epididymides weight $= \dfrac{\text{Epididymides weight}}{\text{Bodyweight}} \times 100$

Table 2 *Sperm quantity and quality after 8-week exposure of male mice to DEHP*

Dose	Time	Sperm count $\times 10^6$ ±SD	Percent of motile spermatozoa ±SD	Percent of abnormal spermatozoa ±SD	Comet score of DNA damage
Control	8 weeks	2.56 ± 0.52	62.00 ± 8.08	9.68 ± 2.67	232.5
1/16 LD_{50} DEHP	8 weeks	2.31 ± 0.54 ns	62.75 ± 10.69 NS	12.00 ± 3.10$^{\#\#}$	217.6
1/4 LD_{50} DEHP	8 weeks	2.84 ± 1.01 ns	67.25 ± 13.70 NS	11.95 ± 3.25$^{\#}$	224.6
Control	8 + 4 weeks	2.69 ± 0.41	77.00 ± 10.75	11.44 ± 1.90	173.6
1/16 LD_{50} DEHP	8 + 4 weeks	2.47 ± 0.60 ns	69.20 ± 20.57 NS	13.10 ± 2.25 NS	197.8
1/4 LD_{50} DEHP	8 + 4 weeks	2.75 ± 1.02 ns	57.63 ± 8.44$^{\#\#}$	17.62 ± 4.73$^{\#\#\#}$	208.7

Notes: χ^2 test: NS not significant, $^{\#}p < 0.05$, $^{\#\#}p < 0.01$, $^{\#\#\#}p < 0.001$; Student t-test: ns not significant.

abnormal foetuses usually confirmed the necroscopy observations. There was a slight increase in the percentage of abnormal skeletons in both the experimental groups, but results were not statistically significantly different compared with the controls.

The results of postnatal bodyweight of pups and percent of growth retardation are shown in Table 5. At birth there was no effect on the bodyweight of pups of exposed and unexposed males. Pups from control and exposed groups did not show any congenital malformations. A statistically significant lower bodyweight was observed during the first week of postnatal life in the offspring of males treated with 2000 mg kg^{-1} bw and 8000 mg kg^{-1} bw of DEHP daily. By the next week, the bodyweight of these pups had increased to normal levels in the group of males exposed to 1/4 LD_{50} of DEHP. The bodyweights of pups of males exposed to 1/16 LD_{50} exceeded the bodyweights of control pups in some cases. At 5 weeks of age mean body weight of animals from both experimental groups were higher than the mean body weight of control. The result of the 1/16 LD_{50} group was significantly different from control group. Then, up to 8 weeks of age the bodyweight of pups from exposed and unexposed males were similar. At birth, we observed growth retardation in 1.5% of pups from males exposed to the higher dose. During the first week of postnatal life, about 23% pups of males exposed to the lower dose and about 18% offspring of males exposed to the higher dose of DEHP were growth retarded. Some pups from the group of males exposed to 8000 mg kg^{-1} bw also showed growth retardation at 5, 6 and 8 weeks of age, but this was not statistically significant.

Table 3 *Effects of 8-week exposure of male mice to DEHP on prenatal development of fetuses*

Dose	% of fertile males	% of pregnant females	No of implantations/ pregnant female ± SD	No of live foetuses ± SD	No of dead foetuses/pregnant female ± SD	% of early deaths	% of late deaths	% DLM
Control	94	67	10.60 ± 1.50	10.14 ± 1.35	0.43 ± 0.76	2.60	1.30	–
1/16 LD 50 DEHP × 3 × 8	100	76	10.09 ± 2.83 ns	9.57 ± 2.94 ns	0.52 ± 0.79 ns	4.31	0.86	6
1/4 LD 50 DEHP × 3 × 8	94	72	8.47 ± 3.85 ns	8.00 ± 3.57*	0.47 ± 0.70 ns	4.97	0.62	21

Notes: Student *t*-test: ns not significant, *$p < 0.05$.

Table 4 Effects of subchronic paternal DEHP exposure on the induction gross and skeletal malformations of the surviving foetuses in mice

Dose	Mean body weight of living foetuses (g)	% abnormal foetuses	Type of gross malformations	% of abnormal skeletons	Type of skeletal malformations
Control	1.30	2.8	2 runts: 63.8%, 61.9%; 2 bent tail	4.23	missing rib – 1; extra rib – 2
1/16 LD$_{50}$ DEHP $\times$ 3 $\times$ 8	1.28	1.9 ns	3 runts: 74.4%, 73.3%, 67.2%; 1 runt (74.4%) with exencephaly + cleft nose	5.08 ns	missing rib – 1; extra rib – 2; dislocations of skull bones[a] – 1; retardation of development of foot bones – 1; rudimentary rib-1
1/4 LD$_{50}$ DEHP $\times$ 3 $\times$ 8	1.23	3.3 ns	2 runts: 68.5%, 50.5%; 2 bumps on the spine; 1 bent hind limb	5.8 ns	extra rib – 4; retardation of development of palm bones-1

Notes: Fisher test: ns not significant.
[a] Runt with exencephaly, cleft nose.

Table 5   *Changes in postnatal bodyweight (g) and percent of growth-retarded pups of males exposed to DEHP*

Time after birth	bw/ % g-r	Control	1/16 LD_{50} DEHP	1/4 LD_{50} DEHP
			Paternal dose	
at birth	bw	1.62 ± 0.25	1.56 ± 0.16ns	1.59 ± 0.23 ns
	%g-r	0	0	1.5 NS
1 week	bw	4.03 ± 0.43	3.64 ± 0.74**	3.69 ± 0.65**
	%g-r	4.35	23.08[#]	17.74 NS
2 week	bw	5.85 ± 1.27	6.03 ± 1.11 ns	5.98 ± 0.98 ns
	%g-r	0	0	0
3 week	bw	7.85 ± 2.50	8.02 ± 2.03 ns	7.67 ± 2.13 ns
	%g-r	0	0	0
4 week	bw	12.52 ± 4.47	13.98 ± 3.43 ns	12.77 ± 4.53 ns
	%g-r	0	0	0
5 week	bw	18.57 ± 4.54	20.74 ± 3.26**	19.26 ± 4.98 ns
	%g-r	0	0	3.33 NS
6 week	bw	23.61 ± 3.37	24.64 ± 3.05 ns	23.71 ± 4.06 ns
	%g-r	0	0	6.67 NS
7 week	bw	26.55 ± 3.71	27.01 ± 3.15 ns	26.51 ± 3.60 ns
	%g-r	0	0	0
8 week	bw	28.53 ± 3.24	28.43 ± 2.65 ns	28.31 ± 3.41 ns
	%g-r	0	0	1.75 NS

Notes: % g-r stands for percent of growth-retarded pups; χ^2 test: NS not significant, [#]$p < 0.05$, Student t-test: ns not significant, **$p < 0.01$.

There were no effects on litter size (Table 6). The percent of postnatal mortality increased with dose but not with statistical significance. There were no significant differences between control and experimental groups in the time of appearance of pinna detachment, fur development and eye opening.

Paternal exposure to DEHP for 8 weeks induced slight delay in the appearance of vaginal opening and a significant delay in testes descent, especially at the lower dose (Table 6).

9.4   Discussion

Phthalates including DEHP are widely used as plasticisers. They constitute 10–60% by weight of many plastics because they impart flexibility, transparency and other desirable physical properties. Phthalates are not covalently bound to the polymers with which they are mixed, so they can leach into the food, beverages or other materials contained in these plastics.[13]

In this paper, effects on spermatozoa are presented following subchronic exposure of all stages of male mice germ cells: spermatogonial stem cells, differentiating spermatogonia, spermatocytes, spermatids and mature spermatozoa. Exposure of mice to DEHP for 8 weeks reduced both testes and relative testes weight, and slightly decreased epididymides weight. Results of the experiments presented in this chapter confirmed the results obtained by other

Table 6 *Postnatal mortality and appearance of physiological markers in pups of males exposed to DEHP*

Paternal dose	Mean litter size	Percent of mortality	Time of appearance, days (mean ± SD)				
			Pinna detachment	Fur development	Eye opening	Vagina opening	Testes descent
Control	9.6	6.67	6.40 ± 0.89	6.60 ± 0.89	15.00 ± 0.00	24.63 ± 2.17	23.82 ± 1.59
1/16 LD$_{50}$ DEHP	9.2	10.00 NS	6.33 ± 0.82 ns	6.50 ± 0.84 ns	15.00 ± 0.00 ns	25.54 ± 2.90 ns	25.33 ± 1.69**
1/4 LD$_{50}$ DEHP	9.6	11.67 NS	6.00 ± 0.00 ns	6.43 ± 0.53 ns	15.67 ± 0.52 ns	25.90 ± 2.79 ns	24.80 ± 1.97 ns

Notes: Student t-test: ns not significant, **$p < 0.01$, χ^2 test: NS not significant.

authors in rodents.[19,34–40] Surprisingly, the reduction in gonadal weights was not correlated with diminished sperm counts. In the articles of other authors, in the reduction in sperm production, a correlation with diminished epididymides and testes weights was observed.[38,41]

In earlier papers, there is no evidence of a direct effect of phthalates on the germ cells. Sertoli cells are primarily affected by treatment with phthalates, and are involved in the regulation of the proliferation and differentiation of all spermatogenic cells. The effect would be detectable in the ejaculate later.[8,42–44] Damage to Sertoli cell function may be very dangerous, because Sertoli cells do not proliferate after puberty.[42,45–48] A lack of normal Sertoli cells' function may affect spermatogenesis.[49] Exposure to phthalates including DEHP induced necrosis of spermatogonia and shedding of spermatocytes and spermatids from the tubules lumen associated with the morphological changes in the supporting Sertoli cells.[39,50] Adult male rats exposed to 1000 mg kg^{-1} and 2000 mg kg^{-1} DEHP for 5 days showed a loss of both spermatogonia and spermatocytes.[51] Loss of some germ cells could be a reason for the diminished testes weight. Exposure to DEHP has been reported to lead to testicular fragmentation and apoptosis in mouse testis.[52] Destruction of Sertoli cells could be the cause of diminished sperm motility at 4 weeks after the end of exposure to 1/4 LD$_{50}$ of DEHP daily and slightly increased DNA damage in male germ cells at 4 weeks after the last treatment; though the mechanism is not clear. Increase in DNA damage at 4 weeks after the end of exposure may suggest an accumulation of damage and lack of repair of DNA damage induced in haploid germ cells.

Disruption of spermatogenesis may be induced in rats by feeding them a diet containing DEHP. The administration of DEHP induced aspermatogenesis accompanied by testicular atrophy.[40,44] Agarwal *et al.*[37] observed reduction in epididymal sperm density and motility after exposure of rats to DEHP. Decreased sperm motility in rats after exposure to other phthalate ester, dibutyl phthalate (DBP) was observed by Tsutsumi *et al.*[53] In our study, a significantly increased frequency of abnormal spermatozoa was observed at 24 h as well as at 4 weeks after the last treatment. Similar results were obtained by Agarwal *et al.*[37] after 60 days of exposure of rats to 20,000 ppm of DEHP. It is a known fact that malformed spermatozoa are usually less motile[54] and they have less ability to fertilise eggs. Our results showed an association between diminished sperm motility and an increase in abnormal spermatozoa at 4 weeks after the last treatment with 1/4 LD$_{50}$ of DEHP.

In our study, an 8-week exposure to DEHP did not decrease the male's fertility or percentage of pregnant females. Similarly, no consistent changes in the fertility of males were observed by Dostal *et al.*[51] and Agarwal *et al.*[37] after the exposure of male rats to doses between 10 mg kg^{-1} and 2000 mg kg^{-1} of DEHP. Subcutaneous exposure of male mice to DEHP just before mating to untreated females resulted in a reduction of the incidence of pregnancies. On the other hand, mating at longer intervals after exposure did not induce effects on the incidence of pregnancy,[55] suggesting that the effect is transient.

Defects in sperm DNA or chromosomes may be associated with effects determined on the viability of the embryo and potential health risk to the newborn.[56] In our study, after daily exposure to 1/4 LD_{50} DEHP, a reduction in total implantations as well as the frequency of live foetuses was observed. Decrease in total implantations could reflect an excess of unfertilized eggs or preimplantation loss caused by damage of spermatogonial cells leading to the death of fertilized eggs before implantation. Preimplantation losses may also be correlated with defects manifested in spermatozoa, which are then not able to fertilize eggs.

The frequency of early deaths increased in a dose-related manner. This may reflect a genetic effect, especially if it correlates with a lower incidence of total implantations. The foetal losses detected in the dominant lethal test are usually caused by numerical and structural chromosome damage or lethal gene mutations derived from the fertilizing spermatozoa.[57] In our study, the frequency of early deaths exceeded the frequency of late deaths and survival of foetuses with gross malformations. Most of the induced mutations lead to death shortly after implantation.[58] The reason for the higher incidence of late deaths in the control group may be connected with the larger litter size. The dead foetuses were found in litters of at least 11.

There is relatively little information about developmental toxicology caused by exposure of males before conception. The majority of articles regarding DEHP describes the effects on the offspring after exposure of pregnant females.

Dostal *et al.*[51] did not observe a decrease in total and live implantations after a 5-day exposure of male rats to 10–2000 mg kg^{-1} DEHP. In contrast, the average litter size was reduced in rats paternally exposed to 20,000 ppm DEHP administered for 60 days.[37] The number of live foetuses per dam after exposure to 1000 mg kg^{-1} day^{-1} of DEHP of pregnant female rats was reduced in some cases.[20,59] Hellwig *et al.*[59] reported that the frequency of resorptions increased 10 times after exposure of dams to 1000 mg kg^{-1} day^{-1} DEHP, and late deaths were more frequent than early deaths. Foetal deaths averaging 20–36% were observed in the group of female rats receiving 1000 mg kg^{-1} DEHP during pregnancy.[60]

We observed similar bodyweights among foetuses of exposed and unexposed males and the percentages of gross and skeletal malformations were not significantly changed. Similar results were obtained by Agarwal *et al.*[37] In other studies, mean foetal bodyweight was reduced, and up to 63% foetuses showed gross and skeletal malformations after maternal exposure of rats to 1000 mg kg^{-1} day^{-1} DEHP.[20,59] Another phthalate ester, butyl benzyl phthalate (BBP) also induced malformations in foetuses after maternal exposure during pregnancy.[61–63]

In our study, we found a slight, but not statistically significant, increase in the frequency of postnatal mortality in the offspring of exposed males. Similarly, DEHP appeared to cause dose-dependent increases in postnatal mortality in the study by Moore *et al.*[13] In contrast, Agarwal *et al.*[37] observed that the rate of neonatal deaths was similar in the control and DEHP treated groups.

Mean litter sizes were not changed in the offspring of treated males in comparison to the control groups, in our experiment. This differs from the results of other authors. Lamb *et al.*[64] observed decreases in the number and proportion of pups born alive after exposure of both sexes of mice to DEHP in diet. Also, the number of rat pups born per dam exposed to 1500 mg kg^{-1} day^{-1} DEHP during pregnancy was significantly reduced.[13]

We observed a significant delay in testicular descent after an 8-week paternal exposure to DEHP. Similarly, increases in reproductive system malformations and induction of abnormalities of sexual development in the male rats exposed in utero and during lactation to DEHP have been shown.[13,20,65,66] Moore *et al.*[13] reported that the male reproductive system is far more sensitive to DEHP early in development than when animals are exposed as juveniles or adults. Exposure of pregnant females to monobenzyl phthalate, the metabolite of DEHP, caused a significant increase in the incidence of undescended testes in male foetuses.[67] Adverse effects on the development of the reproductive system of male offspring (*i.e* undescended testes) were reported also in Wistar rats administered by gavage with BBP on days 15–17 of pregnancy.[68]

Recently, it has been reported that male reproductive tract abnormalities are associated with changes of gene expression in the foetal testis. DEHP and other phthalates administered during pregnancy significantly altered the expression of 391 genes. Gene targets include alpha inhibition, which is essential for Sertoli cells' development, and genes involved with communication between Sertoli cells and gonocytes.[69]

One reason for the differences between our results and those given in other papers may be the various exposure levels, differences in time and route of exposure and also differences in the sensitivity of species.[70–72]

Comparison of results regarding dominant lethality and postnatal development offspring of exposed male mice leads to the conclusion that high and low doses of DEHP might cause different effects in male mice germ cells. The higher dose of DEHP ($1/4$ LD$_{50}$) induces rather stronger damage in male gametes caused in pre- or early post-implantation deaths. The lower dose ($1/16$ LD$_{50}$) leads more often to growth retardation of pups and delay in testes descent in the offspring of exposed males.

Results obtained here suggest that subchronic exposure of male germ cells to DEHP cause genetic defects that might be transmitted to the offspring. This is the first demonstration of the results of the exposure of males to DEHP for the full spermatogenic cycle.

Acknowledgments

This work was supported by Polish Ministry of Scientific Research and Information Technology (2004–2006), Project no 2PO5D 02926.

References

1. Agency for Toxic Substances and Disease Registry (ATSDR), *Toxicological Profile for Diethyl Phthalate*, Atlanta, GA, 1995.
2. Agency for Toxic Substances and Disease Registry (ATSDR), *Toxicological Profile for Di-2-(ethylhexyl) Phthalate*, Atlanta, GA, 2000.

3. Agency for Toxic Substances and Disease Registry (ATSDR), *Toxicological Profile for Di-n-butyl Phthalate*, Atlanta, GA, 2001.

4. United States Environmental Protection Agency (USEPA), Drinking water and health, Office of Ground Water and Drinking Water 1999. Retrieved from http://www.epa.gov/ogwdw000/dwh/c-soc/Phthalat.html.

5. N.H. Kleinsasser, E.R. Kastenbauer, H. Weissacher, R.K. Muenzenrieder and U.A. Harréus, *Environ. Mol. Mutagen.*, 2000, **35**, 9.

6. H.A. Dirven, P.H.H. Van Den Broek and F.J. Jongeneelen, *Int. Arch. Occup. Environ. Health*, 1993, **64**, 549.

7. R. Hauser, S. Duty, L. Goldfrey-Bailey and A.M. Calafat, *Environ. Health Perspect.*, 2004, **6**, 751.

8. T.J.B. Gray, I.R. Rowland and P.M. Foster, *Toxicol. Lett.*, 1982, **11**, 141.

9. L. Hardell, C.G. Ohlson and M. Fredrikson, *Int. J. Cancer*, 1997, **73**, 828.

10. R. Wine, L. Li, L. Barnes, D. Gulati and R. Chapin, *Environ. Health Perspect.*, 1997, **105**, 102.

11. E. Mylchreest, R. Cattley and P. Foster, *Toxicol. Sci.*, 1998, **43**, 47.

12. E. Gray, C. Wolf, P. Lambright, R. Mann, R. Price, R. Cooper and J. Ostby, *Toxicol. Ind. Health*, 1999, **15**, 94.

13. R.W. Moore, T.A. Rudy, T.M. Lin, K. Ko and R.E. Peterson, *Environ. Health Perspect.*, 2001, **109**, 229.

14. H.M. Koch, H. Drexler and J. Angerer, *Int. J. Hyg. Environ. Health*, 2003, **206**, 1.

15. J. Doull, R. Cattley, C. Elcombe, B.G. Lake, J. Swenberg, C. Wilkinson, G. Williams and M.A. van Gemert, *Regul. Toxicol. Pharmacol.*, 1999, **29**, 327.

16. R. Kavlock, K. Boeckelheide, R. Chapin, M. Cunningham, E. Faustman, P. Foster, M. Golub, R. Henderson, I. Hinberg, R. Little, J. Seed, K. Shea, S. Tabacova, R. Tyl, P. Williams and T. Zacharowski, *Reprod. Toxicol.*, 2002, **16**, 529.

17. W.W. Huber, B. Grasl-Kraupp and R. Schulte-Hermann, *Crit. Rev. Toxicol.*, 1996, **26**, 365.

18. M.A. Faouzi, T. Dine, B. Gressier, K. Kambia, M. Luyckx, D. Pagniez, C. Brunet, M. Cazin, A. Belabed and J.C. Cazin, *Int. J. Pharm.*, 1999, **180**, 113.

19. F.A. Arcadi, C. Costa, C. Impertore, A. Marchese, A. Rapisarda, M. Salemi, G.R. Trimarchi and G. Costa, *Food Chem. Toxicol.*, 1998, **36**, 963.

20. L.E. Gray Jr., J. Ostby, J. Furr, M. Price, D.N.R. Veeramachaneni and L. Parks, *Toxicol. Sci.*, 2000, **58**, 350.

21. Agency for Toxic Substances and Disease Registry (ATSDR), *Toxicological Profile for Di(2-ethylhexyl) Phthalate*, Atlanta, GA, 2002.

22. R. Tandon, S.R. Chowdhary, P.K. Seth and S.P. Srivastava, *J. Environ. Biol.*, 1990, **11**, 345.

23. R. Tandon, P.K. Seth and S.P. Srivastava, *Indian. J. Exp. Biol.*, 1991, **29**, 1044.

24. A.G. Searle and C.V. Beechey, *Mutat. Res.*, 1974, **22**, 69.

25. A. Harrison and P.C. Moore, *Health Phys.*, 1980, **39**, 219.

26. P.K. Working, J.S. Bus and T.E. Hamm Jr., *Toxicol. Appl. Pharmacol.*, 1985, **77**, 144.
27. A.J. Wyrobek and W.R. Bruce, *Proc. Natl. Acad. Sci. USA*, 1975, **72**, 4425.
28. N.P. Singh, M. Mc Coy, R.R. Tice and E.L. Schneider, *Exp. Cell. Res.*, 1988, **175**, 184.
29. D. Anderson, T.-W. Yu, B.J. Phillips and P. Schmezer, *Mutat. Res.*, 1994, **307**, 261.
30. I. Knudsen, E.V. Hansen, O.A. Meyer and E. Poulsen, *Mutat. Res.*, 1977, **48**, 267.
31. D. Anderson, A.J. Edwards and M.H. Brinkwort, in *Butadiene and Styrene: Assesment of Health Hazards*, M. Sorsa, K. Peltonen, H. Vainio and K. Hemminki (eds), vol. 127, IARC Scientific Publications, Lyon, 1993, 171.
32. M.K. Kirk and M.F. Lyon, *Mutat. Res.*, 1984, **125**, 75.
33. M. Hossain, P. Uma Devi and K.S. Bisht, *Teratology*, 1999, **59**, 133.
34. T.J. Gray, K.R. Butterworth, I.F. Gaunt, G.P. Grasso and S.D. Gangolli, *Food. Cosmetic. Toxicol.*, 1977, **15**, 389.
35. S. Oishi and K. Hiraga, *Toxicol. Appl. Pharmacol.*, 1980, **53**, 35.
36. S. Oishi, *Arch. Toxicol.*, 1986, **59**, 290.
37. D.K. Agarwal, S. Eustis, J.C. Lamb, J.R. Reel and W.M. Kluwe, *Environ. Health Perspect.*, 1986, **65**, 343.
38. A. Siddiqui and S.P. Srivastava, *Bull. Environ. Contam. Toxicol.*, 1992, **48**, 115.
39. R. Poon, P. Lecavalier, R. Mueller, V.E. Valli, B.G. Procter and I. Chu, *Food Chem. Toxicol.*, 1997, **35**, 225.
40. M. Ishihara, M. Itoh, K. Miyamoto, S. Suna, Y. Takeuchi, I. Takenake and F. Jitsunari, *Int. J. Androl.*, 2000, **23**, 85.
41. D. Parmar, S.P. Srivastava and P.K. Seth, *Toxicology*, 1986, **42**, 47.
42. P.M.D. Foster, J.R. Foster, M.W. Cook, L.V. Thomas and S.D. Gangolli, *Toxicol. Appl. Pharmacol.*, 1982, **63**, 120.
43. D.M. Creasy, L.M. Beech and T.J.B. Gray, *Toxicol. In vitro*, 1988, **2**, 83.
44. M. Ablake, M. Itoh, H. Terayama, S. Hayashi, S. Shoji, M. Naito, K. Takahashi, S. Suna and F. Jitsunari, *Int. J. Androl.*, 2004, **27**, 274.
45. V. Hansson, B. Jégou, H. Attramadal, T. Jahnsen, F. Le Gac, M. Tvermyr, A. Frøysa and R. Horn, in *Recent Advances in Male Reproduction: Molecular Basis and Clinical Implications*, R.D' Agata, M.B. Lipsett, P. Polosa and H.J. van der Molen (eds), Raven Press, New York, 1983, 53.
46. B.M. Sanborn, J.R. Wagle, A. Steinberger and D.J. Lam, in *Male Reproduction: Molecular Basis and Clinical Implications*, R.D' Agata, M.B. Lipsett, P. Polosa and H.J. van der Molen (eds), Raven Press, New York, 1983, 69.
47. J.P. Bonde and A. Giwercman, *Reprod. Med. Rev.*, 1995, **4**, 59.
48. P.M.D. Foster, J.R. Foster, M.W. Cook, L.V. Thomas and S.D. Gangolli, *Toxicol. Appl. Pharmacol.*, 1994, **63**, 120.
49. P. Sjöberg, N.G. Lindqvist and L. Plöen, *Environ. Health Perspect.*, 1986, **65**, 237.

50. H.B. Jones, D.A. Garside, R. Liu and J.C. Roberts, *Exp. Mol. Pathol.*, 1993, **58**, 179.

51. L.A. Dostal, R.E. Chapin, S.A. Stefanski, M.W. Harris and B.A. Schwetz, *Toxicol. Appl. Pharmacol.*, 1988, **95**, 104.

52. T. Ichimura, M. Kawamura and A. Mitani, *Toxicology*, 2003, **194**, 35.

53. T. Tsutsumi, T. Ichihara, M. Kawabe, H. Yoshino, M. Asamoto, S. Suzuki and T. Shirai, *Reprod. Toxicol.*, 2004, **18**, 35.

54. D.F. Katz, *Biol. Reprod.*, 1982, **26**, 566.

55. D.K. Agarwal, W.H. Lawrence and J. Autian, *J. Toxicol. Environ. Health*, 1985, **16**, 71.

56. A.J. Wyrobek, S.M. Schrader, S.D. Perreault, L. Fenster, G. Huszar, D.F. Katz, A.M. Osorio, V. Sublet and D. Evenson, *Reprod. Toxicol.*, 1997, **11**, 243.

57. B.F. Hales and D.G. Cyr, in *Advances in Male Mediated Developmental Toxicity*, B. Robaire and B.F. Hales (eds), Kluver Academic/Plenum Publishers, New York, 2003, 271.

58. A.J. Bateman, in *Handbook of Mutagenicity Test Procedures*, B. Kilbey, M. Legar, W. Nichols and C. Ramel (eds), Elsevier, Amsterdam, 1977, 325.

59. J. Hellwig, J. Freudenberger and R. Jackh, *Food Chem. Toxicol.*, 1997, **35**, 501.

60. M. Shirota, Y. Saito, K. Imai, S. Horiuchi, S. Yoshimura, M. Sato, T. Nagao, H. Ono and M. Katoh, *J. Toxicol. Sci.*, 2005, **30**, 175.

61. M. Ema, T. Itami and H. Kawasaki, *J. Appl. Toxicol.*, 1992, **12**, 179.

62. M. Ema, T. Itami and H. Kawasaki, *J. Appl. Toxicol.*, 1992, **12**, 57.

63. A.-M. Saillenfait, J.-P. Sabaté and F. Gallissot, *Reprod. Toxicol.*, 2003, **17**, 575.

64. J.C. Lamb, R.E. Chapin, J. Teague, A.D. Lawton and J.R. Reel, *Toxicol. Appl. Pharmacol.*, 1987, **88**, 255.

65. L.G. Parks, J.S. Ostby, C.R. Lambright, B.D. Abbott, G.R. Klinefelter, N.J. Barlow and L.E. Gray Jr., *Toxicol. Sci.*, 2000, **58**, 339.

66. J. Borch, O. Ladefoged, U. Hass and A.M. Vinggaard, *Reprod. Toxicol.*, 2004, **18**, 53.

67. M. Ema, E. Miyawaki, A. Hirose and E. Kamata, *Reprod. Toxicol.*, 2003, **17**, 407.

68. M. Ema and E. Miyawaki, *Reprod. Toxicol.*, 2002, **16**, 71.

69. K. Liu, K.P. Lehmann, M. Sar, S.S. Young and K.W. Gaido, *Biol. Reprod.*, 2005, **73**, 180.

70. G.J. Ikeda, P.P. Sapienza, J.L. Couvillion, T.M. Farber and E.J. Van Loon, *Food Cosmetic. Toxicol.*, 1980, **18**, 637.

71. C. Rhodes, T.C. Orton, J.S. Pratt, P.L. Batten, H. Braft, S.J. Jackson and C.R. Elcombe, *Environ. Health Perspect.*, 1986, **65**, 299.

72. P.W. Albro and S.R. Lavenhar, *Drug. Metab. Rev.*, 1989, **21**, 13.

CHAPTER 10

Prevention of Adverse Effects of Cancer Treatment on the Germline

MARVIN L. MEISTRICH, ZHEN ZHANG, KAREN L. PORTER, OLGA U. BOLDEN-TILLER AND GUNAPALA SHETTY

Department of Experimental Radiation Oncology, The University of Texas M.D. Anderson Cancer Center, Unit 066, 1515 Holcombe Blvd., Houston Texas 77030, USA

10.1 Infertility Resulting from Cancer Treatment

For children and young adults who have cancer, the success of treatment with regimens that are toxic to gonadal function has made infertility an important problem. When the cancer is controlled, quality of life then becomes a major issue. To many of the young men who have received chemotherapy or radiation therapy for cancer, a major issue of life quality is the ability to have a normal child. However, concerns about the heritable effects on the sperm produced become secondary for many men who experience prolonged and even permanent sterility. First their fertility must be regained, only then does heritable genetic disease becomes an issue.

Many men are treated for cancer with doses of alkylating agents, platinum drugs, or radiation that are sufficient to induce prolonged azoospermia.[1] Others treated with cyclophosphamide for autoimmune diseases, usually involving the kidney, also become azoospermic.[2]

Of all the cells in the testis, the germ cells appear to be the most sensitive to killing by these cytotoxic agents. Among the germ cells, the differentiating spermatogonia proliferate most actively and are extremely susceptible to these treatments (Table 1). As the later stage germ cells (spermatocytes and especially spermatids) are less sensitive to killing and progress through spermatogenesis, sperm count is maintained for one or two months after cytotoxic treatment. Subsequently, sperm count diminishes when sperm in the ejaculate were spermatocytes at the time of treatment and reaches a minimum when the

Table 1 Comparison of effects of comparable doses of radiation on the spermatogenic cells of mouse, rat, monkey, and human. Chemotherapy effects are included but it was not possible to compare doses

Cell stage: species (strain)	Stem spermatogonia	Differentiating spermatogonia		Spermatocytes		Spermatids
		Immediate Effects	Development during recovery	Immediate effects	Development during recovery	
Mouse	Some survive doses up to 16 Gy. Killed by some alkylating agents, adriamycin, and radiation	Very sensitive to direct killing by radiation and chemotherapy	Develop normally from surviving stem cells	Moderately resistant (more than stem cells) to radiation	Develop normally although efficiency may be reduced	Highly resistant to killing by radiation and many chemotherapeutic drugs
Rat (inbred strains)	Survive doses up to 10 Gy	Very sensitive to direct killing by radiation. Likely to be sensitive to chemotherapy too	Strong block of differentiation after radiation and some chemotherapeutic drugs	Relatively resistant to irradiation	Significant loss of differentiating cells	Relatively resistant to irradiation
Rat (outbred strains)	Survive doses up to 10 Gy	Very sensitive to direct killing by radiation. Likely to be sensitive to chemotherapy too	Variable, intermediate between mice and inbred rats	Resistant (no cytotoxic effects of 3 Gy)	Develop normally after 3 Gy	Resistant (no cytotoxic effects of 3 Gy)
Monkey (macaque)	Most killed by doses of 7 Gy. Killing of A_{dark} is delayed	Very sensitive to direct killing by radiation	Develop from surviving stem cells after transient block ($\sim$100 days)	Resistant (negligible cytotoxic effects of 4 Gy)	Develop normally after 4 Gy	Resistant (negligible cytotoxic effects of 4 Gy)
Human	More sensitive to radiation than rodents. Killing of A_{dark} is delayed. Sensitive to alkylating agents	Very sensitive to direct killing by radiation. Most sensitive stage to chemotherapy	Differentiation is transiently blocked after radiation and chemotherapy	Relatively resistant to radiation	Develop normally after 4 Gy	Relatively resistant to radiation. Intermediate sensitivity to some chemotherapy regimens

differentiating spermatogonia would have become sperm. However, although the later stage cells are more resistant to killing, they are sensitive to the induction of mutagenic damage, and studies in rodents have shown they are more susceptible than the stem cells with respect to transmitting mutations induced in their DNA to the next generation.[3]

The eventual recovery of sperm production depends on the survival of the spermatogonial stem cells and their ability to differentiate. After many radiotherapeutic and chemotherapeutic regimens, the surviving stem cells regenerate spermatogenesis within a matter of months. However, there are many cases of prolonged azoospermia in which there appears to be killing of all of the stem cells by cytotoxic agents.[4] In other instances, however, the stem spermatogonia survive but have a long-term delay before they can again differentiate into sperm, as evidenced by the spontaneous reinitiation of spermatogenesis in some patients after many years of azoospermia.[5] There is evidence of arrest at the spermatogonial[6,7] stage during the azoospermic period caused by cytotoxic agents (Table 1).

In contrast, the somatic cells, including the Leydig and Sertoli cells, which do not proliferate in adults, and the peritubular myoid and vascular cells, which proliferate slowly, survive most cytotoxic therapies. These cells may, however, suffer functional damage.

10.2   Block to Spermatogonial Differentiation

In a rat model, many stem cells survive treatment with certain doses of either radiation[8] or procarbazine,[9] but these cells fail to differentiate (Table 1). The number of stem spermatogonia remains relatively constant for a long time. The cells proliferate actively but, before they reach the stage at which they would normally differentiate into A_1 spermatogonia, they undergo apoptosis.[10] Examination of the hormonal status of these animals reveals that the failure of differentiation of spermatogonia cannot be a result of insufficient stimulation by gonadotropins or testosterone. Follicle-stimulating hormone (FSH) levels were 1.8-fold normal and luteinizing hormone (LH) levels were 2- to 4-fold normal after the rats were treated with radiation[8] or procarbazine.[9] Whereas serum testosterone levels remained unchanged, intratesticular testosterone concentrations reached 2- to 4-fold normal.[9,10] Androgen receptors remained present in the somatic cells, and FSH receptor mRNA expression in the Sertoli cells was unchanged.[11]

It is not clear how widespread is the phenomenon of cytotoxic agents inducing this block in spermatogonial differentiation. It is rarely observed in the mouse[12,13] and to a limited extent in monkeys,[14] but may occur in humans as described above.

10.3   Approaches to Prevention of Adverse Effects on the Germline

A variety of biochemical and biological approaches have been suggested and tested to protect the testes in experimental animal model systems against radiation and chemotherapy. A thiol compound, amifostine (WR-2721), which acts to protect normal tissues against radiation by scavenging reactive oxygen species and against chemotherapy by binding reactive metabolites of cisplatin, was tested in mice. Although it did produce some radiation protection against single doses of radiation, it was directly toxic to spermatogonial stem cells and showed reduced protection when the doses were given as a fractionated regimen.[15] Prostaglandin analogues, such as misoprostol, were shown to protect mouse spermatogonial stem cells against radiation, but no data on its effects on tumors were presented to determine whether the therapeutic benefit of radiation would remain with such a treatment.[16] Growth factors have potential protective roles; although expression of FGF-4 in testes was shown to protect against prolonged testicular damage produced by Adriamycin, it also protected germ cell tumors and thus neutralized the therapeutic benefit of Adriamycin.[17] Although compounds like sphingosine-1-phosphate, which counterbalances the effects of ceramide, protect oocytes against radiation and chemotherapy, it did not offer any significant protection to mouse testes.[18] Physical methods of reducing blood flow to testes by ligation or temporary cryptorchidism have been shown to protect ram testes against damage from Adriamycin.[19] None of these methods has been demonstrated to be promising enough to undergo clinical trials.

In contrast, there has been interest and clinical trials of hormonal modulation in attempts to prevent damage to the germline from radiotherapy and chemotherapy. The mechanism originally proposed was that interruption of the pituitary–gonadal axis would reduce the rate of spermatogenesis and render the resting testis more resistant to the effects of chemotherapy.[20] However the original suggestive studies with mice could not be reproduced,[21] no truly protective effects have ever been observed in mice,[13] and the premise that the kinetics of spermatogenesis in mammals could be altered by suppressing gonadotropins and testosterone was incorrect.[22] Despite this, it has been convincingly and reproducibly shown that suppression of gonadotropins and testosterone in the rat protects the testis so as to enhance the recovery of spermatogenesis after chemotherapy or radiation.[23–25] Our current understanding of this phenomenon and the status of clinical applications are discussed in the following sections.

10.4 Protection or Stimulation of Spermatogenic Recovery by Hormonal Suppression

Despite the enhanced recovery of spermatogenesis observed following suppression of gonadotropins and testosterone before and during cytotoxic therapy, there is no evidence that the hormonal suppression is actually acting to protect stem cells. The numbers of type a spermatogonia surviving after irradiation were unaffected by hormonal suppression.[26] Rather, hormone pretreatment appears to act in some unknown way to preserve the subsequent ability of the testis to support the differentiation of the surviving stem spermatogonia. The enhancement of recovery of spermatogenesis by hormonal suppression appears to only occur when radiation or chemotherapy blocks the subsequent differentiation of surviving spermatogonia, because it protects against the appearance of this block. The failure of hormonal suppression to stimulate recovery of spermatogenesis in mouse[13] is a result of the absence of a radiation- or chemotherapy-induced block in spermatogonial differentiation in this species (Table 1). Hormonal suppression before and immediately after irradiation also failed to enhance the recovery of spermatogenesis after irradiation of rhesus macaques.[27] The above data indicate that protection of spermatogenic recovery by hormonal suppression has only conclusively been demonstrated in rat and appears to act to protect the somatic cells of the testis so that they can continue to support the differentiation of surviving stem cells.

Based on the above observations, we investigated whether suppression of gonadotropins and testosterone *after* the cytotoxic insult could restore the ability of the rat testis environment to support the differentiation of spermatogonia that survive the toxic treatment. Gonadotropin-releasing hormone (GnRH) agonists or antagonists given immediately after irradiation[28] or procarbazine treatment[9] prevented the block in spermatogonial differentiation. Even after the block to spermatogonial differentiation had developed, 10–20 weeks after irradiation, administration of GnRH analogues still overcame this block and restored spermatogenesis.[10,28] Although the spermatogonia differentiated, it should be noted that they could not progress past the round spermatid stage as long as testosterone and FSH levels were suppressed. However, when the suppressive GnRH-analogue treatment was stopped, spermatogenesis went to completion and sperm production and fertility were restored.[28] It has also been shown that giving a GnRH agonist after busulfan,[29] hexanedione,[30] dibromochloropropane,[31] or heat[32] treatment also stimulated the recovery of spermatogenesis in rats, supporting the generality of this phenomenon, at least in rat. However, suppression of gonadotropins and testosterone after irradiation failed to restore spermatogenesis in macaque monkeys.[14] This observation is consistent with the fact that radiation did not induce a prolonged block in differentiation in macaques; instead, there were no spermatogonia remaining in these atrophic tubules after irradiation (Table 1).

10.5 Mechanisms of Hormonal Stimulation of Spermatogenic Recovery in Rat

It is surprising that suppression of the hormones testosterone and/or FSH, without which normal spermatogenesis cannot go to completion, stimulates spermatogonial differentiation after cytotoxic treatment. Indeed we showed that when irradiated, GnRH-analogue-treated rats were also given exogenous testosterone, or for that matter any other androgenic compound, it dose-dependently reduced the GnRH-analogue-stimulated spermatogonial differentiation.[33–35] In these circumstances spermatogonial differentiation becomes sensitive to inhibition even by physiological levels of testosterone. This androgen-induced reduction in spermatogonial differentiation was reversed by the simultaneous treatment with the androgen receptor antagonist, flutamide.[33]

However, these experiments could not unequivocally prove an inhibitory effect of testosterone on spermatogonial differentiation. The GnRH-analogue treatment also reduces FSH levels, addition of testosterone to GnRH-analogue treatment increases FSH levels, and flutamide also partially reverses the effect of testosterone on FSH levels.[33] However in other experiments, we have recently shown that testosterone is indeed the major factor inhibiting spermatogonial differentiation but that FSH also has a minor inhibitory effect (G. Shetty *et al.*, Endocrinology, in press).

The mechanisms by which testosterone and FSH inhibit spermatogonial differentiation are not known. However, it should be noted that the spermatogonia lack androgen receptor[36] and are generally believed to lack FSH receptors, so that the hormones must act on the somatic cells, likely the Sertoli cell, which then affects the spermatogonia by paracrine or juxtacrine interactions. Models have been developed to explain how spermatogonial differentiation might be inhibited by testosterone in rats treated with gonadotoxic agents but not in untreated rats.[37] The specific details of the model depend upon whether the primary cause of the cessation in spermatogonial differentiation is the induction of apoptosis in spermatogonia or a failure in the differentiation-signaling pathway for the spermatogonia.

10.6   Potential for Clinical Application

Although recovery of spermatogenesis can be protected or restored in rat models, the question arises whether the procedures will work in men. There are many similarities in spermatogenesis between rodents and primates and so we would expect that the same principles would apply. In addition, many aspects of the hormonal regulation of spermatogenesis are similar in rodents and primates, with both androgen and FSH supporting normal differentiation.[38] It is also noteworthy that, of all the anticancer agents, radiation and the alkylating agents are generally the most effective in producing prolonged azoospermia in both humans and rodents, implying that similar mechanisms may be involved.[39,40] In addition, the doses required to produce prolonged or permanent azoospermia are quite similar. For example, the doses that produce prolonged azoospermia in humans and rats, are respectively, about 3 and 3.5 Gy for radiation,[7,8] 4 and 3 g m^{-2} for procarbazine,[25,41] and 600 and 200 g m^{-2} for busulfan.[29,42]

However, there are some differences in the processes of spermatogonial proliferation and differentiation and the action of hormones and antineoplastic agents in rodents and primates. Whereas, in the absence of androgens and FSH, spermatogenesis proceeds to the spermatocyte or early spermatid stage in rodents, it appears to be blocked at the spermatogonial stage in primates.[43]

Another difference is that spermatogonia are always present in the testes of rats subjected to doses of radiation that render them azoospermic, but this finding is variable in humans.[4,6]

Six out of seven human studies, in which treatment with GnRH agonists, antiandrogens, and/or steroids was given before and during chemotherapy or radiotherapy for cancer, failed to show any benefit of the treatment with respect to recovery of sperm count.[44–49] The failure of some of these human trials could be due to heterogeneity of the population, too high doses of cytotoxic therapy killing all stem cells, too low doses of cytotoxic therapy allowing recovery in all patients, inadequate controls, the use of medroxyprogesterone acetate that is ineffective at stimulating recovery,[35] and testosterone supplementation of the GnRH-agonist treatment, which in the rat model blunts the stimulatory effects of the GnRH agonist.[33] However, a more recent report, showed that low-dose systemic testosterone given to suppress intratesticular testosterone levels did induce recovery of spermatogenesis in all men treated with cyclophosphamide for nephrotic disorders.[2] Although this study has not been repeated to confirm the results, possible reasons for the success might have been that the dose regimen of the particular chemotherapeutic agent did not kill all of the stem cells but was sufficient to block their differentiation.

There has been only one trial of the use of hormonal suppression after the completion of chemotherapy, and in that case, no recovery was observed.[50] It should be noted that this study used testosterone combined with medroxyprogesterone acetate treatment, a hormone combination that has been shown not to be very effective at restoring spermatogenesis in irradiated rats,[35] and all the patients had been treated before puberty with high doses of procarbazine or radiation and in which cases complete loss of stem cells was likely.

Overall, simple hormonal protection or restoration of spermatogenic function in men treated with chemo- or radiotherapy seems at most to have only very limited application. However, it is important to understand the molecular mechanisms and the specific cells involved in the androgen and FSH inhibition of spermatogonial differentiation after cytotoxic exposure in the rat model. This will allow identification of downstream targets affected by the hormones, which will be useful in formulating treatment regimens that may be successful in humans.

10.7   Application in Conjunction with Spermatogonial Transplantation

As hormonal treatments, even when given before the cytotoxic insult, do not appear to protect the survival of stem cells from killing by drugs or physical agents, alternative strategies need to be developed to protect them. One alternative would be to remove the stem cells in a biopsy before cytotoxic therapy, cryopreserve them, and later transplant them back into the testis.[51] However, this procedure may not be successful if the stromal tissue is damaged and consequently the spermatogonia do not have an appropriate environment to colonize and in which differentiate. Several studies have shown that treatment of recipient mice[52] or rats[53] with GnRH analogues enhances the ability of donor spermatogonial stem cells to colonize and produce differentiated cells in busulfan-treated recipient testes. Hence, suppression of testosterone and/or FSH may be necessary to restore the environment to support differentiation.

There was one study in which irradiated monkey testes were used, without any hormonal treatment, as a host for transplantation of cells from the contralateral testis, but the success of the transplantation was uncertain because of recovery of endogenous cells and the lack of a maker to identify donor germ cells.[54] Clinical trials involving cryopreservation of testicular cells from men before the start of chemotherapy for Hodgkin's disease and non-Hodgkin's lymphoma and reinfusing the cells into the testis after chemotherapy are underway,[55] but no positive results have been reported. Hormonal treatment might be an important, and possibly a necessary, adjunct to these attempts at restoration of spermatogenesis and fertility by spermatogonial transplantation in primates and humans.

Acknowledgments

Much of the research from our laboratory that is included in this review was supported by a National Institutes of Health (USA) Grant ES-08075.

References

1. M.L. Meistrich, R. Vassilopoulou-Sellin and L.I. Lipshultz, in *Principles and Practice of Oncology*, V.T. DeVita, S. Hellman and S.A. Rosenberg (eds), J.B. Lippincott Co., Philadelphia, 2001, 2923.
2. A. Masala, R. Faedda, S. Alagna, A. Satta, G. Chiarelli, P.P. Rovasio, R. Ivaldi, M.S. Taras, E. Lai and E. Bartoli, *Ann. Intern. Med.*, 1997, **126**, 292.
3. M.L. Meistrich, *Hum. Reprod.*, 1993, **8**, 8.
4. D.H. Van Thiel, R.J. Sherins, G.H. Myers and V.T. De Vita, *J. Clin. Invest.*, 1972, **51**, 1009.
5. M.L. Meistrich, G. Wilson, B.W. Brown, M.F. da Cunha and L.I. Lipshultz, *Cancer*, 1992, **70**, 2703.
6. E.D. Kreuser, E. Kurrle, W.D. Hetzel, B. Heymer, R. Porzsolt, R. Hautmann, W. Gaus, U. Schlipf, E.F. Pfeiffer and H. Heimpel, *Klinische Wochenschrift*, 1989, **67**, 367.
7. M.L. Meistrich and M.E.A.B. van Beek, *Adv. Radiat. Biol.*, 1990, **14**, 227.
8. M. Kangasniemi, I. Huhtaniemi and M.L. Meistrich, *Biol. Reprod.*, 1996, **54**, 1200.
9. M.L. Meistrich, G. Wilson and I. Huhtaniemi, *Cancer Res.*, 1999, **59**, 3557.
10. G.A. Shuttlesworth, D.G. de Rooij, I. Huhtaniemi, T. Reissmann, L.D. Russell, G. Shetty, G. Wilson and M.L. Meistrich, *Endocrinology*, 2000, **141**, 37.
11. S. Maiti, M.L. Meistrich, G. Wilson, G. Shetty, M. Marcelli, M.J. McPhaul, P.L. Morris and M.F. Wilkinson, *Endocrinology*, 2001, **142**, 1567.

12. M.E.A.B. van Beek, M.L. Meistrich and D.G. de Rooij, *Cell Tissue Kinet.*, 1990, **23**, 1.

13. M. Kangasniemi, K. Dodge, A.E. Pemberton, I. Huhtaniemi and M.L. Meistrich, *Endocrinology*, 1996, **137**, 949.

14. K. Boekelheide, H. Schoenfeld, S.J. Hall, C.C.Y. Weng, G. Shetty, J. Leith, J. Harper, M. Sigman, D.L. Hess and M.L. Meistrich, *J. Androl.*, 2005, **26**, 222.

15. M.L. Meistrich, M.V. Finch, N. Hunter and L. Milas, *Int. J. Radiat. Oncol. Biol. Phys.*, 1984, **10**, 2099.

16. D.G. de Rooij, M.E. van Beek, D.H. Rutgers, A. van Duyn-Goedhart and P.P. van Buul, *Genet. Res.*, 1998, **72**, 185.

17. H. Yamamoto, T. Ochiya, S. Tamamushi, H. Toriyama-Baba, Y. Takahama, K. Hirai, H. Sasaki, H. Sakamoto, I. Saito, T. Iwamoto, T. Kakizoe and M. Terada, *Oncogene*, 2002, **21**, 899.

18. M. Otala, L. Suomalainen, M.O. Pentikainen, P. Kovanen, M. Tenhunen, K. Erkkila, J. Toppari and L. Dunkel, *Biol. Reprod.*, 2004, **70**, 759.

19. J. van Vliet, D.G. de Rooij, C.J. Wensing and A.L. Bootsma, *Andrologia*, 1993, **25**, 251.

20. L.M. Glode, J. Robinson and S.F. Gould, *Lancet*, 1981, **1**, 1132.

21. M.F. da Cunha, M.L. Meistrich and S. Nader, *Cancer Res.*, 1987, **47** 1093.

22. M.L. Meistrich, G. Wilson, Y. Zhang, B. Kurdoglu and N.H.A. Terry, *Cancer Res.*, 1997, **57**, 1091.

23. J.I. Delic, C. Bush and M.J. Peckham, *Cancer Res.*, 1986, **46**, 1909.

24. B. Jégou, J.F. Velez de la Calle and F. Bauche, *Proc. Natl. Acad. Sci. USA*, 1991, **88**, 8710.

25. N. Parchuri, G. Wilson and M.L. Meistrich, *J. Androl.*, 1993, **14**, 257.

26. M.L. Meistrich, G. Wilson, M. Kangasniemi and I. Huhtaniemi, *J. Androl.*, 2000, **21**, 464.

27. A. Kamischke, M. Kuhlmann, G.F. Weinbauer, M. Luetjens, C.-H. Yeung, H.L. Kronholz and E. Nieschlag, *J. Endocrinol.*, 2003, **179**, 183.

28. M.L. Meistrich, G. Wilson, G. Shuttlesworth, I. Huhtaniemi and T. Reissmann, *J. Androl.*, 2001, **22**, 809.

29. K. Udagawa, T. Ogawa, T. Watanabe, Y. Yumura, M. Takeda and M. Hosaka, *Int. J. Urol.*, 2001, **8**, 615.

30. K.T. Blanchard, J. Lee and K. Boekelheide, *Endocrinology*, 1998, **139** 236.

31. M.L. Meistrich, G. Wilson, K.L. Porter, I. Huhtaniemi, G. Shetty and G. Shuttlesworth, *Toxicol. Sci.*, 2003, **76**, 418.

32. B.P. Setchell, L. Ploen and E.M. Ritzen, *Reproduction*, 2001, **122**, 255.

33. G. Shetty, G. Wilson, I. Huhtaniemi, G.A. Shuttlesworth, T. Reissmann and M.L. Meistrich, *Endocrinology*, 2000, **141**, 1735.

34. G. Shetty, G. Wilson, M.P. Hardy, E. Niu, I. Huhtaniemi and M.L. Meistrich, *Endocrinology*, 2002, **143**, 3385.

35. G. Shetty, C.C.Y. Weng, O.U. Bolden-Tiller, I. Huhtaniemi, D.J. Handelsman and M.L. Meistrich, *Endocrinology*, 2004, **145**, 4461.

36. M.L. Meistrich, G. Shetty, O.U. Bolden-Tiller and K.L. Porter, in *Sertoli Cell Biology*, M.K. Skinner and M.D. Griswold (eds), Elsevier Academic Press, San Diego, 2005, 437.

37. M.L. Meistrich and G. Shetty, *J. Androl.*, 2003, **24**, 135.

38. S. Bhasin, N. Berman and R.S. Swerdloff, *J. Androl.*, 1994, **15**, 386.

39. M.L. Meistrich, *Biomed. Pharmacother.*, 1984, **38**, 137.

40. M.L. Meistrich, R. Vassilopoulou-Sellin and L.I. Lipshultz, in *Cancer: Principles and Practice of Oncology*, V.T. DeVita, S. Hellman and S.A. Rosenberg (eds), Lippincott Williams & Wilkins, Philadelphia, 2005, 2560.

41. M.F. da Cunha, M.L. Meistrich, L.M. Fuller, J.H. Cundiff, F.B. Hagemeister, W.S. Velasquez, P. MacLaughlin, S.A. Riggs, F.F. Cabanillas and P.G. Salvador, *J. Clin. Oncol.*, 1984, **2**, 571.

42. J.E. Sanders, J. Hawley, W. Levy, T. Gooley, C.D. Buckner, H.J. Deeg, K. Doney, R. Storb, K. Sullivan, R. Witherspoon and F.R. Appelbaum, *Blood*, 1996, **87**, 3045.

43. G.F. Weinbauer, E. Gockeler and E. Nieschlag, *J. Clin. Endocrinol. Metab.*, 1988, **67**, 284.

44. D.H. Johnson, R. Linde, J.D. Hainsworth, W. Vale, J. Rivier, R. Stein, J. Flexner, R.V. Welch and F.A. Greco, *Blood*, 1985, **65**, 832.

45. J.H. Waxman, R. Ahmed, D. Smith, P.F.M. Wrigley, W. Gregory, S. Shalet, D. Crowther, L.H. Rees, G.M. Besser, J.S. Malpas and T.A. Lister, *Cancer Chemother. Pharmacol.*, 1987, **19**, 159.

46. J.R. Redman and D.R. Bajorunas, *in Workshop on Psychosexual and Reproductive Issues Affecting Patients with Cancer*, American Cancer Society, New York, 1987, 90.

47. S.D. Fossa, O. Klepp and N. Norman, *Br. J. Urol.*, 1988, **62**, 449.

48. E.D. Kreuser, W.D. Hetzel, R. Hautmann and E.F. Pfeiffer, *Horm. Metab. Res.*, 1990, **22**, 494.

49. W. Brennemann, K.A. Brensing, N. Leipner, I. Boldt and D. Klingmuller, *Clin. Investig.*, 1994, **72**, 838.

50. A.B. Thomson, R.A. Anderson, D.S. Irvine, C.J.H. Kelnar, R.M. Sharpe and W.H.B. Wallace, *Hum. Reprod.*, 2002, **17**, 1715.

51. M.R. Avarbock, C.J. Brinster and R.L. Brinster, *Nat. Med.*, 1996, **2**, 693.

52. M. Ohmura, T. Ogawa, M. Ono, M. Dezawa, M. Hosaka, Y. Kubota and H. Sawada, *Biol. Reprod.*, 2003, **68**, 2304.

53. T. Ogawa, I. Dobrinski and R.L. Brinster, *Tissue Cell*, 1999, **31**, 461.

54. S. Schlatt, L. Foppiani, C. Rolf, G.F. Weinbauer and E. Nieschlag, *Hum. Reprod.*, 2002, **17**, 55.

55. J. Radford, *Horm. Res.*, 2003, **59**(1), 21.

CHAPTER 11

Molecular Changes in Sperm and Early Embryos after Paternal Exposure to a Chemotherapeutic Agent

BERNARD ROBAIRE,[a,b] ALEXIS M. CODRINGTON[a]
AND BARBARA F. HALES[a]

[a] Department of Pharmacology and Therapeutics, McGill University, Montreal, Canada

[b] Department of Obstetrics and Gynecology, McGill University, Montreal, Canada

11.1 Introduction

The consequences of exposure to drugs, radiation, and environmental toxicants on reproduction and development are a growing concern. The extent to which paternal exposures contribute to human infertility and pregnancy loss is unknown. Cyclophosphamide (CPA), a commonly used anticancer drug, remains one of the best studied examples of a male-mediated developmental toxicant with clear, stage-specific effects on male germ cells.[1–3] Spermatogenesis is a highly ordered and regulated process. Rat spermatogonia (stem cells) undergo five mitotic cell divisions to become spermatocytes; as spermatocytes they undergo two meiotic cell divisions to form spermatids (spermacytogenesis). Spermatids differentiate into spermatozoa, primarily by condensing nuclear elements, developing a propulsion mechanism, and shedding most of their cytoplasm; this process is known as spermiogenesis.[4]

One can deduce the stage specificity of the susceptibility of germ cells during spermatogenesis from the timing between toxicant exposure and the effect on offspring.[1,5] Two weeks of chronic low dose CPA treatment of male rats increased post-implantation loss; this post-implantation loss rose dramatically to plateau at a level dependent on drug dose by 4 weeks of treatment. Thus, CPA-induced post-implantation loss was associated primarily with germ cell exposure during spermiogenesis. Post-meiotic germ cells were also most

124

susceptible to the induction of learning abnormalities in the progeny after paternal exposure to CPA.[6,7] In mice, heritable translocations were found after exposure of spermatids and spermatozoa to CPA.[8] Exposure of spermatocytes undergoing meiosis, to CPA resulted in synaptic failure, fragmentation of the synaptonemal complex, altered centromeric DNA sequences, induction of frame-shift mutations, and gene conversions as well as increased pre-implantation loss.[2,9,10] An increase in malformed (hydrocephaly, edema, micrognathia) and growth retarded fetuses was observed among progeny sired by germ cells first exposed to CPA as spermatogonia, prior to meiosis.[2] Chronic treatment with CPA resulted in an overall increase in chromosomal aneuploidy.[11] Thus, both the increase in malformations and in numerical chromosome abnormalities after paternal CPA exposure occurred in germ cells exposed prior to pachynema. It is noteworthy that the malformations produced by exposure of male mice to urethane or X-rays, including dwarfism, open eyelids, and tail anomalies, are similar to those observed in the progeny of male rats exposed to CPA.[12] Significantly, the increases in post-implantation loss and malformations persisted to the F_2 generation.[13]

To ascertain the molecular mechanisms underlying the male mediated developmental toxicity induced as a result of paternal exposure to CPA, complementary approaches were pursued. In the first, we tested the hypothesis that CPA induces a stage specific response in male germ cells. In the second, we tested the hypothesis that paternal CPA exposure disrupts epigenetic programing in the early embryo. Disturbances in epigenetic programing may contribute to heritable instabilities later in development, emphasizing the importance of considering the effects of chemotherapeutics on the epigenome in risk assessment.

11.2 Cyclophosphamide Induces a Stage-Specific Response in Male Germ Cells

11.2.1 Altered Gene Expression during Spermatogenesis

Male-mediated developmental toxicity is influenced by the stage at which germ cells are exposed. The balance between damage incurred and the ability of the cell to cope with such damage, either by repair or apoptosis, may determine the fate of germ cells exposed to insult. Post-meiotic germ cells (spermatids and mature spermatozoa) do not undergo apoptosis following CPA, etoposide, or doxorubicin exposure;[14–16] however, the ability to repair DNA lesions has been detected in germ cells up to the mid-spermatid stage.[17,18]

Gene products involved in the cellular response to stress, such as DNA repair, antioxidant defense, and heat shock proteins, are differentially regulated during germ cell development.[19] Exposure to CPA alters the expression of stress response genes in male germ cells. We have found that acute CPA treatment primarily affected gene expression in round spermatids;[20] this may permit these cells to mount a response to the damaging effects of CPA. Chronic

CPA treatment resulted in a dramatic decrease in gene expression in pachytene spermatocytes and round spermatids. In contrast, fewer genes were expressed in elongating spermatids, correlating with the transcriptional inactivation that takes place during mid-spermiogenesis; interestingly, 20% of expressed genes were up-regulated.[21] The down-regulation of genes involved in different stress response mechanisms following chronic CPA exposure may decrease the ability of germ cells to respond to insult, thereby allowing damage to accumulate as cells progress through spermatogenesis and become mature spermatozoa.

11.2.2   Increased Chromosomal Aberrations

Post-meiotic germ cells are most susceptible to the effects of alkylating agents.[2,22,23] However, mitotic and meiotic cells are also vulnerable to damage. Chromosome synapsis is completed and recombination takes place during the pachytene stage of meiotic prophase; cells then enter the diplotene and diakinesis substages of prophase before eventually reaching metaphase I.[24] This transition from G_2 (prophase) to metaphase I requires proper chromosome alignment, synaptonemal complex formation, and complete recombination.[25,26] CPA has been reported to damage the synaptonemal complex during prophase, disrupt chromosomal synapsis and alter centromeric sequences, induce gene conversions and frameshift mutations, and induce structural and numerical chromosomal aberrations in mice.[9,10,27] Using fluorescence *in situ* hybridization (FISH), studies on spermatozoa from men treated with cisplatin, etoposide, and bleomycin have revealed an increased rate of aneuploidy.[28] We have used the rat sperm Y-4 FISH assay to assess the incidence of numerical chromosomal abnormalities.[11] Chronic CPA treatment for 9 weeks, but not 6 weeks, significantly increased the overall frequency of spermatozoa with chromosome 4 disomy and nullisomy about 2-fold. Thus, CPA induced aneuploidy prior to pachynema, further emphasizing the ability of exposed germ cells to continue to develop to mature spermatozoa and alluding to the lack of capacity of these cells to mount a response to damage following chronic drug exposure.

11.2.3   Effects on Cell-Cycle Progression

In response to genetic damage, eukaryotic cells arrest or delay cell-cycle progression at certain checkpoints to activate DNA repair mechanisms or cell death pathways.[29] This prevents the transmission of damage as cells divide. In male germ cells, arrest at different points of meiotic prophase has been reported for a number of mice with null mutations in genes involved in monitoring chromosome structure and synapsis, as well as repairing DNA damage.[30–35] We have evaluated the response of pachytene spermatocytes to damage caused by acute and chronic CPA exposure *in vivo* by assessing the ability of these cells to undergo the G_2/MI transition induced *in vitro* by okadaic acid.[36] Following acute CPA treatment, the transition from G_2 to MI was impaired; the number of metaphase I cells decreased with corresponding increases in the number of cells at the diakinesis and diplotene substages of late meiotic prophase. This impairment correlated with extensive DNA double strand breaks, as indicated by the immunocytochemical detection of phosphorylated histone H2AX (γH2AX). Significant levels of DNA damage were detected also following chronic CPA treatment; however, meiotic progression was not impaired.

Transient arrest induced by acute CPA exposure may be due to a G_2/MI checkpoint, activated in response to the damaging effects of the drug. This delay would provide the cell with the opportunity to repair damage or to initiate cell death. Fewer γH2AX foci were detected in spermatocytes cultured with okadaic acid, indicative of possible DNA repair. However, chronic exposure still resulted in a greater proportion of cells with higher numbers of γH2AX foci compared to acute exposure. In accordance with decreased transcript levels for genes involved in checkpoint response regulation following chronic CPA exposure,[19] activation of the G_2/MI checkpoint may be blocked, and DNA repair and apoptotic mechanisms bypassed. The absence of an arrest after chronic treatment raises concern about the functionality of defense mechanisms in male germ cells after repeated exposure to low doses of genotoxic agents. Faulty activation of surveillance mechanisms may result in accumulation of unrepaired genetic damage and consequent genomic instability, affecting both germ cell quality and embryo development.[37]

11.2.4 Disturbances in Chromatin Structure

Increased DNA damage in human spermatozoa is associated with both infertility[38] and exposure to chemotherapeutic agents.[39] Mature spermatozoa contain highly packaged chromatin organized in a specific manner,[40–42] presumably to allow access to genetic information required for embryogenesis. Abnormal sperm chromatin condensation has been correlated with the presence of DNA strand breaks.[43–46]

Using the Comet assay, spermatozoa were analyzed for DNA strand breaks following CPA exposure.[47] Acute exposure to a high dose of the drug did not result in dramatic increases in DNA damage in cells exposed to the drug as Step 9 or 15 spermatids. However, subchronic exposure revealed a dose-related increase in DNA damage with maximal damage seen in spermatozoa exposed to CPA as Steps 9–14 elongating spermatids. Chronic low dose administration resulted in an accumulation of damage over time. Damage reached a plateau after 21 days of drug treatment, a time when germ cells progress from elongating spermatids (Steps 9–19) to mature spermatozoa. CPA may have a maximal effect on elongating spermatids and spermatozoa since these cells cannot repair damage or undergo apoptosis. Thus, DNA damage induced by CPA is germ cell phase-specific.

During spermiogenesis, chromatin remodeling occurs as chromosomal histones are acetylated and ubiquitinated in mid-spermiogenic spermatids to allow transition proteins to bind to DNA.[48] Transition proteins facilitate the preferential binding of protamines in late spermatids to fully condense the chromatin.[49,50] The most damaging effects of CPA occurred during a key point

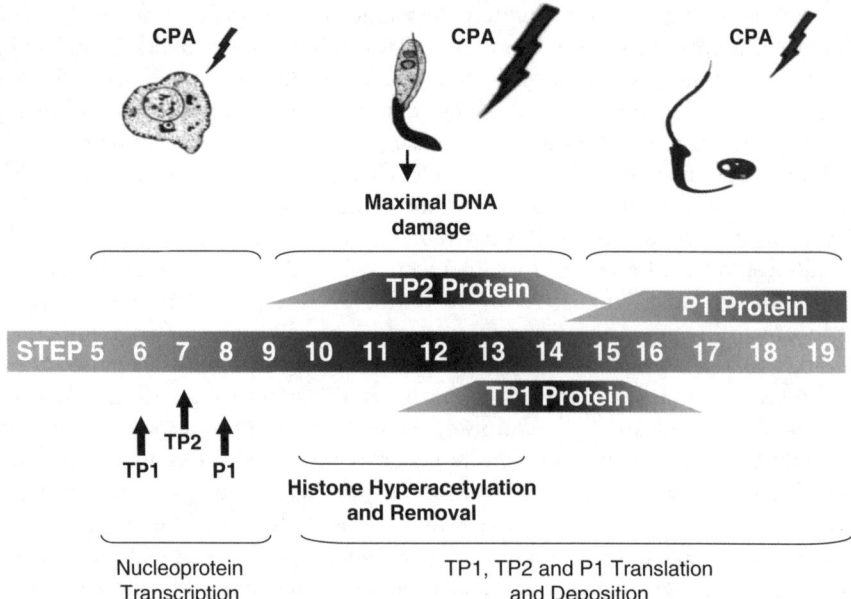

Figure 1   *Spermiogenic germ cell phase-specific susceptibility to cyclophosphamide (CPA) damage. Effects were maximal in mid-spermiogenic Steps 9–14 spermatids during the chromatin remodeling histone-protamine exchange. TP indicates transition protein and P1 protamine 1*

of sperm chromatin remodeling (histone hyperacetylation and transition protein deposition) (Figure 1). Increased DNA damage in mid-spermiogenic cells could alter the binding of protamines to DNA, thus affecting chromatin condensation. Indeed, we have reported that CPA disturbs sperm chromatin decondensation both *in vitro*[51] and in denuded hamster oocytes.[52] In addition, the formation of the male pronucleus was earlier in rat oocytes fertilized by drug treated males.[52] CPA-induced chromatin damage may alter condensation during spermiogenesis and hence result in more rapid decondensation after fertilization.

Decreased iodoacetamide binding, indicative of a reduced sulfhydryl content in sperm nuclei, was observed in CPA-exposed germ cells, suggesting that protamines are affected in these cells.[51] Effects on reduced sulfhydryl content may be due to either incomplete protamine binding or an increase in alkylation of the sulfhydryl groups in protamine. The fluorochrome chromomycin A3 (CMA3) has been used as an indirect tool to assess sperm chromatin packaging, as it is indicative of underprotaminated cells.[53] Preliminary studies from our laboratory show increased CMA3 binding in sperm chronically exposed to CPA for 4 weeks, therefore targeting spermatids as they develop from Step 1 round spermatids to mature spermatozoa. However, this binding is dependent on the decondensation or thiol status of the cell. Protamines are especially susceptible to alkylation, resulting in blockage of normal disulfide bond formation.[54] Further studies are necessary to determine whether CPA does indeed affect protamination and hence sperm chromatin packaging.

In addition to packaging sperm DNA with protamines, the model for sperm DNA packaging proposes that there are DNA loop domains attached to the protein scaffold of the cell, called the nuclear matrix, in a sequence-specific manner.[40–42] Changes in nuclear matrix protein composition occur during spermiogenesis as the synthesis of some proteins ceases and others appear.[55] In somatic cells, DNA-loop domain organization is involved in DNA replication and transcriptional regulation.[56] It has been shown that alkylating agents preferentially damage DNA regions that are in close proximity to matrix-bound replication and transcription sites and affect DNA attachment to the nuclear matrix.[56] Although very little is known about the organization of DNA in sperm or the functional significance of this organization, it is likely that a unique chromatin architecture is required to facilitate scheduled transcription after fertilization. Using *Xenopus* oocyte extracts, Sawyer and Brown[57] reported a decrease in DNA synthesis in sperm exposed to CPA for 6 weeks compared to control rat sperm. This is consistent with our previous report that the *in vitro* template function of spermatozoal DNA was markedly affected by CPA treatment.[58] It is tempting to speculate that the damaging effects of CPA may disrupt the organization of DNA on the nuclear matrix by targeting the DNA or nuclear matrix proteins, thus affecting the participation of the paternal genome in embryo development.

11.3 Paternal Cyclophosphamide Exposure Disrupts Programming in the Early Embryo

It is clear that events in the early embryo are disturbed by damage to the paternal genome. Studies from our lab[11,47] and that of Marchetti[59] have shown that sperm with DNA damage can fertilize; the early embryos that result frequently can proceed through cleavage stages of development.[59,60] Gene expression is affected in the embryos sired by CPA exposed-spermatozoa.[52,61,62] Total RNA synthesis ($[^{32}P]$-UTP incorporation) was constant in 1–8 cell embryos sired by drug-treated fathers, while in control embryos RNA synthesis increased 4-fold to peak at the 4-cell stage.[62] Moreover, both BrUTP incorporation into RNA and Sp1 transcription factor immunostaining were increased and spread over the cytoplasmic and nuclear compartments in 2-cell embryos sired by CPA-treated males. In contrast, in embryos fertilized by control spermatozoa, BrUTP incorporation and Sp1 immunostaining were confined to the nucleus.[62]

The profile of expression of specific genes was altered in embryos sired by drug-treated males even at the 1-cell stage.[52,61,62] By the 2-cell stage, the relative abundance of transcripts for candidate imprinted, growth factor, and cell adhesion genes was elevated significantly above control in embryos sired by CPA-treated males; a peak in the expression of many of these genes was not observed until the 8-cell stage in control embryos. Thus, paternal drug exposure temporally and spatially dysregulated rat zygotic gene activation, altering the developmental clock. In addition to altered mRNA profiles for cell adhesion molecules, embryos from litters sired by CPA-treated males had altered E-cadherin immmunoreactivity, lower cell numbers and decreased cell–cell contacts. Therefore, paternal genomic, drug exposure leads to decreased cell interactions during early embryo development.[62]

Studies with human populations and mice have provided evidence that paternal irradiation exposures can result in elevated mutation rates in progeny.[63,64] Germ cell mutagen tests have demonstrated that paternal exposures, to a plethora of chemicals, induce dominant lethality or specific locus mutations.[65–68] However, there is a large gap between the generally low rate of genetic "damage" induced by most of these chemicals, as measured by the incidence of chromosomal aberrations or specific locus mutations, and the high rate of adverse progeny outcomes, such as the dominant lethality. This gap has led us to hypothesize that a key mechanism by which toxin perturbation of sperm may lead to heritable alterations in the genome is by epigenetic modifications.

Epigenetic programing of the parental genomes is crucial for embryogenesis; such programing can tightly control DNA replication and transcription. Histone acetylation and DNA methylation constitute intricate regulatory mechanisms that play essential roles in DNA packaging and programing of epigenetic information during pre-implantation development.[69,70] To elucidate whether paternal pre-conceptional exposure to CPA disrupts epigenetic programing in the early embryo, we examined the temporal patterns of histone H4 acetylation at lysine 5 (H4-K5) and 5-methylcytosine immunostaining in zygotes sired by CPA-exposed and control sperm.[71] We found that both the male and female pronuclei in embryos sired by CPA-treated males were significantly hyperacetylated beginning in G_1 and lasting into S-phase; in later stages of zygotic development, the extent of H4 acetylation was not different between drug-exposed and control embryos. DNA methylation reprograming was also remarkably dysregulated. In control zygotes, the male pronuclei underwent a gradual process of active genome-wide demethylation, while the female pronuclei remained hypermethylated. In contrast, the male pronuclei in zygotes fertilized by drug-exposed spermatozoa were dramatically hypomethylated. To the best of our knowledge, this is the first time that a genotoxic exposure has been shown to manifest an epigenetic effect in the early post-fertilization conceptus. Deregulation of parental pronuclear programing may be responsible for altered expression of genes and impaired post-implantation development. Genetic/epigenetic damage persisted to the 2-cell stage.[71]

Paternal drug exposure affected not only the epigenetic programing of the male pronucleus, but also the female pronucleus, demonstrating the presence of pronuclear cross talk during zygotic development. This unique destabilization of maternal chromatin architecture in response to the introduction of a damaged paternal genome indicates that common regulatory factors controlling differential gene activity between parental chromosomes are likely to be affected. Other studies have shown that the damage-free maternal genome has the ability to initiate a p53-dependent checkpoint in response to DNA damage delivered by irradiated spermatozoa.[72] Elucidation of the mechanisms underlying this unique phenomenon, focusing on the DNA damage/repair response in early embryos, is needed to establish endpoints for the assessment of the quality of early embryos obtained using *in vitro* fertilization (IVF) or intracytoplasmic sperm injection (ICSI) technologies.

A mechanistic understanding of how paternal chemical exposure alters male germ cell quality in a manner that impacts on the next generation will permit the assessment and potentially the prevention of male mediated developmental toxicity.

Acknowledgments

Studies from our laboratories which are discussed in this review were funded by the Canadian Institutes of Health Research.

References

1. J.M. Trasler, B.F. Hales and B. Robaire, *Nature*, 1985, **316**, 144.
2. J.M. Trasler, B.F. Hales and B. Robaire, *Biol. Reprod.*, 1986, **34**, 275.
3. D. Anderson, J.B. Bishop, R.C. Garner, P. Ostrosky-Wegman and P.B. Selby, *Mutat. Res.*, 1995, **330**, 115.
4. Y. Clermont, *Physiol. Rev.*, 1972, **52**, 198.
5. L.B. Russell, in *Male-Mediated Developmental Toxicity*, A.F. Olshan and D.R. Mattison (eds), Plenum Press, New York, 1994, 37.
6. M. Auroux, E. Dulioust, J. Selva and P. Rince, *Mutat. Res.*, 1990, **229**, 189.
7. J.D. Fabricant, M.S. Legator and P.M. Adams, *Mutat. Res.*, 1983, **119**, 185.
8. W.M. Generoso, B. Cattanach and A.M. Malashenko, in *Comparative Chemical Mutagenesis*, F.J. de Serres and M.D. Shelby (eds), Plenum Press, New York, 1981, 681.
9. L.C. Backer, M.J. Gibson, M.J. Moses and J.W. Allen, *Mutat. Res.*, 1988, **203**, 317.
10. K.J. Schimenti, W.H. Hanneman and J.C. Schimenti, *Toxicol. Appl. Pharmacol.*, 1997, **147**, 343.
11. T.S. Barton, A.J. Wyrobek, F.S. Hill, B. Robaire and B.F. Hales, *Biol. Reprod.*, 2003, **69**, 1150.
12. T. Nomura, in *Male-Mediated Developmental Toxicity*, A.F. Olshan and D.R. Mattison (eds), Plenum Press, New York, 1994, 117.
13. B.F. Hales, K. Crosman and B. Robaire, *Teratology*, 1992, **45**, 671.
14. L. Cai, B.F. Hales and B. Robaire, *Biol. Reprod.*, 1997, **56**, 1490.
15. M.H. Brinkworth and E. Nieschlag, *Mutat. Res.*, 2000, **447**, 149.
16. T. Sjoblom T, A. West and J. Lahdetie, *Environ. Mol. Mutagen.*, 1998, **31**, 133.
17. R.E. Sotomayor and G.A. Sega, *Environ. Mol. Mutagen.*, 2000, **36**, 255.
18. R.E. Sotomayor, G.A. Sega and R.B. Cumming, *Mutat. Res.*, 1978, **50**, 229.
19. A. Aguilar-Mahecha, B.F. Hales and B. Robaire, *Biol. Reprod.*, 2001, **65**, 119.

20. A. Aguilar-Mahecha, B.F. Hales and B. Robaire, *Mol. Reprod. Dev.*, 2001, **60**, 302.
21. A. Aguilar-Mahecha, B.F. Hales and B. Robaire, *Biol. Reprod.*, 2002, **66**, 1024.
22. U.H. Ehling, R.B. Cumming and H.V. Malling, *Mutat. Res.*, 1968, **5**, 417.
23. H. Jackson, *Br. Med. Bull.*, 1964, **20**, 107.
24. J. Cobb and M.A. Handel, *Semin. Cell. Dev. Biol.*, 1998, **9**, 445.
25. M.A. Handel, *Theriogenology*, 1998, **49**, 423.
26. M.A. Handel, J. Cobb and S. Eaker, *J. Exp. Zool.*, 1999, **285**, 243.
27. F. Pacchierotti, D. Bellincampi and D. Civitareale, *Mutat. Res.*, 1983, **119**, 177.
28. P. De Mas, M. Daudin, M.C. Vincent, G. Bourrouillou, P. Calvas, R. Mieusset and L. Bujan, *Hum. Reprod.*, 2001, **16**, 1204.
29. A. Sancar, L.A. Lindsey-Boltz, K. Unsal-Kacmaz and S. Linn, *Annu. Rev. Biochem.*, 2004, **73**, 39.
30. S.K. Sharan, A. Pyle, V. Coppola, J. Babus, S. Swaminathan, J. Benedict, D. Swing, B.K. Martin, L. Tessarollo, J.P. Evans, J.A. Flaws and M.A. Handel, *Development*, 2004, **131**, 131.
31. B. Kneitz, P.E. Cohen, E. Avdievich, L. Zhu, M.F. Kane, H. Hou Jr., R.D. Kolodner, R. Kucherlapati, J.W. Pollard and W. Edelmann, *Genes Dev.*, 2000, **14**, 1085.
32. C. Barlow, M. Liyanage, P.B. Moens, M. Tarsounas, K. Nagashima, K. Brown, S. Rottinghaus, S.P. Jackson, D. Tagle, T. Ried and A. Wynshaw-Boris, *Development*, 1998, **125**, 4007.
33. S. Eaker, J. Cobb, A. Pyle and M.A. Handel, *Dev. Biol.*, 2002, **249**, 85.
34. X. Xu, O. Aprelikova, P. Moens, C.X. Deng and P.A. Furth, *Development*, 2003, **130**, 2001.
35. D.L. Pittman, J. Cobb, K.J. Schimenti, L.A. Wilson, D.M. Cooper, E. Brignull, M.A. Handel and J.C. Schimenti, *Mol. Cell.*, 1998, **1**, 697.
36. A. Aguilar-Mahecha, B.F. Hales and B. Robaire, *Biol. Reprod.*, 2005, **72**, 1297.
37. O. Niwa, *Oncogene*, 2003, **22**, 7078.
38. D.S. Irvine, J.P. Twigg, E.L. Gordon, N. Fulton, P.A. Milne and R.J. Aitken, *J. Androl.*, 2000, **21**, 33.
39. R. Chatterjee, G.A. Haines, D.M. Perera, A. Goldstone and I.D. Morris, *Hum. Reprod.*, 2000, **15**, 762.
40. R. Balhorn, *J. Cell. Biol.*, 1982, **93**, 298.
41. W.S. Ward and D.S. Coffey, *Biol. Reprod.*, 1991, **44**, 569.
42. W.S. Ward, *Biol. Reprod.*, 1993, **48**, 1193.
43. W. Gorczyca, F. Traganos, H. Jesionowska and Z. Darzynkiewicz, *Exp. Cell Res.*, 1993, **207**, 202.
44. G.C. Manicardi, P.G. Bianchi, S. Pantano, P. Azzoni, D. Bizzaro, U. Bianchi and D. Sakkas, *Biol. Reprod.*, 1995, **52**, 864.
45. G.C. Manicardi, A. Tombacco, D. Bizzaro, U. Bianchi, P.G. Bianchi and D. Sakkas, *Histochem. J.*, 1998, **30**, 33.
46. B.L. Sailer, L.K. Jost and D.P. Evenson, *J. Androl.*, 1995, **16**, 80.

47. A.M. Codrington, B.F. Hales and B. Robaire, *J. Androl.*, 2004, **25**, 354.
48. P. Sassone-Corsi, *Science*, 2002, **296**, 2176.
49. D. Poccia, *Int. Rev. Cytol.*, 1986, **105**, 1.
50. D. Wouters-Tyrou, A. Martinage, P. Chevaillier and P. Sautiere, *Biochimie*, 1998, **80**, 117.
51. J. Qiu, B.F. Hales and B. Robaire, *Biol. Reprod.*, 1995, **52**, 33.
52. W. Harrouk, S. Khatabaksh, B. Robaire and B.F. Hales, *Mol. Reprod. Dev.*, 2000, **57**, 214.
53. P.G. Bianchi, G.C. Manicardi, D. Bizzaro, U. Bianchi and D. Sakkas, *Biol. Reprod.*, 1993, **49**, 1083.
54. G.A. Sega and J.G. Owens, *Mutat. Res.*, 1983, **111**, 227.
55. J.L. Chen, S.H. Guo and F.H. Gao, *Mol. Reprod. Dev.*, 2001, **59**, 314.
56. H.J. Muenchen and K.J. Pienta, *Crit. Rev. Eukaryot. Gene Expr.*, 1999 **9**, 337.
57. D.E. Sawyer and D.B. Brown, *Toxicol. Lett.*, 2000, **114**, 19.
58. J. Qiu, B.F. Hales and B. Robaire, *Biol. Reprod.*, 1995, **53**, 1465.
59. F. Marchetti, J.B. Bishop, L. Cosentino, D. Moore II and A.J. Wyrobek, *Biol. Reprod.*, 2004, **70**, 616.
60. S.M. Austin (Kelly), B. Robaire and B.F. Hales, *Biol. Reprod.*, 1994 **50**, 55.
61. W. Harrouk, A. Codrington, R. Vinson, B. Robaire and B.F. Hales, *Mutat. Res.*, 2000, **461**, 229.
62. W. Harrouk, B. Robaire and B.F. Hales, *Biol. Reprod.*, 2000, **63**, 74.
63. Y.E. Dubrova, *Adv. Exp. Med. Biol.*, 2003, **518**, 115.
64. Y.E. Dubrova, *Radiat. Res.*, 2005, **163**, 200.
65. P.G. Odeigah, *Mutat. Res.*, 1997, **389**, 141.
66. D. Anderson, A.J. Edwards, M.H. Brinkworth and J.A. Hughes, *Toxicology*, 1996, **113**, 120.
67. S.A. Narod, G.R. Douglas, E.R. Nestmann and D.H. Blakey, *Environ. Mol. Mutagen.*, 1988, **11**, 401.
68. B.I. Ghanayem, K.L. Witt, L. El-Hadri, U. Hoffler, G.E. Kissling, M.D. Shelby and J.B. Bishop, *Biol. Reprod.*, 2005, **72**, 157.
69. F. Santos, V. Zakhartchenko, M. Stojkovic, A. Peters, T. Jenuwein, E. Wolf, W. Reik and W. Dean, *Curr. Biol.*, 2003, **12**, 1116.
70. T. Hashimshony, J.M. Zhang, I. Keshet, M. Bustin and H. Cedar, *Nat. Genet.*, 2003, **34**, 187.
71. T.S. Barton, B. Robaire and B.F. Hales, *Proc. Natl. Acad. Sci. USA*, 2005, **102**, 7865.
72. T. Shimura, M. Inoue, M. Taga, K. Shiraishi, N. Uematsu, N. Takei, Z.M. Yuan, T. Shinohara and O. Niwa, *Mol. Cell Biol.*, 2002, **22**, 2220.

CHAPTER 12

Transmissible Genetic Risk Causing Tumours in Mice and Humans

TAISEI NOMURA

Department of Radiation Biology and Medical Genetics, Graduate School of Medicine, Osaka University, B4 2-2, Yamada-oka, Suita, Osaka 565-0871, Japan

12.1   Introduction

Germinal exposure to radiation and chemicals may cause various adverse effects in the offspring. We conducted the first and largest experiments with ICR mice between 1967 and 1981 and these were subsequently extended to N5 and other strains of mice. The main aim of all these studies was to discover whether parental exposure to radiation and chemicals induces tumours and birth defects in the offspring derived from exposed germ cells and identify the underlying mechanisms. We found that radiation and chemical mutagens induced germ cell alterations causing tumours, malformations, and embryonic deaths in the offspring.[1-11] These studies were referred to as "Transgenerational Carcinogenesis", "Paternal Toxicology", or "Male-mediated Developmental Toxicology",[6,12,13] though preconceptional exposure of females also induced such defects in the offspring.[1,3]

In humans, a higher risk of leukaemia and congenital malformations has been reported in the children of fathers who had been exposed to radionuclides at the nuclear reprocessing plants and to diagnostic radiations.[14-20] However, no increases of adverse effects (mutation, abortion, malformation, cancer, *etc.*) have been proven in the children of A-bomb survivors in Hiroshima and Nagasaki, who had been exposed to higher doses (about 0.4 Sv) from atomic radiations.[21,22]

This article presents an overview of our experimental results and subsequent molecular studies aimed at gaining insights into the possible mechanisms underlying transgenerational tumourigenesis. Using these as a framework, transmissible genetic risk of tumours is discussed.

12.2 Experimental Evidences

The first experiments were carried out with ICR mice and with X-rays. In the initial experiment, adult male ICR mice were exposed to X-rays and then mated with untreated oestrous females at various intervals between exposure and conception, and vaginal plug was checked to determine the date of conception and the exposed stages of germ cells. Some mice were euthanized on the 18th day of gestation and F_1 foetuses were examined for embryonic deaths, congenital malformations, *etc.* by Caesarian operation.[1–3] Other parental mice delivered live offspring which were submitted to examination for congenital malformations, tumours, other chronic diseases, and related molecular changes.[1–11]

12.2.1 Embryonic Deaths and Malformations

X-ray exposure at postmeiotic stages resulted in a high incidence of dead embryos, indicating that paternal exposure to radiation kills their offspring (dominant lethals). The incidence of dominant lethals increased with paternal doses of X-rays, for example, about 60% after 5 Gy exposure at the spermatid stage.[1–3,6] Spermatogonial exposure, however, hardly ever induced dominant lethals, because dominant lethals are caused by large chromosomal aberrations and spermatogonia with large chromosomal aberrations are killed or eliminated during meiosis so that only a sperm without large chromosomal changes can survive and be ejaculated. Consequently, there would be no increases of embryonic deaths in the F_1 of A-bomb survivors; most of them had been exposed at the spermatogonial stage.

In the surviving foetuses, varieties of congenital malformations (cleft palate, hydrocephalus, gastroschisis, buphthalmus, diaphragm hernia, dwarfism, tail anomalies, *etc.*) were observed.[1–3,6] Most of the induced malformations are types commonly observed in humans, except for tail anomaly which corresponds to abnormality of vertebral bones in humans. Postmeiotic stages were two times more sensitive than spermatogonial stages. Similar results were obtained from female exposure.[1–3,6] In general, the rate of congenital anomalies detected prenatally was 3-fold higher than that detected after birth, indicating that about 70% of the malformations were lethal after birth.[2,3,6] Some viable anomalies such as dwarfs,[2,3] open eyelids, and tail anomalies were transmitted to further generations with varying degrees of penetrance and expressivity.[2,3,6,9] Spermatogonial exposure induces only a 0.2% increase of congenital malformations per Gy in mice.[9] Considering that the background incidence of congenital malformations is about 5% in Japanese populations, and that the average dose to the A-bomb survivors was about 0.4 Sv, a 0.1% increase in the children of A-bomb survivors cannot be identified. X-ray induction of congenital anomalies in mice and their transmission have been confirmed by Lyon and her colleagues using the same doses and the same treated stages as in our studies but with a different strain.[23–26] Müller *et al.*[27] showed that in the mouse strain, namely, 'the Heiligenberger Stamm' there was an increase in the frequency of one particular malformation, gastroschisis.

12.2.2   Induction of Tumours in the Offspring

Cancer incidences are influenced and/or modified by postnatal environments. Consequently, the investigation has to be carried out carefully using well-controlled procedures such as the use of high quality animal and animal facilities and a strict coding system.[11,28,29] Randomly selected mice were exposed to radiation or chemicals, and unexposed litter mates were used for the concurrent controls. Live offspring were nursed in the barrier (SPF) section with or without postnatal treatments, and the presence or absence of tumours was determined macroscopically and microscopically at autopsy. All processes between treatment and autopsy were kept blind. After the autopsy, mice were matched to their parents retrospectively, and classified as to the irradiated or unirradiated group of the parents. Their descendants (F_2, F_3, *etc.*) were also classified as to the retrospectively determined tumour-type of their parental mice (F_1, F_2, *etc.*).[3,5,7,28,29]

12.2.2.1   Dose Response and Stage Sensitivity

Initial studies were carried out with ICR mice with X-rays (dose rate; 0.72 Gy/min). Male ICR mice were exposed to a single or fractionated doses (0.36 Gy of X-rays given 2 h apart) of X-rays at spermatozoa, spermatids, and spermatogonia stages[1-3]. Females were also exposed to X-rays at various intervals. Incidences of tumours in F_1 offspring increased with X-ray dose to the parents in the dose range of 0.36–5.04 Gy. For male treatment, postmeiotic stages were more sensitive than the spermatogonial stage. No differences were observed in the incidence of tumours between single and fractionated doses after postmeiotic exposure. However, there were apparent fractionation effects after spermatogonial exposure, that is, dose fractionation showed a large reduction of tumour incidence in F_1 offspring. Similar results were obtained after the exposure of females at mature oocyte stages. Oocytes were resistant to single dose of X-rays up to 1 Gy for tumour induction in F_1, but large increases were observed at higher doses. Furthermore, an apparent fractionation effect – a large reduction of tumour incidence was observed after mature oocyte exposure, that is, large reduction of tumour incidence by dose fractionation. These suggest some repair activities in spermatogonia and oocytes. Eighty-seven percent of the induced tumours in ICR mice were in the lung (papillary adenoma), the remainder being of various types – ovarian tumour, lymphocytic leukaemia, stomach tumour, lipoma, granulosa cell tumour, thyroid tumour, liver hemangioma and hepatoma. The spectrum of tumour incidence in the offspring of treated mice is essentially the same as in controls (90% in the lung). The germ cell sensitivity for tumour induction in the offspring is very similar to that of specific locus mutations by radiation.[30-34] However, the most serious reservation was that the tumour frequency is more than two orders of magnitude higher than that of specific locus mutations.

Confirmation studies were carried out with two other inbred strains of mice, LT and N5.[5] F_1 progeny of LT males irradiated at postmeiotic stages with 5.04 Gy of X-rays developed significantly higher incidence of lung tumours and lymphocytic leukaemias. The induced incidence of lung tumours was similar to that in ICR mice.[5,28] However, a more than 4-fold higher induced incidence of leukaemia was observed in LT than in ICR mice.[2–3,5] Postmeiotic treatment of N5 males also induced very high incidences of tumours. Spermatogonial treatment of the N5 strain increased the incidence of various tumours such as lung tumours, ovarian tumours, and leukaemias in the F_1 offspring. Induced incidences of lung tumours and ovarian tumours in N5 were also similar to those in ICR. However, the induced incidence of lymphocytic leukaemias in N5 was higher than that in ICR. Hepatomas were found in N5, but rarely in ICR and LT. The incidence of leukaemia in the offspring of irradiated LT and N5 parents was about 5 and 10 times higher than that of unexposed controls in each strain, respectively. Consequently, there are strain differences in the incidence of lymphocytic leukaemias among these three strains (reviewed by Nomura[28]) (Table 1).

12.2.2.2 The Heritable Nature of Tumours

To confirm that the induced tumours are heritable, the F_1 progeny of treated parents were mated and their progeny examined. The ICR mice were mated as young adults and the presence or absence of tumours was determined at 8 months of age. The progeny were then classified as to the retrospectively determined type of the parent. A significantly higher incidence of lung tumours was observed in the F_2, when their parental F_1 had tumours. In each case, the original treated mouse was a male, as a precaution against the possible transmission of chemical or cytoplasmic factors to the progeny. The results were confirmed by continuing the X-ray treated group into the F_3 generation,

Table 1 *Induced rate ($\times 10^3$) of tumours in the offspring per parental dose (Gy) of radiation in different strains of mice*

Strain	Radiation	TBA	Hepatoma	Leukaemia and lymphoma	Lung tumour	Reference
ICR	X-rays	24.6	0	1.9	22.7	2,3
N5[a]	X-rays	30.0	5.8	3.0	21.5	–[a]
LT	X-rays	24.3	0	8.5	21.0	5
B6-C3F$_1$	^{60}Co	12.0	6.7	6.4	1.0	–[b]
B6-C3F$_1$	^{252}Cf	–	800	–	–	40,41
C3H/HeH	X-rays	–	–	–	0	42
BALB/cJ	X-rays	–	–	–	0	43
N5	X-rays	–	–	6.8	–	45
N5	^{3}H β-rays	–	–	13.2	–	45
CBA/JNCrj	X-rays	–	90.0	65.0	22.5	39

Note: TBA indicates tumour-bearing animal.
Source: Reproduced from T. Nomura, *Mutat. Res.*, 2003, **544**, 425–432, with permission.
[a] Spermatogonial exposure. Average of two doses (5.04 and 2.16 Gy). Others are postmeiotic exposure.
[b] K. Kamiya, Personal communication.

that is, F_2 males were mated with unirradiated females. Significantly higher incidence of lung tumours was observed in the F_3, when the F_2 male parent had tumours.[2–4] The pattern of inheritance is that of an autosomal dominant with about 40% penetrance. Reduced penetrance was also found for dominant skeletal mutations[35] and congenital malformations.[2–3,6,9] Similar results were also obtained with urethane and 4-nitroquinoline-1-oxide (4-NQO).[2,3] Such a tendency was confirmed with other strains of mice N5 and LT[5] and in further generations[5,11,28] (Figure 1).

Studies with N5 mice, which develop various types of tumours, suggest the inheritance of tumour susceptibility; for example, mating of preconceptionally irradiated lung tumour-bearing F_1 offspring with their litter-mates yields a variety of tumors, such as lung tumours, ovarian tumours, multiple embryonic tumours, myxoma, and thymic lymphomas (lymphocytic leukaemias) in the F_2 offspring.[11,28,29] Observation continued for further generations: F_3 progeny of F_2 parental mice with multiple tumours and lung tumour developed not only multiple tumours and lung tumour, but also other types of tumours (ovarian tumour, hepatoma, leukaemia, *etc.*), and congenital anomalies. The pattern is similar to that of Li–Fraumeni syndrome, suggesting the inheritance of tumour susceptibility. Germ cell exposure may have induced a transmissible hyper susceptibility to tumour induction in the offspring. In further studies, induced tumours in N5 were used for molecular analyses.

12.2.2.3 *Manifestation of Germ-Line Alteration Causing Tumours by Postnatal Environments*

If a germ-line mutation can lead to cancer, all cells composing that organ must be mutated and have an equal likelihood to form tumours.[4] However, only one tumour nodule was induced in the organ. Presumably, such changes induced in the offspring by parental exposure to X-rays must be weakly carcinogenic by themselves, and their expression will be influenced by ageing and by naturally existing carcinogenic and promoting agents in the diet and environment. This hypothesis was proven by the fact that unusually large clusters of tumour nodules developed in the lung after postnatal treatment with small amounts of urethane.[4] This study was confirmed in the lung of the offspring of parentally X-irradiated outbred SHR mice by postnatal treatment with urethane,[36] in the skin by postnatal treatment with 12-*O*-tetradecanoylphorbol-13-acetate (TPA),[37] and in the incidence of leukaemia by preconceptional [239]Pu irradiation and postnatal methylnitrosourea treatment.[38] However, such enhancing effects were not observed when CBA/J male mice were irradiated and their offspring were postnatally treated with urethane.[39]

To confirm the persistent hypersensitivity induced by germinal exposure to radiation, N5 male mice were treated with 2.16 Gy of [60]Co γ-rays (0.52 Gy/min), at the spermatogonial stage and mated with untreated N5 females. Offspring (6 weeks old) were treated twice a week with TPA for 18 weeks. A significant increase in skin cancer and leukaemia was observed in the offspring by preconceptional irradiation and postnatal TPA treatment, and a higher

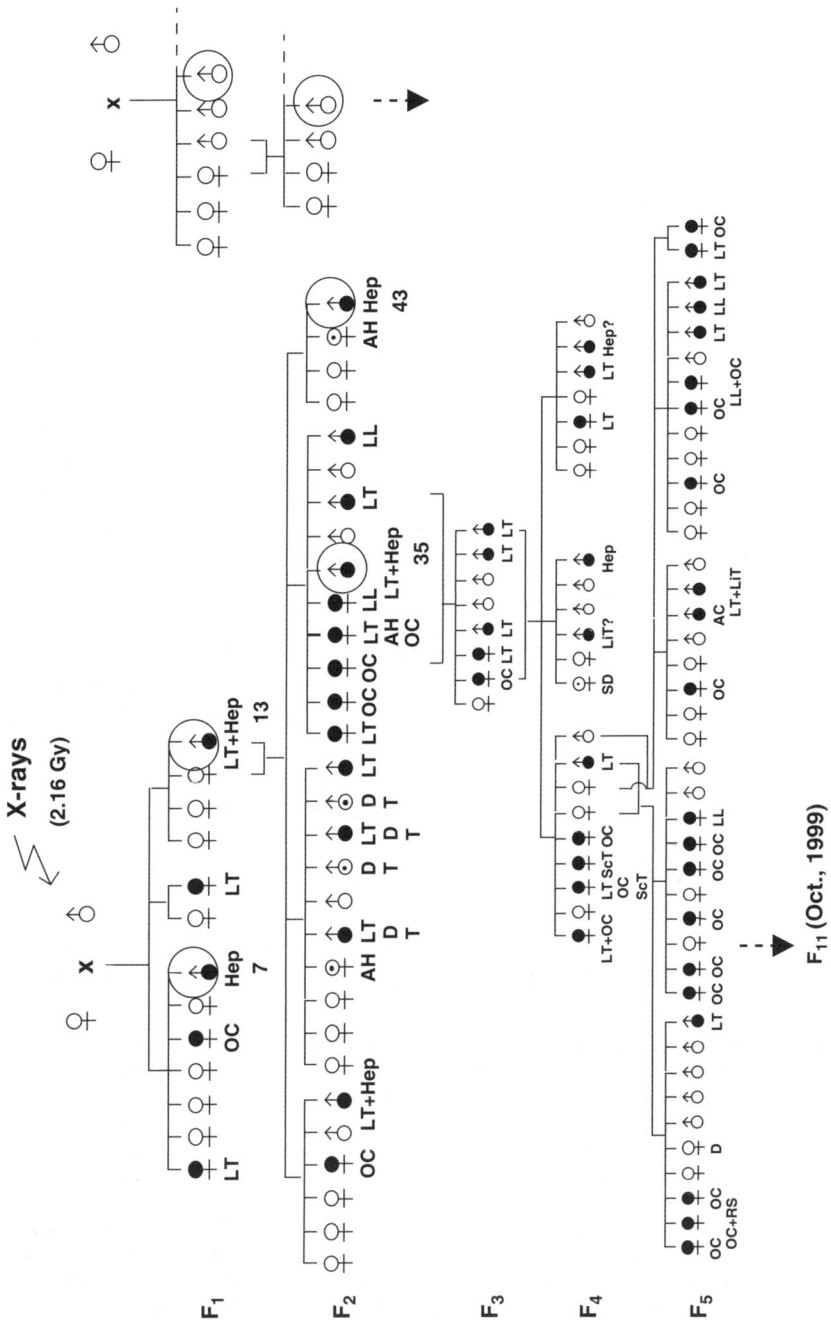

incidence of skin tumours developed in F_2 and further generations.[11] However, it was not seen without the postnatal TPA treatment. Thus, germ cell exposure to radiation appears to induce transmissible hypersensitivity to tumour induction in the next generation, although it is very weakly carcinogenic by itself.[4,11]

12.2.2.4   Further Mouse Studies

After Gardner's report in 1990[17] on the higher risk of leukaemia in the children of fathers who had been exposed to nuclear radiations, experimental studies were carried out with various strains of mice in several countries, to reconcile the differences between our experimental results and epidemiological findings. High incidence of liver tumours was observed in the F_1 offspring of C3H male mice which had been exposed to 0.5 Gy of ^{252}Cf (66% neutron) and mated with C57BL/6 females.[40,41] A slight increase in tumours was also observed in B6C3F_1 offspring after paternal exposure to ^{60}Co-γ-rays (Kamiya, personal communication). Cattanach *et al.*[42,43] observed no significant increase but a seasonal change in the incidence of lung tumours in the offspring of BALB/cJ and C3H/HeH mice exposed to X-rays following the experimental protocol of Nomura.[2,3] We believe that a seasonal change is observed, when experiments are carried out in insufficient animal facilities and experimental conditions, for example, change of light–dark intervals significantly influenced tumour frequencies in mice.[44]

Another study was carried out in Canada with N5 mice provided by Nomura. Male N5 mice were irradiated in conditions close to those used by Nomura[2,3] with 5 Gy of X-rays.[2,3] In this study, the probabilities of dying from leukaemia and overall survival were statistically different between the offspring of X-ray-treated males and unirradiated controls. Earlier occurrence of leukaemia was also observed in the F_1 offspring after the treatment of male N5 mice with tritiated water.[45]

A lifetime experiment showed a trend towards a higher incidence of tumours of the hematopoietic system and broncho-alveolar adenocarcinomas in the offspring of male CBA/JNCrj mice exposed to X-rays 1 week before mating (spermatozoa irradiation), although no increase in tumour incidence was

---

Figure 1   *Tumours and congenital anomalies in the offspring of male N5 mice exposed to 2.16 Gy of X-rays at the spermatogonial stage. This is one of the pedigrees of tumour-susceptible offspring of N5 mice exposed to X-rays. A part of the pedigree of concurrent (untreated) controls is shown on the right side. Hepatoma lesions and/or normal liver tissues of six circled male offspring in the figure were used for Gene Chip analysis. This figure was modified from T. Nomura, Cong. Anomalies, 2000, 40, S54–S67 with permission. Closed symbols for males and females indicate tumour-bearing offspring, and dotted symbols indicate offspring with congenital anomalies. Open symbols indicate offspring without tumours and anomalies. The abbreviations used are: Hep, hepatoma; LT, lung tumour; OC, ovarian tumour; LL, lymphocytic leukaemia; ScT, subcutaneous tumours (fibrosarcoma, rhabdomyosarcoma); LiT, liver tumour; RS, reticulum cell neoplasia; AH, atresia hymenalis; D, dwarf; T, tail anomalies; SD, diverticulum of stomach*

observed in the offspring of males irradiated 9 weeks before conception (spermatogonial irradiation).[39]

In general, there are apparent strain differences in the types of induced tumours and tumour incidences in the preconceptionally irradiated offspring (Table 1), indicating that pre-existing genetic predisposition, that is strain difference, is essential and important in transgenerational carcinogenesis.[28] In other words, preconceptional irradiation may enhance background incidences of tumours in each strain. In fact, most of the induced tumours were commonly observed types of tumours in each strain, and only a few were very rare types.

12.2.2.5   *Malignancy of Induced Tumours*

To examine the malignancy of tumours, we tested transplantability of induced tumours in the offspring of X-irradiated N5 mice.[5] Twenty-six tumours (11 lung tumours, 5 leukaemias, 2 fibrosarcomas, 3 undifferentiated tumours, 2 hepatomas, *etc.*), were transplated subcutaneously in N5 mice. Among them, only three (one lung tumour and two hepatomas) failed to grow, but others grew rapidly, metastasized and killed the recipient N5 mice. Even very small lung tumour nodules (1–2 mm diameter) could grow and kill the animals.

12.2.3   Possible Mechanism and Molecular Studies

12.2.3.1   *Transgenerational Chromosomal Changes*

To study the genesis of the high incidence of tumours and congenital anomalies in the progeny, first, we examined chromosomal aberrations. There is a considerable amount of literature on the induction of translocations by radiation or chemicals in the germ cells of mice (reviewed in UNSCEAR).[46] We studied radiation- and urethane-induced translocations cytogenetically in the germ cells of treated adult ICR males as well as in the F_1 male progeny sired by them to examine whether there was any correlation between induced translocations and tumours in the F_1 progeny.[2,5,29] However, tumours occurred no more frequently in the offspring with translocations than in those without translocations, that is, there is no correlation between the induction of translocations and the occurrence of tumours.[2,3,29]

Then we examined whether any relationship could be discerned between visible chromosomal changes (in bone-marrow preparations using G and CQ-band analysis) in 36 tumour-bearing offspring in the N5 strain (from post radiation generations F_1–F_5) and 53 irradiated but non-tumour bearing controls. The results do not provide any evidence for a cytogenetically detectable chromosomal abnormality in these animals.[5,29] The above sets of data considered together suggest that induced germ cell alterations causing tumours are not related to gross chromosomal changes detectable with the cytogenetic techniques employed. However, they do not exclude the possibility that smaller genetic changes may be involved.[2,3,5,29] These findings were supported by the fact that urethane, an intragenic mutagen,[47] induced tumours, but neither translocations nor dominant lethals in the offspring.[2,3]

12.2.3.2   Oncogene Activation

To determine the potential contribution of oncogenes to multigeneration carcinogenesis, activation of 17 known oncogenes was examined in DNA extracted from transplantable tumours induced in the offspring of X-irradiated N5 parents (undifferentiated tumours, fibrosarcomas, lung tumours, lymphocytic leukaemias, etc.).[7] Ras, mos, and/or abl were amplified in two rare tumours (undifferentiated multiple tumours in Figure 1) and lymphocytic leukaemia.[7] Furthermore, two of three tumour DNAs which did not hybridize with any of the oncogenes so far tested had transforming ability on hamster cells, and were found to contain mos[48] and cot[49] oncogenes. These were the first reports of detection of activated mos and cot genes by transfection assay. P53 mutation was also detected in brain tumour (glioma) of X-irradiated N5 progeny (data not shown). Thus, known and new oncogenes were activated in the tumours which were not observed spontaneously in the N5 strain, but induced in the descendants of X-irradiated N5 parents. In commonly observed tumours, however, we rarely found oncogene activation and p53 tumour suppressor gene mutation.

Consequently, the majority of tumours induced in the progeny were commonly observed types in the strain used in which oncogene activations were rarely detected, although germ-line alteration could produce some specific tumours and molecular changes.

12.2.3.3   Genomic Instability

Considerable numbers of papers have been published in the literature on germline mutations at the expanded simple tandem repeat (ESTR) loci induced by both low and high LET (neutrons from ^{252}Cf) irradiation[50–56] (reviewed in UNSCEAR[57]). In the experiments of Dubrova et al., however, spermatogonial stages are more sensitive than postmeiotic stages and there were no dose rate effects. Their results were different from those published by Niwa and colleagues[55,56] and also different from those of specific locus mutations.[30–34]

To assess whether genomic instability induced by irradiation of germ cells could show some association with induced tumours, similar studies are being carried out in our laboratory using the offspring of ^{60}Co-irradiated (2.16 Gy) and unirradiated N5 males (postmeiotic irradiation). Mice were euthanized at 12 months of age, and liver and tumour tissues were examined for Pc3 mutations by analysing the PCR products by Gene Scan (ABI PRISM 3100 Genetic Analyzer: Applied Biosystems, Foster City, CA).[29,58] The results obtained thus far show no increase of outliers either in 65 offspring of irradiated males or in 13 tumour tissues. Further studies are currently underway using microsatellite probes in mice and humans.[59]

12.2.3.4   Changes in Gene Expression in Cancer-Prone Progeny

Characteristics of germ-line alterations causing tumours that we have observed: (a) 100-fold higher incidence of tumours in the offspring than ordinary mouse mutations; (b) radiation-induced genetic changes are weakly carcinogenic by themselves with manifestation by postnatal treatment; (c) inheritance of tumour susceptibility; (d) increases over the background or strain dependent incidence of tumours (strain differences), are suggesting that the accumulation of changes in functional genes underlying immunological, biochemical, and physiological functions may slightly elevate or enhance tumour incidences in each strain.

The question of alterations in gene expression (that are known to occur in many genes in tumours) is now being addressed using the Gene Chip technology (Affymetrix, Inc., Santa Clara, CA). In these experiments, the affected F_1 offspring (with tumours and malformations) of N5 male mice exposed to 2.16 Gy of X-rays at the spermatogonial stage were used as the starting material. These animals were mated to their litter mates as young adults and their offspring were examined for the presence or absence of tumours at 12 months of age and classified as to the retrospectively-determined type of the parent.

A total of 6500 genes examined in the F_3 offspring of ^{60}Co-γ-ray irradiated male N5 mice, 254 and 75 functioning genes showed more than a 4-fold differences in the expression level (both suppression and over-expression) in the skin cancer lesion and surrounding normal skin area of preconceptionally irradiated F_3 offspring, respectively, in comparison with the skin of unirradiated concurrent control F_3. In the non-tumour area, for instance, macrophage inflamatory protein, osteopontin precurser, SV40 induced 24p3 mRNA and a variety of genes were 50-fold over-expressed or suppressed.[29]

Gene Chip expression analysis has also been carried out for the hepatoma and normal liver tissue of the affected offspring and compared with the liver tissue of concurrent controls (Figure 1).[29] Progeny No. 7 developed only one hepatoma. However, No. 13 developed three tumour nodules and both No. 35 and No. 43 developed two nodules. Among 12,000 genes examined, there were more than a 4-fold differences in 30 and 110 functioning genes in the normal liver tissue and hepatomas of irradiated F_1 and F_2 offspring, respectively (Table 2). The average numbers of abnormal expression of oncogenes per individual liver tissue and hepatoma were 4.3 and 20, respectively. Many genes are altered and pre-exist in the liver tissue of cancer-prone progeny, and the majority of the abnormally expressed genes were those involved in normal physiological, biochemical, and immunological functions. Consequently, changes in gene expression seem to occur in various normal functional genes rather than oncogenes *per se* in irradiated cancer-prone or tumour susceptible descendants, and their progressive accumulation may contribute to cancer; this has been our hypothesis.[2–3,7] In fact, the numbers of such oncogene-related genes and tumour suppressor-related genes have increased enormously during the last 3 years in literature with advance of biomedical research and technology (NetAffx Analysis Center, Affymetrix, Inc.) (Table 2, A, B).[28,29].

About 60% of genes showing altered expression were different among three hepatoma nodules in No. 13 and two nodules in No. 35 and 43. These observations suggest that each tumour nodule is derived from a different liver cell in the individual mouse. More precise analyses on the genes involved, tumour types, *etc.* are necessary, before definitive conclusions can be reached.

Table 2 *Changes in expression of functioning genes in the hepatoma and adjacent normal liver tissue of the descendants of N5 male mice exposed to 2.16 Gy of spermatogonial X-irradiation*

Liver	Increase ≥ × 4				Decrease ≤ × 1:4			
	Total	A	B	C	Total	A	B	C
Total numbers of genes	12,000	1,888	390	725	12,000	1,888	390	725
Average numbers of genes in abnormal expression								
Normal tissue	22.8	3	0.8	1.8	7	1.3	0	0.8
Hepatoma	83.9	16.4	3.3	6.1	26	4.4	0.1	3.6

Source: Data derived from NetAffx Analysis Center (Affymetrix, Inc.) on July 29, 2005.
Eight hepatomas and four adjacent normal liver tissues of four male offspring of X-irradiated N5 male mice and normal liver tissues of two concurrent controls (Figure 1) were used for Gene Chip analyses (U74Av.2 Array, Affymetrix, Inc., Santa Clara, CA). Gene expression in the hepatoma and normal liver tissues of the offspring of X-irradiated male N5 mice was compared to those of concurrent controls (offspring of unirradiated N5 mice). Numbers of functioning genes showing 4-fold increases or decreases were scored. The average values of eight hepatomas and four normal liver tissues are shown in the table. A: oncogenes and oncogene-related genes, B: tumour suppressor genes and tumour suppressor-related genes, C: immune-related genes.

12.3 Relevant Human Studies

In 1990, Gardner *et al.*[17] reported that there was about 6–8 fold higher risk of leukaemia in the children of fathers who were employed at Sellafield nuclear reprocessing plant and had been exposed to 10–100 mSv of radiation before conception. As a possible cause of leukaemia induction, the sperm damage by fathers' exposure to radiation was put forth on the basis of our mouse experiments.[2–5,8] However, his report has not been supported by the epidemiological study on the children of atomic bomb survivors in Hiroshima and Nagasaki who were exposed to an average dose of 435 mSv,[21,22] although some epidemiological studies have reported (but have not proved) the increase of leukaemia in the children of fathers who had been exposed to low doses of diagnostic radiation.[14–16]

The discrepancies between epidemiological studies and also between humans and animals can be partly reconciled by the following considerations:[8] (1) Much higher doses were used in animal experiments (0.36–5 Gy) in comparison to those experienced in epidemiological studies, (2) human populations were exposed to radiation at the spermatogonial stage (mostly) which is known to be less sensitive to radiation than spermatozoa and spermatid stages, (3) possible existence of differences in the genetic predisposition may exist in both human populations and mouse strains, (4) postnatal exposure by radiation and/or chemically contaminated environment may enhance tumour incidence in the offspring as seen in mice, (5) epidemiology focuses on childhood cancer and leukaemia which develop at younger ages (below 20 years), but not on adult diseases. Although the direct link of cancer to sperm exposure has not been confirmed,[19] adult-types of cancer and other adult (chronic) diseases may increase at cancer-prone ages in the human population, as anticipated by the animal experiments,[8,11,18] because induced rate of solid tumours in the offspring of mice exposed to radiation is much higher than that of leukaemia. In Japan, an epidemiological survey has been underway since 2002 to investigate the possibility of transgenerational transmission of chronic diseases in the offspring of A-bomb survivors.

12.4 Summary

Parental exposure of mice to radiation and also chemicals causes a variety of adverse effects (*e.g.*, tumours, congenital malformations, mutations, and embryonic deaths) in the progeny. Tumour-induction in the progeny shows apparent dose-rate effects in experiments involving spermatogonial and oocyte exposures, probably due to the repair of induced DNA damage in these germ cells. The tumour phenotype is transmissible beyond the first postradiation generation and the transmission pattern is consistent with inheritance of tumour susceptibility, that is, increased incidence of common types of tumours that are observed in the background of the used strain. There are strain differences in tumour types. Germ-line alteration causing tumours is very weakly tumourigenic by itself, but induces persistent hypersensitivity to tumour induction in the offspring for the subsequent development of tumours by the postnatal exposure to tumour promoting/carcinogenic agents.

Cytogenetic and molecular analyses provided no evidence for the presence of chromosomal anomalies in tumour-bearing offspring, although smaller changes beyond the resolution of the method could not be excluded. Mutations of oncogenes and tumour suppressor genes such as *p53, ras, mos, abl, cot, etc.* were detected in some specific tumours induced in cancer prone descendants. However, the majority of tumours observed in the progeny were those commonly observed in the strain used. Cumulative alterations in many normal gene loci concerning immunological, biochemical, and physiological function may slightly elevate or enhance tumour incidence. In fact, Gene Chip analyses show significant suppression and/or over-expression of many functional genes in the cancer-prone lines of mice. Potential links of mini- and micro-satellite mutations to transmissible adverse effects have not been seen in our mouse system or in humans.

Acknowledgments

The work was supported by JSPS and MEXT of Japan, Nissan, Showa-Shell, and Heiwa Nakajima Foundations, and Osaka, Inoue, and Kihara Science Awards. I wish to thank all the past and/or present members of the former Institute for Cancer Research, Department of Radiation Biology and Medical Genetics, Graduate School of Medicine, Osaka University, and Medical

Genetics, and University of Wisconsin for their advice and help over the years. I am grateful to Professors J.F. Crow (Wisconsin) and K. Sankaranarayanan (Leiden) for their constructive comments and editorial suggestions for this article.

References

1. T. Nomura, *Cancer Res.*, 1975, **35**, 264–266.
2. T. Nomura, in *Tumors of Early Life in Man and Animals*, L. Severi (ed), Perugia University Press, Perugia, 1978, 873–891.
3. T. Nomura, *Nature*, 1982, **296**, 575–577.
4. T. Nomura, *Mutat. Res.*, 1983, **121**, 59–65.
5. T. Nomura, in *Genetic Toxicology of Environmental Chemicals, Part B: Genetic Effects and Applied Mutagenesis*, C. Ramel (ed), Alan R. Liss, New York, 1986, 13–20.
6. T. Nomura, *Mutat. Res.*, 1988, **198**, 309–320.
7. T. Nomura, in *Perinatal and Multigeneration Carcinogenesis*, N.P. Napalkov, J.M. Rice, L. Tomatis and H. Yamasaki (eds), IARC Sci. Publ. No. 96, IARC, Lyon, 1989, 375–387.
8. T. Nomura, *Nature*, 1990, **345**, 671.
9. T. Nomura, in *Male-Mediated Developmental Toxicity*, D.R. Mattison and A.F. Olshan (eds), Plenum Press, New York, 1994, 117–127.
10. IARC, *IARC Monograph on the Evaluation of Carcinogenic Risks to Humans. Vol. 75: Physical Agents: Ionizing Radiation, Part I, X-rays, γ-rays and neutrons*, IARC, Lyon, 1999.
11. T. Nomura, *Cong. Anomalies*, 2000, **40**, S54–S67.
12. H.J. Evans, *Nature*, 1982, **296**, 488–489.
13. N.A. Brown, *Nature*, 1985, **316**, 110.
14. S. Graham, M.L. Levin, A.M. Lilienfield, L.M. Schuman, R. Gibson, J.E. Dowd and L. Hemplemann, *Natl. Cancer Inst. Monogr.*, 1966, **19**, 347–371.
15. P.H. Shiono, C.S. Chang and N.C. Myrianthopoulos, *J. Natl. Cancer Inst.*, 1980, **65**, 681–686.
16. X.O. Shu, Y.T. Gao, L.A. Brinton, M.S. Linet, J.T. Tu, W. Zheng and J.F. Fraumeni Jr., *Cancer*, 1988, **62**, 635–644.
17. M.J. Gardner, M.P. Snee, A.J. Hall, C.A. Powell, S. Downes and J.D. Terrell, *Br. Med. J.*, 1990, **300**, 423–429.
18. T. Nomura, *Br. Med. J.*, 1993, **306**, 1412.
19. COMARE, Committee on Medical Aspects of Radiation in the Environment. Fourth Report. Department of Health, London, 1996.
20. L. Parker, M.S. Pearce, H.O. Dickinson, M. Aikin and A.W. Coft, *Lancet*, 1999, **354**, 1407–1414.
21. Y. Yoshimoto and K. Mabuchi, *J. Radiat. Res.*, 1991, **32**, 294–300.
22. Y. Yoshimoto, H. Kato and W.J. Schull, *J. Radiat. Res.*, 1991, **32** 231–238.
23. K.M. Kirk and M.F. Lyon, *Mutat. Res.*, 1982, **106**, 73–83.
24. K.M. Kirk and M.F. Lyon, *Mutat. Res.*, 1984, **125**, 75–85.

25. M.F. Lyon and R. Renshaw, *Mutat. Res.*, 1988, **198**, 277–283.
26. A.G. Searle and C.V. Beechey, in *Genetic Toxicology of Environmental Chemicals Part B: Genetic Effects and Applied Mutagenesis*, C. Ramel (ed), Alan R. Liss, New York, 1986, 511–518.
27. W.U. Müller, C. Streffer, A. Wojcik and F. Niedereichholz, *Mutat. Res.*, 1999, **425**, 99–106.
28. T. Nomura, *Mutat. Res.*, 2003, **544**, 425–432.
29. T. Nomura, H. Nakajima, H. Ryo, L.Y. Li, Y. Fukushima, S. Adachi, H. Gotoh and H. Tanaka, *Cytogenet. Genome. Res.*, 2004, **104**, 252–260.
30. W.L. Russell, L.B. Russell and E.M. Kelly, *Science*, 1958, **128**, 1546–1550.
31. M.F. Lyon, D.G. Papworth and R.J.S. Phillips, *Nature (London), New Biol.*, 1972, **238**, 101–104.
32. W.L. Russell and E.M. Kelly, *Proc. Natl. Acad. Sci. USA*, 1982a, **79** 539–541.
33. W.L. Russell and E.M. Kelly, *Proc. Natl. Acad. Sci. USA*, 1982b, **79** 542–544.
34. U.H. Ehling, in *Mutations in Man*, G. Obe (ed), Springer-Verlag, Berlin, 1984, 292–318.
35. M. Bartsch-Sandhoff, *Humangenetik*, 1974, **25**, 93–100.
36. I.E. Vorobstova and E.M. Kitaev, *Carcinogenesis*, 1988, **9**, 1931–1934.
37. I.E. Vorobstova, L.M. Aliyakparova and V.N. Anisimov, *Mutat. Res.*, 1993, **287**, 207–216.
38. B.I. Lord, L.B. Woodford, L. Wang, V.A. Stones, D. McDonald, S.A. Lorimore, D.P. Papworth, E.G. Wright and D. Scott, *Br. J. Cancer*, 1998, **78**, 301–311.
39. U. Mohr, C. Dasenbrock, T. Tillmann, M. Kohler, K. Kamino, G. Hagemann, G. Morawietz, E. Campo, M. Cazorla, P. Fernandez, L. Hernandez, A. Cardesa and L. Tomatis, *Carcinogenesis*, 1999, **20**, 325–332.
40. T. Takahashi, H. Watanabe, K. Dohi and A. Ito, *Cancer Res.*, 1992, **52**, 1948–1953.
41. H. Watanabe, T. Takahashi, J.Y. Lee, M. Ohtaki, G. Roy, Y. Ando, K. Yamada, T. Gotoh, K. Kurisy, N. Fujimoto, Y. Satow and A. Ito, *Jpn. J. Cancer Res.*, 1996, **87**, 51–57.
42. B.M. Cattanach, G. Patrick, D. Papworth, D.T. Goodhead, T. Hacker, L. Cobb and E. Whitehill, *Int. J. Radiat. Biol.*, 1995, **67**, 607–615.
43. B.M. Cattanach, D. Papworth, G. Patrick, D.T. Goodhead, T. Hacker, L.E. Cobb and L. Whitehill, *Mutat. Res.*, 1988, **403**, 1–12.
44. H. Nakajima, I. Narama, T. Matsuura and T. Nomura, *Cancer Lett.*, 1994, **78**, 127–131.
45. A. Daher, M. Varin, Y. Lamontagne and D. Oth, *Carcinogenesis*, 1998, **19**, 1553–1558.
46. UNSCEAR, United Nations Scientific Committee on the Effects of Atomic Radiation. Ionizing Radiation: Levels and Effects. Vol. II, Effects, A Report to the General Assembly with annexes. United Nations, New York, 1972.
47. T. Nomura and N. Kurokawa, *Jpn. J. Cancer Res.*, 1997, **88**, 461–467.

48. Y. Ohuchi, Y. Kinuta, H. Sasai, J. Miyoshi, T. Nomura and K. Toyoshima, *Oncogene*, 1992, **7**, 331–338.
49. H. Sasai, T. Higashi, S. Nakamori, J. Miyoshi, F. Suzuki, T. Nomura and T. Kakunaga, *Br. J. Cancer*, 1993, **67**, 262–267.
50. Y.E. Dubrova, A.J. Jeffreys and A.M. Malashenko, *Nat. Genet.*, 1993, **5**, 92–94.
51. Y.E. Dubrova, M. Plumb, J. Brown, J. Fennelly, P. Bois, D. Goodhead and A.J. Jeffreys, *Proc. Natl. Acad. Sci. USA*, 1998a, **95**, 6251–6255.
52. Y.E. Dubrova, M. Plumb, J. Brown and A.J. Jeffreys, *Int. J. Radiat. Biol.*, 1998b, **74**, 689–696.
53. Y.E. Dubrova, M. Plumb, J. Brown, E. Boulton, D. Goodhead and A.J. Jeffreys, *Mutat. Res.*, 2000a, **453**, 17–24.
54. Y.E. Dubrova, M. Plumb, B. Gutierrez, E. Boulton and A.J. Jeffreys, *Nature*, 2000b, **405**, 37.
55. S. Sadamoto, S. Suzuki, K. Kamiya, R. Kominami, K. Dohi and O. Niwa, *Intl. J. Radiat. Biol.*, 1994, **65**, 549–557.
56. O. Niwa, Y.J. Fan, M. Numoto, K. Kamiiya and R. Kominami, *J. Radiat. Res.*, 1996, **37**, 217–224.
57. UNSCEAR, United Nations Scientific Committee on the Effects of Atomic Radiation. Hereditary Effects of Radiation. Report to the General Assembly with Scientific Annexes. United Nations, New York, 2001.
58. T. Nomura, H. Nakajima, L.Y. Li, Y. Fukudome, R. Baskar, H. Ryo, J.Y. Koo and K. Mori, *Environ. Mutagen. Res.*, 1999, **21**, 207–211 (in Japanese).
59. K. Furitsu, H. Ryo, K.G. Yeliseeva, L.T.T. Thuy, H. Kawabata, E.V. Krupnova, V.D. Trusova, V.A. Rzheutsky, H. Nakajima, N. Kartel and T. Nomura, *Mutat. Res.*, 2005, **581**, 69–82.

CHAPTER 13

Heritable Effects on DNA Damage Following Paternal F_0 Germline Irradiation

MING-WEN LI AND JANET E. BAULCH

Center for Health and the Environment, University of California, Davis, CA

13.1 Introduction

The production of genetically healthy sperm is essential to reproduction since sperm deliver the paternal genome to oocytes. Spermatogenesis is an extremely complex process involving many rounds of cellular proliferation, meiosis, and highly specialized chromatin packaging. Chemical or radiation exposures can disrupt spermatogenesis and can damage the paternal genetic contribution to the offspring. If not properly repaired, damage to sperm DNA can lead to reduced fertility, spontaneous abortion, and impaired or abnormal fetal development. In addition to these outcomes, molecular studies now show that there may be other subtle, potentially harmful effects observed in offspring conceived of sperm whose DNA has been damaged by chemical or radiation exposure.

Sperm morphology, motility, and counts are classic tests used to predict spermatogenic dysfunction and heritable mutations.[1,2] These tests, along with the 7-specific locus test[3], are primarily thought to be indicators of targeted mutational events following a germline chemical or radiation exposure. These endpoints may be predictive of effects at moderately low doses, but require large numbers of mice to obtain statistical significance. More recent animal studies using other methods, such as analysis of expanded simple tandem repeat (ESTR) DNA instability, or analysis of p^{un} reversion events, are more sensitive to relatively low dose exposures and demonstrate the transmission of heritable effects through the male germline following premeiotic spermatogonial irradiation or chemical exposures.[4-9]

Studies from our laboratory have evaluated heritable effects of paternal spermatogonial irradiation (0.1–1.0 Gy) on cell proliferation and protein endpoints. Using outbred CD1 mice, we demonstrated heritable effects of ^{137}Cs γ irradiation of the paternal F_0 Type B spermatogonia through the F_2 generation using embryo cell proliferation rate as the endpoint in

149

preimplantation embryo chimera assays.[10] These results were surprising because there was no degradation in the frequency or severity of the decreased cell proliferation rate among the preimplantation embryos with a history of paternal Type B spermatogonial irradiation between the F_1 and F_2 generation of embryo chimeras.

In subsequent studies, adult F_2 chimeric male mice were generated with both a component from paternal F_0 Type B spermatogonial irradiation and a control component.[11,12] Since these chimeric males produced both sperm with paternal F_0 irradiation history and control sperm, litters that included F_3 offspring with paternal F_0 irradiation history and offspring with no irradiation history were obtained from a single dam. The basal hepatic activities of cytosolic receptor tyrosine kinase (RTK), protein kinase C (PKC), and mitogen activated protein kinase (MAPK) were altered in the juvenile F_3 offspring with an irradiation history in comparison with the littermate controls, as were nuclear protein levels of Trp53 and $p21^{waf1}$.[11] PKC and MAPK enzyme activities and Trp53 and $p21^{waf1}$ protein levels were selected for evaluation in these studies because of their involvement in cellular proliferation, radiation response, and genomic instability.

Longitudinal evaluation of the F_2 chimeric males revealed evidence of germline chimeric drift with selection against the component of the germline having a history of Type B spermatogonial irradiation.[12] The observed germline drift suggested the possibility that interactions between the germlines in chimeric male mice could have contributed to the biochemical differences between offspring arising from the irradiation history and from the control cell lineages. However, our findings on protein endpoints in the chimera studies have been supported by studies of conventionally bred mice.[13-16]

The results of those studies as well as the results of germline and heritable ESTR mutation analyses suggest that the observed radiation-induced transgenerational phenotype has an underlying mechanism involving a form of inherited genomic instability.[5,6,8,9] Studies of transgenerational effects of paternal germline irradiation demonstrate a higher incidence of affected individuals than would be predicted by Mendelian genetics, suggesting that the heritable phenotype is a result of non-targeted or epigenetic effects of germline irradiation. Additionally, studies have demonstrated that the offspring conceived of mature sperm that were premeiotic spermatogonia at the time of paternal germline irradiation respond differently from acute radiation or chemical exposure, indicating that this inherited phenotype is functional.[16,17] While these heritable effects have been shown to affect *in vitro* fertilization capacity of first generation male mice from sires that received a moderate dose of ionizing radiation,[18] there has been no demonstrable effect on the *in vivo* fertility of normal, wild type mice.[19]

The comet assay, or single cell gel electrophoresis (SCGE), has been used to measure DNA damage endpoints including single-strand breaks (SSBs), double-strand breaks (DSBs), and alkaline labile sites in the DNA of somatic cells following exposure to genotoxins. The comet assay has also been modified to evaluate DNA damage in human and mouse sperm.[20-25] In a previous study, we demonstrated altered basal levels of DNA damage in kidney cells from third-generation offspring with a paternal F_0 germline irradiation history using the alkaline pH comet assay.[16] The aim of the present study was to use the comet assay to evaluate the sperm of offspring from irradiated male mice for heritable changes in basal levels of DNA damage or chromatin electrophoretic mobility as a result of the paternal irradiation history. There is a significantly increased frequency of alkali-labile sites in normal sperm DNA compared with the DNA of most somatic cell types.[26] These alkali-sensitive sites are manifest as DNA breaks in alkaline sperm comet assays. To avoid the possibility that this increased background level of DNA breaks would confound our results, we used a neutral pH sperm comet assay.

In this study, 129S6/SvEv mice were used to evaluate inherited effects of paternal exposure to ionizing radiation on the germline of F_2 and F_3 generation male offspring conceived from mature sperm that were premeiotic Type B spermatogonia at the time of paternal F_0 germline irradiation. These offspring were evaluated for conventional spermatogenic endpoints, including epididymides and testes weights, sperm motility, sperm morphology, and testicular sperm counts. They were also evaluated for sperm DNA damage using the comet assay. The hypothesis being tested is that irradiation of paternal F_0 spermatogonia prior to conception of the first generation initiates a form of transgenerational genomic instability that results in the inheritance of germline DNA damage or chromatin effects in male offspring of irradiated male mice without inherited gross effects on spermatogenesis or *in vivo* fertility.

13.2   Materials and Methods

13.2.1   Animals

Male mice from a single cohort of 8–10-week-old 129S6/SvEv mice (Jackson Laboratory, Bar Harbor, ME) were stratified by body weight and then randomly assigned to the 0.1 Gy irradiated group (17 mice) and the sham-irradiated concurrent control (16 mice) group of F_0 males. In this manner, we obtained dose and control groups of mice with comparable mean body weight (±SE). Male mice from two different single cohorts were grouped for the 0.5 Gy irradiation experiments using the same protocol (16 irradiated and 16 sham-irradiated concurrent control for initial sperm analysis and transgenerational study; 10 irradiated and 10 sham-irradiated for the subsequent analysis of F_0 male epididymal weight and sperm counts).

For each breeding event, female 129S6/SvEv mice (13–14 weeks old; Taconic, Germantown, NY) were also from single cohorts and assigned to males among the irradiated and the control groups. Mice were maintained under a 12 h light/ 12 h dark photoperiod.

All animal experimentation was carried out in accordance with the principles of the American Association for Accreditation of Laboratory Animal Care (AAALAC) in fully accredited facilities.

13.2.2 Irradiation and Dosimetry

The experimental males received an absorbed dose of 0.1 or 0.5 Gy from acute whole-body attenuated ^{137}Cs γ irradiation. Mice were irradiated using a J.L. Shepherd & Associates (San Fernando, CA) Mark I Model 30 calibrated ^{137}Cs γ ray (0.662 MeV) irradiator at a dose rate of 0.017 Gy min^{-1} to obtain a dose of 0.1 Gy and at a dose rate of 0.037 Gy min^{-1} to obtain a dose of 0.5 Gy. For each group, control mice were sham irradiated within the exposure chamber with the door closed but without γ rays. The exposure for each group of mice was measured using the average of three or more commercially supplied thermoluminescent dosimeters (TLD-100 LiF powder, Englehard Corp., Harshaw, OH), supplied and read by Radiation Detection Co. (Gilroy, CA). Air dose dosimeter measurements were converted to tissue absorbed dose estimates using the ratio of the energy mass absorption coefficients of γ photons for tissue and air.

At postirradiation week 6 (42 days postirradiation), each male was mated to two females. Following this mating, at 45 days postirradiation and approximately 15 weeks of age for the male mice, the breeding trios were separated and F_0 males were sacrificed for analysis of delayed effects of acute irradiation of Type B spermatogonia as observed in the subsequent mature sperm 45 days later.

The females delivered their F_1 litters. For each litter, the F_1 male offspring were weighed at 19 days of age and a mean male offspring body weight was obtained for the litter. The F_1 male closest in body weight to this mean body weight was selected from each F_1 litter. Beginning at 15–19 weeks of age, the selected F_1 male was mated to a different pair of females to provide the F_2 generation of offspring. The same breeding protocol was used to obtain the F_3 generation of offspring. The non-breeding F_2 generation male mice from the litters were sacrificed to obtain sperm at 18–19 weeks of age. F_3 generation males were sacrificed at 15–16 weeks of age for sperm assays. The F_3 generation males were not bred.

New female mice were obtained from the vendor for each generation's matings to preclude inbreeding or closed colony effects. As with the F_0 dams, the F_1 and F_2 dams were each from a different, single cohort, stratified based on body weight and randomly assigned to either the irradiation history or concurrent control experimental groups. Throughout the study F_1, F_2, and F_3 offspring were tracked with respect to the F_0 sires. For each given generation, both experimental groups of mice were housed together within the same room.

13.2.3 Sample Collection

Male mice were sacrificed by cervical dislocation and then weighed. Both left and right epididymides and testes were removed, weighed and then frozen at -20 °C for later homogenization and sperm counts. To obtain cauda epididymal sperm for assessment of sperm motility, morphology, and DNA integrity, both cauda epididymides were placed in a Petri dish containing 4 mL

Dulbecco's phosphate buffered saline (DPBS) with 3 mg mL^{-1} bovine serum albumin (BSA) and gently punched and pressed with a 30-G needle to release sperm. For F_0 sires on day 45 postirradiation, sperm were collected from both vas deferens by using a teasing action with forceps under dissecting microscope. The sperm were placed in a dark box to protect them from light, and incubated at room temperature for 5 min before sperm cell concentrations were determined using a hemocytometer.

13.2.4   Assessment of Sperm Motility and Morphology

After adjusting sperm concentration to approximately 10^5 sperm mL^{-1}, 6 µL of sperm suspension was mounted onto a microscope slide and sperm motility was analyzed under phase-contrast microscope at 200 × magnification. Sperm motility was calculated as the percentage of motile sperm at room temperature out of 200 sperm per animal.

After motility analysis, 50 µL of sperm suspension from each animal was transferred into a microcentrifuge tube and kept on ice for the comet assay. Concentrated EDTA solution was added to the sperm suspension (final concentration, 10 mM EDTA) to protect the DNA from DNase degradation until comet slides were prepared.

Two-hundred air-dried sperm were scored under a phase-contrast microscope at 400 × magnification to obtain the percentage of sperm with normal head morphology. Sperm smears were prepared for each animal by evenly spreading 5 µL of the sperm suspension along the length of a microscope slide and air-drying. Sperm tail defects were not considered in this study. Normal sperm head morphology was judged as described by Burruel *et al.*[27]

For each animal, the remaining sperm suspension with cauda epididymides tissue was frozen at −20 °C together with the caput and corpus epididymides for later homogenization and epididymal sperm count.

13.2.5   Testicular and Epididymal Sperm Counts

For each animal, both testes were thawed and homogenized together in DPBS containing 3 mg mL^{-1} BSA using a manual glass homogenizer for 16 strokes. The homogenate was mixed thoroughly by vortexing and the elongated heads of spermatids and sperm were counted using a hemocytometer. An average of four independent counts was used to calculate testicular sperm numbers. Sperm counts were expressed as number of sperm per milligram of testes weight. The same method was used to obtain epididymal sperm counts for each animal, using both epididymides plus any reserved epididymal sperm suspension and tissues from the same animal that had been used in other sperm assays. Sperm counts were expressed as number of sperm per mg epididymides weight.

13.2.6   Sperm Comet Assay

Comet slides were prepared from fresh sperm samples on the same day that the samples were collected from the mice. The following procedure was performed under yellow light to minimize light-induced damage to sperm DNA using the techniques of Haines *et al.*[20], with some modifications. Briefly, normal microscope slides were coated with agarose in advance by dipping slides into 1% normal melting-point agarose (NMA) in Ca^{2+}, Mg^{2+}-free PBS and allowing the slides to air-dry at room temperature. A mixture of sperm suspension and 0.7% low melting point agarose (LMA) in Ca^{2+}, Mg^{2+}-free PBS (9:1 v/v), was prepared at 37 °C. Eighty-five μL of this mixture was pipetted onto the surface of a precoated slide and spread by covering with a coverslip. The slides were placed flat and incubated for 10 min at 4 °C. The coverslips were then gently removed, and a 85 μL layer of LMA was pipetted onto each slide and spread with a coverslip. After 10 min at 4 °C, the coverslips were gently removed, and the slides were placed into coplin jars containing freshly prepared, cold lysing solution (10 mM Tris-HCl, pH 10.0 containing 2.5 M NaCl, 100 mM EDTA, 1% Triton X-100, and 10 mM dithiothreitol; 4 °C overnight). The next day, lysing solution was drained off and replaced with freshly made lysing solution containing 0.1 mg mL^{-1} DNase-free Proteinase K. After 4 h at 37 °C, the slides were removed from the coplin jars, drained of any remaining liquid and placed in a 20 × 40 cm horizontal electrophoresis tank side by side to equilibrate in TAE buffer (100 mM Tris-HCl, pH 8.5, 100 mM sodium acetate, and 50 mM EDTA). Slides were subjected to electrophoresis for 30 min at 25 V and 35 mA. Following electrophoresis, the slides were drained and transferred into coplin jars containing distilled water to remove salts and detergents, then fixed in cold 100% ethanol and air-dried at room temperature until scored. The slides were coded randomly, stained with 20 μg mL^{-1} EtBr in Ca^{2+}, Mg^{2+}-free PBS, and viewed using a Zeiss Photoscope fluorescence microscope. Using Komet 5 software (Kinetic Imaging Ltd, Liverpool, UK) 50 sperm per slide were scored. Two slides were analyzed for a total of 100 sperm per animal. Comet tail parameters evaluated include comet tail length (the maximum distance [μm] the damaged DNA migrates from the centre of the cell nucleus), the percent DNA in the tail (the percent of the total DNA that migrates from the nucleus into the comet tail), and tail extent moment (TEM; TEM= (tail length × % tail DNA)/100), for an integrated measurement of overall DNA damage in the cell.

For each dose and generation concurrent control and experimental slides were processed and electrophoresed together within 24 h of slide preparation to minimize variation among slides and animals (*e.g.*, 0.1 Gy and concurrent control F_0 male mice or 0.5 Gy irradiation history F_3 offspring and concurrent controls). This allowed the best possible comparison of experimental animal data to concurrent control data. Because some months elapsed between groups and generations of animals there is variation in comet assay experimental data. This is due to changes in laboratory conditions (temperature and humidity), chemical lots, *etc.* As a result, percent change relative to concurrent control values is the appropriate comparison among groups, rather than the raw comet data values as they are reported in Tables 2 and 3.

13.2.7 Statistical Analyses

Comet data were log transformed and evaluated and contrasted by treatment group with respect to geometric mean and geometric standard error (SE_g).[28] Statistical evaluations of comet data were performed using the non-parametric Mann–Whitney U-test. All other sperm measurements were analyzed by standard ANOVA. Statistical significance was taken as $\alpha \leq 0.05$.

13.3   Results

13.3.1 F_0 Males

Analysis of the F_0 male mouse data demonstrated that there was no effect of acute irradiation using either 0.1 or 0.5 Gy on whole animal body weight, testis weight, sperm motility, sperm morphology, or testis sperm counts at 45 days postirradiation. Although no effect on epididymides weights was observed in the male mice that had been exposed to the lower, 0.1 Gy dose of gamma rays, a small decrease in epididymal sperm counts to 89% of the control level was observed ($p = 0.04$, Table 1). These epididymal sperm correspond to mature sperm that were irradiated 45 days earlier as premeiotic spermatogonia. The 0.5 Gy dose group of F_0 males had significantly decreased epididymides weights and a correlative significant decrease in epididymal sperm counts to approximately 48% of concurrent control counts ($p = 0.01$ and $p = 0.0001$, respectively; Table 1).

Results of neutral pH sperm comet assays to mature sperm that had been acutely irradiated 45 days earlier are shown in Table 2 and Figure 1. Many of these results are statistically significant by Mann–Whitney U-test, and we believe that they are important indicators of the effects of germline irradiation. However, we are only considering those significant changes greater than 5% to be biologically relevant.

Comet analysis of sperm collected from vas deferens of the acutely irradiated F_0 male mice 45 days after irradiation demonstrated significant effects on all endpoints evaluated. The magnitude of the delayed effect of acute irradiation on all comet measurements was larger in the 0.5 Gy dose group relative to concurrent controls than in the 0.1 Gy dose group relative to their concurrent controls. For the 0.1 Gy group all comet measurements were significantly decreased relative to concurrent controls. However, for the 0.5 Gy group all comet measurements were significantly increased relative to concurrent controls. TEM measurements were 16% lower and 36% higher for the 0.1 Gy and the 0.5 Gy groups, respectively, relative to their concurrent controls. Comet tail length measurements were 12% lower and 25% higher for the 0.1 and the 0.5 Gy groups, respectively, relative to their concurrent controls, and measurements of the percent DNA in the comet tail were 5% lower and 7% higher for the 0.1 Gy and the 0.5 Gy groups, respectively, relative to their concurrent controls.

Table 1 *Epididymal weight and sperm counts providing mean values ($\bar{X}$), standard errors of the mean (SE), and number of animals (N) for mice acutely irradiated using ^{137}Cs γ rays. Assays performed at 45 days postirradiation, corresponding to mature sperm that were irradiated as Type B spermatogonia[a]*

Experimental group		Epididymal weight (g)	Epididymal sperm count ($\times 10^6$ mg^{-1})
Sham-irradiated concurrent controls	$\bar{X}$	0.076	84.0
	SE	0.002	2.7
	N	10	10
0.1 Gy acute irradiation	$\bar{X}$	0.078	74.4
	SE	0.002	3.3
	N	10	10
p value		0.32	**0.04**
Sham-irradiated concurrent controls	$\bar{X}$	0.078	86.3
	SE	0.002	7.1
	N	10	10
0.5 Gy acute irradiation	$\bar{X}$	0.069	41.6
	SE	0.002	3.3
	N	10	10
p value		**0.01**	**0.0001**

[a] Statistical significance obtained using ANOVA with α < 0.05.

13.3.2 Breeding Results for Acutely Irradiated F_0 Males

Mice were mated at 42 days postirradiation, so that the offspring were conceived of mature sperm that were premeiotic Type B spermatogonia at the time of irradiation. From the 34 female mice mated to the 17 F_0 males irradiated using 0.1 Gy, 10 F_1 litters with male F_1 offspring were obtained from eight different F_0 sires (Figure 2). From the 32 female mice mated to the 16 sham-exposed F_0 males, eight F_1 litters were obtained from seven different F_0 sires. From the 32 female mice mated to the 16 F_0 males irradiated using 0.5 Gy, seven F_1 litters with male F_1 offspring were obtained from seven different F_0 sires. Six of these litters had male F_1 offspring. From the 32 female mice mated to the 16 sham-exposed F_0 males, eight F_1 litters were obtained from seven different F_0 sires.

Based on the results of these various mating events, no effect of acute spermatogonial irradiation was observed on F_1 generation litter size or *in vivo* fertility of the F_0 male mice 42–45 days after exposure. F_1 male mice obtained from these litters were bred to provide the F_2 (and F_3) generation of mice with 0.1 and 0.5 Gy irradiation history and respective concurrent control animals for the transgenerational evaluation of sperm endpoints.

Table 2   *Effects of acute ^{137}Cs γ irradiation on comet measurements in F_0 mouse mature sperm that were Type B spermatogonia 45 days earlier at the time of irradiation, providing geometric mean values ($\bar{X}_g$), geometric standard error (SE_g), and number of animals (N)[a]*

Experimental group		TEM (50–100 cells per animal)	Tail length (μm)	Tail DNA (%)
Sham-irradiated concurrent	$\bar{X}_g$	3.73	27.73	13.43
controls	SE_g	1.03	1.02	1.02
	N	10	10	10
0.1 Gy acute irradiation	$\bar{X}_g$	3.12	24.49	12.74
	SE_g	1.03	1.01	1.016
	N	10	10	10
p value		**<0.0001**	**<0.0001**	**0.02**
Sham-irradiated concurrent	$\bar{X}_g$	3.01	22.49	12.82
controls	SE_g	1.05	1.05	1.02
	N	7	7	7
0.5 Gy acute irradiation	$\bar{X}_g$	4.10	28.18	13.74
	SE_g	1.06	1.06	1.03
	N	7	7	7
p value		**<0.0001**	**<0.0001**	**0.04**

[a] Levels of significance are obtained with Mann–Whitney *U*-tests.

13.3.3 F_2 Generation Male Offspring

Analysis of whole animal body weight, epididymides weight, testes weight, sperm motility, sperm morphology, and testis sperm counts for F_2 male offspring demonstrated that there was no inherited effect of paternal F_0 spermatogonial irradiation using 0.1 or 0.5 Gy on any of these endpoints relative to concurrent control male offspring. Additionally, there was no inherited effect of paternal irradiation history on the size of F_2 litters sired by F_1 males relative to concurrent controls or on the size of F_3 litters sired by F_2 males (Figure 2). While no inherited effect was observed on sperm comet endpoints for the mice in the 0.1 Gy group, sperm comet endpoint evaluations for the 0.5 Gy dose group demonstrated that there were statistically significant inherited effects of the higher dose paternal F_0 spermatogonial irradiation on DNA damage (Table 3; Figure 1). Sperm comets from F_2 offspring with a paternal F_0 germline irradiation history using 0.5 Gy showed no significant effect on TEM (+4%), but increased tail lengths and decreased percent DNA in the tails relative to concurrent control F_2 offspring were observed (+10%, $p < 0.0001$ and −7%, $p = 0.0004$, respectively).

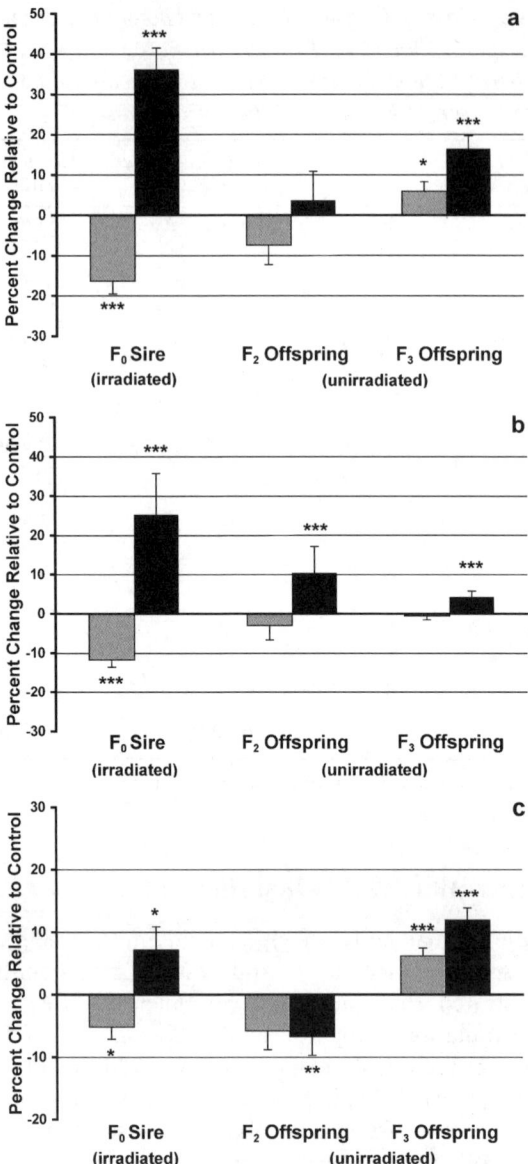

Figure 1   *Transgenerational effects of paternal F_0 germline irradiation on DNA damage as measured by evaluation of comet endpoints in epididymal sperm 45 days after acute irradiation of the F_0 male at 0.1 Gy (gray bars) and at 0.5 Gy (black bars) and also in the unirradiated sperm of second- and third-generation male offspring. Data shown represent the percent change of the geometric mean values (with geometric standard error) for a given endpoint in the acutely irradiated animal or unirradiated F_2 and F_3 offspring with an ancestral history of acute germline irradiation relative to concurrent control animals: (**a**) TEM, (**b**) tail length, and (**c**) percent tail DNA. Geometric mean, geometric standard errors (SE_g), number of animals (N), and statistical probability values (p) are shown for the F_2 and F_3 generations in Table 3. Levels of significance are obtained with Mann–Whitney U-test of comet data (*0.001 > p > 0.05, **0.0001 > p > 0.001, ***p < 0.0001)*

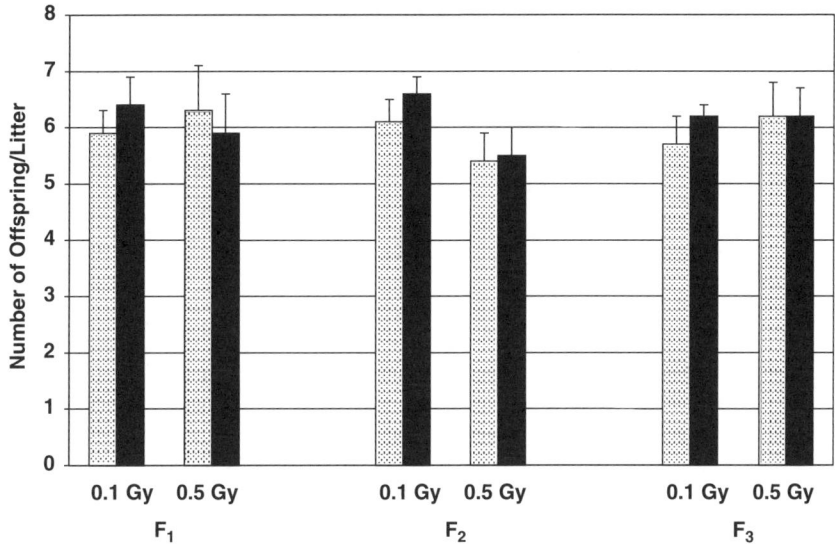

Figure 2 *Breeding results for acutely irradiated male mice and their first- and second-generation male offspring (black bars) and their respective concurrent controls (gray bars) as measured by numbers of offspring per litter*

13.3.4 F_3 Generation Male Offspring

As with the F_2 generation, no effects were observed on animal body weight, epididymides weight, testes weight, and sperm motility for the F_3 male offspring. Testis sperm counts were not evaluated for the F_3 generation. Relative to concurrent controls, the 0.1 Gy dose group of F_3 offspring showed a statistically significant increase in normal sperm morphology, however. This 5% increase in normal morphology for the F_3 experimental group is not expected to be of biological relevance in relation to their paternal F_0 irradiation history. Importantly, evaluation of DNA damage as measured by sperm comet endpoint evaluations demonstrated inherited effects of paternal F_0 spermatogonial irradiation in the F_3 generation (Table 3; Figure 1). For the 0.1 and the 0.5 Gy groups of F_3 offspring from paternal F_0 germline irradiation TEMs were increased by 6 and 16%, respectively (Table 3; $p = 0.01$ and $p < 0.0001$, respectively). For the 0.5 Gy group, the tail length was also significantly increased relative to concurrent controls ($+4\%$, $p < 0.0001$). Percent DNA in the comet tail was significantly increased by 6 and 12%, respectively for the 0.1 and 0.5 Gy dose groups ($p < 0.0001$).

13.4 Discussion

13.4.1 Overview

Mouse studies suggest that exposure of premeiotic spermatogonia to even low doses of DNA damaging agents such as ionizing radiation increases the

Table 3 *Effects of F_0 ^{137}Cs γ irradiation on sperm comet measurements in unirradiated F_2 and F_3 generation offspring, providing geometric mean values ($\bar{X}_g$), geometric standard error (SE_g), and number of animals (N)[a]*

Generation	Experimental group		TEM (100 cells per animal)	Tail length (μm)	Tail DNA (%)
F_2	Sham-irradiated concurrent control	$\bar{X}_g$	3.25	24.72	13.12
		SE_g	1.04	1.02	1.02
		N	6	6	6
	0.1 Gy irradiation history	$\bar{X}_g$	3.01	23.99	12.36
		SE_g	1.04	1.03	1.02
		N	7	7	7
	p value		0.40	0.75	0.11
	Sham-irradiated concurrent control	$\bar{X}_g$	4.04	27.23	14.59
		SE_g	1.05	1.04	1.03
		N	6	6	6
	0.5 Gy irradiation history	$\bar{X}_g$	4.19	30.06	13.61
		SE_g	1.05	1.05	1.02
		N	5	5	5
	p value		0.75	**<0.0001**	**0.0004**
F_3	Sham-irradiated concurrent control	$\bar{X}_g$	8.53	42.95	19.86
		SE_g	1.02	1.01	1.01
		N	7	7	7
	0.1 Gy irradiation history	$\bar{X}_g$	9.04	42.85	21.09
		SE_g	1.01	1.01	1.01
		N	8	8	8
	p value		**0.013**	0.44	**<0.0001**
	Sham-irradiated concurrent control	$\bar{X}_g$	5.71	36.06	15.85
		SE_g	1.02	1.01	1.01
		N	6	6	6
	0.5 Gy irradiation history	$\bar{X}_g$	6.65	37.58	17.74
		SE_g	1.02	1.01	1.01
		N	6	6	6
	p value		**<0.0001**	**<0.0001**	**<0.0001**

[a] Levels of significance are obtained with Mann–Whitney *U*-tests.

frequency of germline DNA damage, predisposing offspring to increased mutation rates and health risks including genetic diseases and cancer.[29–32] *In vitro* and *in vivo* studies have demonstrated dose-related induction of DNA strand breaks in sperm using alkaline and neutral pH comet assays and moderate to high doses of ionizing radiation.[20–23] The *in vivo* studies demonstrate the sensitivity of proliferating spermatogonia to x-irradiation induced, DNA damage and cell killing.[21–23] In our study, we evaluated both the acutely exposed male mouse for spermatogenic effects of irradiation 45 days later, but we also evaluated the offspring conceived of the mature sperm. While the lowest dose used in either of the other studies was 0.25 Gy, we evaluated mice for germline effects of 0.1 Gy from γ irradiation.

13.4.2   Delayed Effects of Acute Spermatogonial Irradiation on Sperm from F₀ Male Mice

In acutely exposed male mice, we observed significant cell killing following irradiation of the Type B spermatogonia using both 0.1 and 0.5 Gy of γ rays. This outcome demonstrates the radiation sensitivity of the premeiotic spermatogonia and is consistent with literature.[33,34] Despite the fact that the epididymal sperm counts were decreased to 48% of concurrent control counts for the 0.5 Gy experimental group, litter sizes for these males were not affected. This result demonstrates that there is no significant effect on *in vivo* fertility rates for these irradiated mice.

The increased DNA eletrophoretic mobility that we observed in sperm 45 days postirradiation using 0.5 Gy agrees with the results of other studies.[22,23] There are differences in the magnitude of the increase among these studies, but these may be a result of strain differences in radiation sensitivity, dose rate effects or comet assay techniques. The cause for the delayed DNA damage effect that we and others observed 45 days postirradiation is unclear. It is possible that the small amount of DNA damage induced by the acute irradiation was repaired, but infidelity of the repair or radiation-induced alterations to the chromatin made it more susceptible to later breakage during spermatogenesis or during the cell processing involved in the comet assay. Some chromatin alterations could also contribute to altered electrophoretic mobility without actual DNA strand breaks. Another likely explanation for the delayed sperm DNA damage effect, however, is that the initial radiation insult resulted in a delayed wave of, or persistent, oxidative stress that caused the chromatin to be susceptible to later breakage during spermatogenesis (see ref 35). Either mechanism for chromatin alteration could underlie a heritable, epigenetic effect on chromatin damage that could be observed in the offspring of the irradiated male.

The results of comet assays of sperm from the F₀ male mice that were irradiated using 0.1 Gy are interesting and represent a dose lower than previously evaluated for these DNA damage endpoints. The 0.1 Gy group had significant decreases in all DNA electrophoretic mobility relative to concurrent control measurements. This result was not predicted by the other studies using higher doses of radiation. The lowest X-ray dose used by Haines *et al.*[22] was 0.25 Gy. At 0.25 Gy, they observed no significant effects on comet endpoints. Again, these differences between studies may be as a result of strain differences, differences in comet methods, or they may be real differences in DNA repair related to dose. These results may also suggest that we have approached the detection limit for the sperm comet assay. We have consistently picked up small but statistically significant effects on DNA damage in both sperm and somatic cell types for 0.1 Gy dose groups and/or their offspring however (unpublished data).

The results for our 0.1 Gy group suggest the possibility that DNA damage was repaired following spermatogonial irradiation and that the low dose irradiation may have even stimulated DNA repair. We hypothesize that the 0.1 Gy dose of radiation could stimulate DNA repair in the spermatogonia, but that the germ cell irradiation still induces chromatin alterations that predispose DNA to increased instability, transgenerationally. The alternative answer to decreases in electrophoretic mobility in comet assays is DNA–protein or DNA–DNA crosslinking.[36] However, this outcome seems less likely since crosslinking is not observed at higher radiation doses. Subsequent experiments will be designed to test these hypotheses regarding DNA damage, repair, and chromatin modifications following acute germline irradiation at doses below 0.5 Gy.

13.4.3   Selection *vs.* Genome Modification

A concern regarding the higher radiation dose heritable effects studies has always been that, rather than inducing genetic or epigenetic changes that were inherited by the offspring of the irradiated sire, we were actually selecting for a small subpopulation of radiation-resistant spermatogonia that carried a particular phenotype. The results of the 0.1 Gy experiment address this concern. Epididymal sperm counts for these irradiated males were decreased by only 11% relative to controls and yet we observe significant transgenerational effects of paternal germline irradiation on comet endpoints. This outcome demonstrates that heritable effects of paternal germline irradiation are not artefacts of selection, but are in fact representative of heritable changes to the genome.

13.4.4   Heritable Effects of Paternal F_0 Irradiation on F_2 and F_3 Generation Sperm

No effect on epididymides weight, testes weight, sperm motility, sperm counts, or litter size was observed in offspring from paternal F_0 germline irradiation for either the 0.1 or 0.5 Gy dose group. This outcome demonstrates that there are no inherited gross effects on spermatogenesis or germline apoptosis related to the paternal irradiation history. Together, the results of the F_2 and F_3 generation sperm comet analyses indicate that there may be important effects of paternal irradiation history on chromatin quality in the offspring of the irradiated male, however. Alone these heritable effects may be of minor biological consequence. Alternatively, though, these effects may be indicators radiation-induced heritable DNA or chromatin effects. These effects may also be indicators of increased health risks and/or a predisposition for increased health risks if the offspring are exposed to genotoxins.[16,17]

13.5   Conclusion

In genomic instability models, it is notably the progeny of the irradiated cell that show delayed effects of the radiation exposure.[37,38] In our study, we observe an effect on sperm DNA in the progeny of the irradiated male mouse, suggesting that spermatogonial irradiation is an initiating event that results in inherited chromatin alterations. While we do not fully understand the implications of this heritable effect, this study suggests the importance of further study to understand the underlying DNA repair mechanisms and the potential risks to offspring of the father whose spermatogonia have been irradiated. Like other biomarkers for transgenerational effects, we do not know the relationship between the mouse and the human for heritable effects on DNA damage, but these studies have implications for people at risk for occupational exposures and those exposed to therapeutic and diagnostic chemical or radiation exposures.

Acknowledgments

The authors would like to thank Dr. Otto G. Raabe for the many productive discussions and critical review of this manuscript. This work was supported by Biological and Environmental Research Program (BER), United States Department of Energy Grant No. DE-FG03-01ER63225 to Dr. J.E. Baulch and NIH RO1 ES06516 to Dr. J.W. Overstreet and Dr. J.E. Baulch.

References

1. W.R. Bruce, R. Furrer and A.J. Wyrobek, *Mutat. Res.*, 1974, **23**, 381–386.
2. A.J. Wyrobek, L.A. Gordon, J.G. Berkhart, M.W. Francis, R.W. Kapp Jr, G. Letz, H.V. Malling, J.C. Topham and M.D. Whorton, *Mutat. Res.*, 1983, **115**, 1–72.
3. W.L. Russell and E.M. Kelly, *Proc. Natl. Acad Sci., USA*, 1982, **79**, 542–544.
4. N. Carls and R.H. Schiestl, *Carcinogenesis*, 1999, **20**, 2351–2354.
5. Y.E. Dubrova, M. Plumb, B. Gutierrez, E. Boulton and A.J. Jeffreys, *Nature*, 2000, **405**, 37.
6. R. Barber, M.A. Plumb, E. Boulton, I. Roux and Y.E. Dubrova, *Proc. Natl. Acad. Sci., USA*, 2002, **99**, 6877–6882.
7. K. Shiraishi, T. Shimura, M. Taga, N. Uematsu, Y. Gondo, M. Ohtaki, R. Kominami and O. Niwa, *Radiat. Res.*, 2002, **157**, 661–667.
8. C.M. Somers, C.L. Yauk, P.A. White, C.L. Parfett and J.S. Quinn, *Proc. Natl. Acad. Sci. USA*, 2002, **99**, 15904–15907.
9. C.M. Somers, B.E. McCarry, F. Malek and J.S. Quinn, *Science*, 2004, **304**, 1008–1010.

10. L.M. Wiley, J.E. Baulch, O.G. Raabe and T. Straume, *Radiat. Res.*, 1997, **148**, 145–151.
11. J.E. Baulch, O.G. Raabe and L.M. Wiley, *Mutagenesis*, 2001, **16**, 17–23.
12. J.E. Baulch, O.G. Raabe, L.M. Wiley and J.W. Overstreet, *Mutagenesis*, 2002, **17**, 9–13.
13. C.S. Giometti, M.A. Gemmell, S.L. Nance, S.L. Tollaksen and J. Taylor, *J. Biol. Chem.*, 1987, **262**, 12764–12767.
14. C.S. Giometti, S.L. Tollaksen and D. Grahn, *Mutat. Res.*, 1994, **320**, 75–85.
15. J.E. Baulch and O.G. Raabe, *Mutagenesis*, 2005, Epub ahead of print.
16. M.M. Vance, J.E. Baulch, O.G. Raabe, L.M. Wiley and J.W. Overstreet, *Int. J. Radiat. Biol.*, 2002, **78**, 513–526.
17. B.I. Lord, L.B. Woolford, L. Wang, D. McDonald, S.A. Lorimore, V.A. Stones, E.G. Wright and D. Scott, *Int. J. Radiat. Biol.*, 1998, **74**, 721–728.
18. V.R. Burruel, O.G. Raabe and L.M. Wiley, *Mutat. Res.*, 1997, **38**, 59–66.
19. P. Warner, L.M. Wiley, D.J. Oudiz, J.W. Overstreet and O.G. Raabe, *Radiat. Res.*, 1991, **128**, 48–58.
20. G. Haines, G. Marples, P. Daniel and I. Morris, *Adv. Exp. Med. Biol.*, 1998, **444**, 79–93.
21. G.A. Haines, J.H. Hendry, C.P. Daniel and I.D. Morris, *Mutat. Res.*, 2001, **495**, 21–32.
22. G.A. Haines, J.H. Hendry, C.P. Daniel and I.D. Morris, *Biol. Reprod.*, 2002, **67**, 854–861.
23. E. Cordelli, A.M. Fresegna, G. Leter, P. Eleuteri, E.M. Spano and P. Villani, *Radiat. Res.*, 2003, **160**, 443–451.
24. N.P. Singh, C.H. Muller and R.E. Berger, *Fertil. Steril.*, 2003, **80**, 1420–1430.
25. R.J. Van Kooij, P. de Boer, J.M. De Vreeden-Elbertse, N.A. Ganga, N. Singh and E.R. Te Velde, *Int. J. Andr.*, 2004, **27**, 140–146.
26. N.P. Singh, D.B. Danner, R.R. Tice, M.T. McCoy, G.D. Collins and E.L. Schneider, *Exp. Cell Res.*, 1989, **184**, 461–470.
27. V.R. Burruel, R. Yanagimachi and W.K. Whitten, *Biol. Reprod.*, 1996, **55**, 709–714.
28. S.J. Wiklund and E. Agurell, *Mutagenesis*, 2003, **18**, 167–175.
29. W.L. Russell, J.W. Bangham and L.B. Russell, *Genetics*, 1998, **148**, 1567–1578.
30. B.I. Lord, *Int. J. Radiat. Biol.*, 1999, **75**, 801–810.
31. M.H. Brinkworth, *Int. J. Andr.*, 2000, **23**, 123–135.
32. Y.E. Dubrova, *Oncogene*, 2003, **22**, 7087–7093.
33. E.F. Oakberg, *Radiat. Res.*, 1955, **2**, 369–391.
34. A.G. Searle and C.V. Beechey, *Mutat. Res.*, 1974, **22**, 63–72.
35. S.A. Lorimore, P.J. Coates and E.G. Wright, *Oncogene*, 2003, **22**, 7058–7069.
36. M.S. Marty, N.P. Singh, M.P. Holsapple and B.B. Gollapudi, *Mutat. Res.*, 1999, **427**, 39–45.
37. W.F. Morgan, J.P. Day, M.I. Kaplan, E.M. McGhee and C.L. Limoli, *Radiat. Res.*, 1996, **146**, 247–258.
38. W.F. Morgan, *Radiat. Res.*, 2003, **159**, 581–596.

CHAPTER 14

Influence of DNA Methylation and Genomic Imprinting in the Male Germ Line on Pregnancy Outcome

JACQUETTA M. TRASLER

McGill University – Montreal Children's Hospital Research Institute, Departments of Pediatrics, Human Genetics and Pharmacology & Therapeutics, McGill University, Montreal, QC H3H 1P3, Canada

14.1 Introduction

The term 'epigenetics' refers to the heritable nonsequence based mechanisms that regulate gene activity. To date, the most well-studied deoxyribonucleic acid (DNA) modification associated with the modulation of gene activity is the methylation of cytosine residues within CpG dinucleotides; CpG methylation occurs at about 30 million sites throughout the genome. DNA methylation plays a role in regulating genes during development and has been implicated in gene regulation, genomic imprinting (variation in the expression of a gene according to its maternal or paternal origin), and X inactivation. Abnormalities in DNA methylation have been linked to cancer as well as growth and behavioural defects. DNA methylation is catalysed by DNA (cytosine-5)-methyltransferases (DNMTs), is initiated in the germ line and then further modified during early embryo development. In the field of male-mediated developmental toxicity, adverse effects on the offspring in animal studies often occur at levels too high to be accounted for by mutagenesis, leading to the suggestion that alternative mechanisms, including epigenetic processes, may be affected. In addition, a number of recent studies have linked the use of assisted reproductive technologies with growth and genomic imprinting disorders in children, implicating epigenetic processes; the imprinting disorders were asso-ciated with DNA methylation abnormalities.[1] However, in the latter studies it was unclear whether the birth defects were related to the underlying infertility or the treatments (*i.e.* intra cytoplasmic sperm injection (ICSI), superovulation, culture conditions) being used. Since DNA methylation events and enzymes are

165

well conserved across mammals, the rat and mouse have served as excellent models relevant to human studies. Using the rodent model, our results and those of others indicate that DNA methylation is highly regulated in the male germ line and implicate alterations in the enzymes involved in DNA methylation, the DNMTs, and DNA methylation, with abnormalities in germ cell as well as embryo development. Three models relevant to male-mediated effects will be reviewed here, DNMT deficiency, effects of cytosine analogues and defects in folate (methyl donor) pathways.

14.2   DNA Methylation and DNA Methyltransferases: Roles and Importance

DNA methylation occurs at the 5-position of cytosine residues for the most part within CpG dinucleotides such that 60–80% of the CpGs within the mammalian genome are methylated. It is a postreplication process catalyzed by a family of DNMTs. Methylation of CpG sites within promoter regions of genes almost invariably silences transcription and provides a means of compartmentalizing large genomes into expressed and unexpressed sequences. Once set down (*de novo* methylation), methylation patterns are clonally inherited (maintenance methylation) or lost through a postulated active demethylation process or when methylation does not occur following replication. The process of maintenance methylation allows gene expression information to be transmitted at the time of DNA replication and cell division. In addition, and critical for normal postnatal development, DNA methylation fulfils the four requirements of the biochemical modification of DNA or chromatin that account for genomic imprinting – the modification must (1) be made before fertilization, (2) be able to confer transcriptional silencing, (3) be stably transmitted through mitosis in somatic cells and, (4) be reversible on passage through the opposite parental germ line.[1] Almost all imprinted genes till date show differences in methylation between maternal and paternal alleles.[2,3] A role for DNA methylation in mammalian genome defence has also been proposed, related to genome defence in bacteria, whereby invading viral sequences are methylated and thereby inactivated.[4] In the large mammalian genome, it is postulated that methylation might serve to block the expression of repeated sequences and retrotransposons that make up an estimated 35% of the genome.[4]

In mammals, five DNMTs have been characterized and classified according to similarities found in their C-terminal catalytic domain: DNMT1, DNMT2, DNMT3a, DNMT3b and DNMT3L (for DNMT3-like).[5–7] Of these, only DNMT1, DNMT3a and DNMT3b are known to be catalytically active *in vivo*; however, DNMT3L is postulated to interact with DNMT3a and DNMT3b. DNMT1, the major methyltransferase in somatic tissues, has a preference for hemimethylated DNA and is critical for the maintenance of methylation patterns during replication of DNA.[4,8–10] DNMT3a and DNMT3b are encoded by essential genes[11] that are expressed at high levels in mouse embryonic stem cells and during embryonic development and have been postulated to function predominantly in *de novo* methylation of DNA.[11–13]

Gene-targeting experiments have helped define the roles of the different DNMTs. Homozygous DNMT1-deficient embryos have <5% of the DNA methylation levels found in normal embryos and such embryos show biallelic expression of most imprinted genes, inactivation of all X chromosomes, activation of retrotransposons, and apoptotic death before mid-gestation.[9,10,14,15] DNMT3a-deficient mice are underdeveloped and die 3–4 weeks after birth, while DNMT3b deficiency results in a more severe mid-gestation embryo lethal phenotype. DNMT3b appears to be specifically required for the methylation of centromeric minor satellite repeats.[11,16] DNMT3L is particularly interesting since its expression appears to be restricted to male and female germ cells and deficiency results in germ line defects.[17,18] Since DNMT3L has not been shown to possess DNA methyltransferase activity, it is postulated to be involved in the acquisition of germ cell methylation through interactions with other DNMTs or other factors. In support of this suggestion, DNMT3L has been shown to stimulate *de novo* methylation through DNMT3a, but not DNMT3b at some imprinted loci *in vitro*.[19] The role of DNMT2 is unknown as deficient mice are normal.[6]

14.3 Acquisition of DNA Methylation Patterns During Spermatogenesis

DNA methylation patterns are erased and reset differentially in the developing male and female gametes, further modified in the early embryo and become relatively stable by late embryogenesis. In both germ lines, DNA methylation patterns on most sequences appear to be erased around the time when primordial germ cells enter the gonad, at approximately 10.5–12.5 days of gestation in the mouse.[20–24] Early studies indicated that genomic methylation patterns are acquired in the germ line and differ markedly for male and female gametes.[25–28] More recent studies suggest that gametic methylation differences are especially striking at imprinted loci where they have important implications for allele-specific gene expression in the offspring.[1,29] It is predicted that most gametic methylation differences will be removed during the genome-wide demethylation that occurs during preimplantation development;[26,30,31] the blastocyst is the other time in development when DNA methylation levels are low.[26,32] However, during the preimplantation wave of demethylation, methylation is maintained on some sequences, as has been shown for imprinted loci and repeat sequences such as *IAP*.[33–35] Following the blastocyst stage, the genome is remethylated to adult levels soon after implantation, between 5.5–7.5 days of gestation in the mouse. Accurate reprogramming is therefore required with every reproductive cycle to ensure proper erasure, acquisition and maintenance of methylation marks.

Evidence to date indicates that DNA methylation timing differs greatly between the two germ lines. A number of different types of experiments including nuclear transplantation studies and bisulfite sequencing of the DNA of oocytes at different stages of development have indicated that methylation in the female germ line is acquired during the postnatal oocyte growth phase.[36]

In contrast to the female, in the male, genomic methylation on imprinted (*e.g. H19, Gtl2/Dlk, Rasgrf*) and repeat (*e.g. LINE1, IAP*, minor satellite) sequences begins to be acquired before birth between 15.5 and 18.5 days of gestation in the mouse and is completed after birth before the end of pachytene.[31,37–42] Certain single copy testis-specific genes (*e.g.* phosphoglycerate kinase-2, transition protein 1) become demethylated prior to their expression during spermatogenesis.[43,44]

14.4 DNA Methyltransferase Enzymes and their Regulation in the Male Germ Line

In rodent models DNA methylation is highly regulated in the male germline and the DNMT1 and DNMT3 (DNMTs 3a, 3b and 3L) enzymes are differentially expressed in germ cell type specific patterns in both prenatal and postnatal germ cells. DNMT expression in the prenatal and postnatal male gonad has been examined in order to correlate specific DNMTs with methylation events. In the prenatal male gonad, DNMT1 is unlikely to play a role in the prenatal acquisition of germ line methylation patterns since the protein levels were down regulated in germ cells between 14.5 and 18.5 days of gestation and were absent at the time of initiation of methylation.[45] In contrast, real-time RT-PCR in whole prenatal gonads provided evidence that DNMT3l and DNMT3a are likely involved in prenatal gonad methylation events.[45] In early work on postnatal male gametogenesis, we showed that DNMT1 protein is localized to the nuclei of all male germ cells up to the pachytene spermatocyte stage.[46] Real-time RT-PCR studies on the DNMT3s in postnatal germ cells also show expression in both mitotic and meiotic cells.[47] DNMTs are predicted to be needed in mitotic cells (spermatogonia) not only for maintenance methylation, but also possibly for *de novo* methylation. The finding of DNMTs in postmitotic germ cells was surprising; we predict that *de novo* methylation may take place in these cells during meiotic prophase. *De novo* methylation in these cells could be involved in genomic imprinting, germ cell-specific methylation or post-DNA repair methylation.

14.5 Perturbing DNA Methylation in the Male Germline: Consequences for the Progeny

Disruption of the DNMTs by gene targeting has underscored the essential nature of DNA methylation for normal development as well as male germ cell development. As described below, a number of germline specific knockout studies have been performed, and have resulted in both male and female infertility. To address whether decreases in DNA methylation affect spermato-genesis, we have used pharmacological and genetic approaches. The latter studies have important clinical relevance since 'epigenetic drugs,' including agents that alter DNA methylation such as demethylating cytidine analogues (*e.g.* 5-aza-2′-deoxycytidine) and histone deacetylase inhibitors have been developed as anticancer agents.[48] In addition, there has been increasing interest in dietary effects on DNA methylation; at least one study has shown that nutrition-based alterations in DNA methylation can be heritable. Since the DNMT knockout mice, including the germ-cell specific knockouts, either have severe systemic effects or are associated with complete infertility, the genetic and pharmacological approaches offer alternative ways to modify DNA methylation in the male germ line and are still able to assess effects on the offspring.

14.5.1 DNA Methyltransferase Deficiency

Recent information from gene-targeting experiments is helping to define better the roles of the specific DNMT enzymes in the male germ line. In males, DNMT3L deficiency results in infertility associated with meiotic defects and decreased methylation of imprinted genes (*H19*) and some dispersed repeat sequences such as the retrotransposon, *IAP*.[17,18,49,50] Seminiferous cords look normal histologically at 1 week after birth, however, there are few differentiated spermatocytes in the testes of *Dnmt3L*-deficient mice by 4 weeks of age. It is postulated that reduced methylation of *IAP* leads to high expression levels in early germ cells and causes failure to complete meiosis.[18] Kaneda[51] deleted the *Dnmt3a* gene in male germ cells; the mice were infertile (similar to the DNMT3L-deficient males) and showed demethylation of some (*H19*, *Gtl2/Dlk1*) imprinted sequences; unfortunately, repeat sequences were not examined. A recent examination of spermatogenesis in mice with DNMT3a deficiency in all cells (DNMT3a-null mice) revealed that, unlike DNMT3L-deficient mice, spermatogenesis proceeded further, with evidence of some, albeit quantitatively reduced as compared to control, synaptonemal complex formation, meiosis and the detection of markers of round spermatids; in addition methylation of the *IAP* and *LINE-1* elements was apparently unaffected in germ cells of the DNMT3a-null mice.[52] The authors of the latter study postulate that DNMT3a may be required for critical DNA methylation events allowing meiosis to proceed. Together, the results of DNMT3a and DNMT3L deficiency suggest that the two enzymes play distinct roles in male germ cell development. Male germ cell depletion of DNMT3b apparently did not cause methylation abnor-malities,[51] perhaps because minor satellite sequences are normally unmethylated in the germline.[53]

14.5.2 Cytidine Analogues

Several drugs, including the cytidine analogues 5-azacytidine and 5-aza-2′-deoxycytidine are known to inhibit DNA methylation.[48] Both 5-azacytidine and 5-aza-2′-deoxycytidine are among the best characterized and most studied of the known methylation inhibitors and are commonly used to manipulate DNA methylation; they are incorporated into DNA during DNA replication but, due to the presence of nitrogen at the C5 position, they are unable to accept a methyl group, DNMTs remains bound as covalent adducts and DNA methylation is inhibited.[48] In addition to directly preventing methylation at drug-substituted sites, 5-azacytidine and 5-aza-2′-deoxycytidine cause indirect hypomethylation and rapid loss of DNMT activity through covalent trapping of the DNMTs.[54] Treatment with the cytidine analogues results in active transcription of previously silent cellular genes and alters the differentiated state of cultured cells.

We have administered 5-azacytidine and 5-aza-2′-deoxycytidine to rats and mice with the goal of perturbing but not completely eradicating DNA methylation patterns in the male germ line in order to permit assessment of effects on male germ cell development and function postfertilization. It was predicted that the induction of DNA methylation aberrations during spermatogenesis would result in ectopic gene expression or altered chromatin structure with lethal consequences for some but not all male germ cells, depending on the degree of DNA demethylation or the specific sequences affected. Those spermatozoa with altered methylation patterns that survived could then be incapable of fertilization or could contribute to abnormalities in embryo, foetal or postnatal development.

In the initial studies, chronic *in vivo* administration of 5-azacytidine to Sprague-Dawley rats, at doses that did not affect the general health of the animals and exposed both mitotic and meiotic populations of developing male germ cells, was associated with dose-dependent decreases in sperm counts and testis weight as well as significant increases in preimplantation loss.[55] Because exposure to the nonhypomethylating agent 6-azacytidine resulted in no differences from saline controls, these studies suggested that the observed effects of 5-azacytidine were mediated by changes in DNA methylation.[55] A follow-up study was designed to examine the mechanisms underlying the effects of 5-azacytidine on sperm function.[56] Male rats were treated chronically for 6 weeks (to expose meiotic and postmeiotic germ cells) or 11 weeks (to expose mitotic, meiotic and postmeiotic germ cells). Effects were most severe after 11 weeks of treatment. After 11 weeks of treatment, doses of 5-azacytidine that caused preimplantation loss also resulted in severe morphological abnormalities in the seminiferous epithelium as well as a 22–29% decrease in genomic methylation levels in epididymal spermatozoa. Both 6 and 11 weeks of treatment resulted in an increase as compared with saline treated controls in the number of apoptotic germ cells in the seminiferous epithelium. When DNA methylation levels were analyzed in isolated germ cells from the treated males, the results indicated that spermatogonia were more sensitive to the hypomethylating effects of 5-azacytidine than were spermatocytes. The latter result was consistent with results from another group that showed that 5-azacytidine and 5-aza-2′-deoxycytidine treatment of 5-day-old neonatal mice inhibited the differentiation of spermatogonia to spermatocytes.[57] The rat studies provided some of the earliest evidence of a link between abnormal DNA methylation, altered male fertility and abnormal embryo development.

In follow-up studies, mice were treated with the more DNA-selective cytidine analogue 5-aza-2′-deoxycytidine.[58] Treatment of mice for 7 weeks, exposing germ cells throughout their development from spermatogonia to spermatozoa resulted in dose-dependent decreases in testicular weight, an increase in histological abnormalities, a decline in sperm counts and up to 29% decreases in sperm DNA methylation. Interestingly, males deficient in DNMT1 appeared to be more resistant to the cytotoxic effects of 5-aza-2′-deoxycytidine. Both *Dnmt1+/+* (control levels of DNMT1) and *Dnmt1+/−* (50% of levels of DNMT1) mice responded similarly to the drug with respect to global sperm DNA methylation and preimplantation loss. However, the treated *Dnmt1+/−* males showed smaller decreases in testis weight and fewer testicular histological abnormalities, suggesting less germ cell toxicity as compared to that seen in the treated *Dnmt1+/+* mice. Juttermann *et al.*[59] demonstrated that the toxicity of cytidine analogues may be mediated *via* the formation of 'toxic' covalent adducts with DNMT1. We postulated that the germ cells of *Dnmt1+/−* males may be more resistant to the cytotoxic effects of 5-aza-2′-deoxycytidine since they have only 50% of the wild-type level of DNMT1, thus reducing the chances of adduct formation. The combination of 5-aza-2′-deoxycytidine and DNMT1 deficiency offers a promising way to alter methylation in male germ cells while minimizing toxicity. Because of well-characterized genetics, the availability of mouse models of DNMT deficiency and well-developed early embryo culture systems, the mechanisms of 5-aza-2′-deoxycytidine's effects on male germ cells and fertility are being further pursued in the mouse. The methylation of individual genes and a detailed examination of the progeny are currently being carried out.

14.5.3 Folate Pathway Enzymes

An alternative way to inhibit methylation in the male germ line is to approach the DNA methylation pathway from the availability of methyl groups necessary for DNA methylation. The folate pathway plays a key role in the synthesis of purines and pyrimidines and the production of methionine; methionine in turn provides methyl groups for numerous biochemical reactions, including DNA methylation. The two pathways are linked by the enzyme methionine synthase, catalysing the reduction of 5-methyltetrahydrofolate to tetrahydrofolate and the remethylation of homocysteine to methionine. Methylenetetrahydrofolate reductase (MTHFR) is a key enzyme in the folate pathway and is involved in a number of biochemical processes including the remethylation of homocysteine to produce methionine. Methionine ultimately provides the methyl groups necessary to form S-adenosylmethionine (SAM), a ubiquitous methyl donor in numerous cellular reactions including DNA methylation. Deficiency in MTHFR results in lowered levels of methionine, SAM and DNA methylation.[60] Moreover, MTHFR was postulated to have a role in spermatogenesis since enzyme activity is higher in testis than in other organs. In support of this proposal, *Mthfr*–/– mice have smaller testes, histological evidence of abnormal germ cell development, and are infertile.[61] Interestingly, supplementation with betaine, an alternative methyl donor serving in the remethylation of homocysteine to methionine, resulted in improvements in testicular histology and fertility in some mice. It was suggested that the defect in the *Mthfr*–/– mice may in part be due to DNA methylation abnormalities in germ cells and this possibility is currently being examined. In summary, the study indicates an important role for MTHFR in germ cell development and suggests that research on other enzymes in the folate pathway, in relation to spermatogenesis, be pursued.

14.6   Conclusions and Implications

The dynamics of DNA methylation and the expression of the different DNMT enzymes appear to be tightly regulated during spermatogenesis. Mouse models of DNMT3L and DNMT3a deficiency have clearly demonstrated the functional importance of DNA methylation to normal male germ cell development; however, the phenotypes in each case were quite severe, resulting in infertility. Treatment of rats and mice with the cytidine analogues 5-azacytidine and 5-aza-2′-deoxycytidine have linked less severe alterations in male germ cell DNA methylation levels with abnormalities in germ cell development, fertility and abnormalities in early embryos. While the provision of methyl donors such as folate are known to be important for normal embryo and foetal development, their role and importance in male germ cell development, and potentially DNA methylation, are just beginning to be studied.

The results to date from animal models are likely to have important implications for human spermatogenesis. Many agents capable of modifying DNA methylation patterns, including 5-aza-2′-deoxycytidine, are currently being tested in clinical trials.[48] Based on the animal studies with 5-azacytidine and 5-aza-2′-deoxycytidine discussed in this review, where the doses were similar to the doses used in human clinical trials, it is likely these drugs will have adverse effects on human spermatozoa. So far, there are only a few studies examining alterations in DNA methylation levels in human sperm. In one, the ability of spermatozoa to fertilize oocytes in vitro was correlated with global sperm methylation.[62] There is a single report linking an imprinting defect to disruptive spermatogenesis. As compared with controls, Marques *et al.*[63] reported lower levels of *H19* methylation in men with moderate-to-severe oligospermia. A number of studies have suggested a possible link between the incidence of imprinting disorders and human assisted reproductive technologies,[1] although it is unclear whether the imprinting disorders are due to underlying infertility or the techniques used. Interestingly, a recent study reported an increased incidence of imprinting defects associated with altered DNA methylation in patients with Angelman syndrome born to subfertile couples with suggestive evidence of an interaction between subfertility and assisted reproductive technologies.[64] These examples emphasize the need not only to understand better the precise roles of DNA methylation in germ cell function but also to study further the existence of epigenetic causes for infertility in males as well as the potential that epigenetic abnormalities in sperm may be one of the mechanisms underlying male-mediated effects of drugs.

Acknowledgments

Jacquetta Trasler is a William Dawson Scholar of McGill University and a Scholar of the Fonds de la recherché en santé du Québec. This work was supported by a grant to J.T. from the Canadian Institutes of Health Research (CIHR).

References

1. D. Lucifero, J.R. Chaillet and J.M. Trasler, *Hum. Reprod. Update*, 2004, **10**, 3–18.
2. M.S. Bartolomei and S.M. Tilghman, *Annu. Rev. Genet.*, 1997, **31** 493–525.
3. R.I. Verona, M.R. Mann and M.S. Bartolomei, *Annu. Rev. Cell Dev. Biol.*, 2003, **19**, 237–259.
4. J.A. Yoder, C.P. Walsh and T.H. Bestor, *Trends Genet.*, 1997, **13**, 335–340.
5. T.H. Bestor, *Hum. Mol. Genet.*, 2000, **9**, 2395–2402.
6. A. Hermann, H. Gowher and A. Jeltsch, *Cell Mol. Life Sci.*, 2004, **61**, 2571–2587.
7. M.G. Goll and T.H. Bestor, *Annu. Rev. Biochem.*, 2004, **74**, 481–514.
8. T.H. Bestor, *EMBO J.*, 1992, **11**, 2611–2617.
9. H. Lei, S.P. Oh, M. Okano, R. Juttermann, K.A. Goss, R. Jaenisch and E. Li, *Development*, 1996, **122**, 3195–3205.
10. E. Li, T.H. Bestor and R. Jaenisch, *Cell*, 1992, **69**, 915–926.
11. M. Okano, D.W. Bell, D.A. Haber and E. Li, *Cell*, 1999, **99**, 247–257.
12. T. Chen, Y. Ueda, S. Xie and E. Li, *J. Biol. Chem.*, 2002, **277**, 38746–38754.
13. M. Okano, S. Xie and E. Li, *Nat. Genet.*, 1998, **19**, 219–220.
14. B. Panning and R. Jaenisch, *Genes Dev.*, 1996, **10**, 1991–2002.
15. C.P. Walsh, J.R. Chaillet and T.H. Bestor, *Nat. Genet.*, 1998, **20**, 116–117.
16. G.L. Xu, T.H. Bestor, D. Bourc'his, C.-H. Hsieh, N. Tommerup, M. Bugge, M. Hulten, X. Qu, J.J. Russo and E. Viegas-Péquignot, *Nature*, 1999, **402**, 187–191.
17. D. Bourc'his, G.L. Xu, C.S. Lin, B. Bollman and T.H. Bestor, *Science*, 2001, **294**, 2536–2539.
18. D. Bourc'his and T.H. Bestor, *Nature*, 2004, **431**, 96–99.
19. F. Chedin, M.R. Lieber and C.L. Hsieh, *Proc. Natl. Acad. Sci. U S A*, 2002, **99**, 16916–16921.
20. P. Hajkova, S. Erhardt, N. Lane, T. Haaf, O. El-Maarri, W. Reik, J. Walter and M.A. Surani, *Mech. Dev.*, 2002, **117**, 15–23.
21. Y. Kato, W.M. Rideout III, K. Hilton, S.C. Barton, Y. Tsunoda and M.A. Surani, *Development*, 1999, **126**, 1823–1832.

22. J. Lee, K. Inoue, R. Ono, N. Ogonuki, T. Kohda, T. Kaneko-Ishino, A. Ogura and F. Ishino, *Development*, 2002, **129**, 1807–1817.
23. P.E. Szabo and J.R. Mann, *Genes Dev.*, 1995, **9**, 1857–1868.
24. P.E. Szabo, K. Hubner, H. Scholer and J.R. Mann, *Mech. Dev.*, 2002, **115**, 157–160.
25. D.J. Driscoll and B.R. Migeon, *Somat. Cell Mol. Genet.*, 1990, **16**, 267–282.
26. M. Monk, M. Boubelik and S. Lehnert, *Development*, 1987, **99**, 371–382.
27. W. Reik, W. Dean and J. Walter, *Science*, 2001, **293**, 1089–1093.
28. J.P. Sanford, H.J. Clark, V.M. Chapman and J. Rossant, *Genes & Development*, 1987, **1**, 1039–1046.
29. W. Reik and J. Walter, *Nat. Rev. Genet.*, 2001, **2**, 21–32.
30. S.K. Howlett and W. Reik, *Development*, 1991, **113**, 119–127.
31. T. Kafri, M. Ariel, M. Brandeis, R. Shemer, L. Urven, J. McCarrey, H. Cedar and A. Razin, *Genes Dev.*, 1992, **6**, 705–714.
32. J.R. Chaillet, T.F. Vogt, D.R. Beier and P. Leder, *Cell*, 1991, **66**, 77–83.
33. N. Lane, W. Dean, S. Erhardt, P. Hajkova, A. Surani, J. Walter and W. Reik, *Genesis*, 2003, **35**, 88–93.
34. A. Olek and J. Walter, *Nat. Genet.*, 1997, **17**, 275–276.
35. K.D. Tremblay, K.L. Duran and M.S. Bartolomei, *Mol. Cell Biol.*, 1997, **17**, 4322–4329.
36. T.L.J. Kelly and J.M. Trasler, *Clin. Genet.*, 2004, **65**, 247–260.
37. T.L. Davis, J.M. Trasler, S.B. Moss, G.J. Yang and M.S. Bartolomei, *Genomics*, 1999, **58**, 18–28.
38. T.L. Davis, G.J. Yang, J.R. McCarrey and M.S. Bartolomei, *Hum. Mol. Genet.*, 2000, **9**, 2885–2894.
39. D.J. Lees-Murdock, M. De Felici and C.P. Walsh, *Genomics*, 2003, **82**, 230–237.
40. J.Y. Li, D.J. Lees-Murdock, G.L. Xu and C.P. Walsh, *Genomics*, 2004, **84**, 952–960.
41. T. Ueda, K. Abe, A. Miura, M. Yuzuriha, M. Zubair, M. Noguchi, K. Niwa, Y. Kawase, T. Kono, Y. Matsuda, H. Fujimoto, H. Shibata, Y. Hayashizaki and H. Sasaki, *Genes Cells*, 2000, **5**, 649–659.
42. C.P. Walsh, J.R. Chaillet and T.H. Bestor, *Nat. Genet.*, 1998, **20**, 116–117.
43. C.B. Geyer, C.M. Kiefer, T.P. Yang and J.R. McCarrey, *Biol. Reprod.*, 2004, **71**(3), 837–844.
44. J.M. Trasler, L.E. Hake, P.A. Johnson, A.A. Alcivar, C.F. Millette and N.B. Hecht, *Mol. Cell. Biol.*, 1990, **10**(4), 1828–1834.
45. S. La Salle, C. Mertineit, T. Taketo, P.B. Moens, T.H. Bestor and J.M. Trasler, *Dev. Biol.*, 2004, **268**, 403–415.
46. K. Jue, T.H. Bestor and J.M. Trasler, *Biol. Reprod.*, 1995, **53**, 561–569.
47. S. La Salle and J.M. Trasler, Proceedings of the 2005 Sex and Gene Expression Conference, March, Winston-Salem, NC, 2005.
48. G. Egger, G. Liang, A. Aparicio and P.A. Jones, *Nature*, 2004, **429**, 457–463.
49. K. Hata, M. Okano, H. Lei and E. Li, *Development*, 2002, **129**, 1983–1993.
50. K.E. Webster, M.K. O'Bryan, S. Fletcher, P.E. Crewther, U. Aapola, J. Craig, D.K. Harrison, H. Aung, N. Phutikanit, R. Lyle, S.J. Meachem,

S.E. Antonarakis, D.M. de Kretser, M.P. Hedger, P. Peterson, B.J. Carroll and H.S. Scott, *Proc. Natl. Acad. Sci. USA*, 2005, **102**, 4068–4073.

51. M. Kaneda, M. Okano, K. Hata, T. Sado, N. Tsujimoto, E. Li and H. Sasaki, *Nature*, 2004, **429**, 900–903.
52. R. Yamen and V. Grandjean, *Mol. Reprod. Dev.*, 2006, **73**, 390–397.
53. J.M. Trasler, *Semin. Cell. Dev. Biol.*, 1998, **9**(4), 467–474.
54. J.K. Christman, *Oncogene*, 2002, **21**, 5483–5495.
55. T. Doerksen and J.M. Trasler, *Biol. Reprod.*, 1996, **55**, 1155–1162.
56. T. Doerksen, G. Benoit and J.M. Trasler, *Endocrinology*, 2000, **141**, 3235–3244.
57. R. Raman and G. Narayan, *Mol. Reprod. Dev.*, 1995, **42**, 284–290.
58. T.L. Kelly, E. Li and J.M. Trasler, *J. Androl.*, 2003, **24**, 822–830.
59. R. Juttermann, E. Li and R. Jaenisch, *Proc. Natl. Acad. Sci. USA*, 1994, **91**, 11797–11801.
60. Z. Chen, A.C. Karaplis, S.L. Ackerman, I.P. Pogribny, S. Melnyk, S. Lussier-Cacan, M.F. Chen, A. Pai, S.W.M. John, R.S. Smith, T. Bottiglieri, P. Bagley, J. Selhub, M.A. Rudnicki, S.J. James and R. Rozen, *Hum. Mol. Genet.*, 2001, **10**, 433–443.
61. T.L. Kelly, O.R. Neaga, B.C. Schwahn, R. Rozen and J.M. Trasler, *Biol. Reprod.*, 2005, **72**(3), 667–677.
62. M. Benchaib, M. Ajina, J. Lornage, A. Niveleau, P. Durand and J.F. Guérin, *Fertil. Steril.*, 2003, **80**, 947–953.
63. C.J. Marques, F. Carvalho, M. Sousa and A. Barros, *Lancet*, 2004, **363**, 1700–1702.
64. M. Ludwig, A. Katalinic, S. Groβ, A. Sutcliffe, R. Varon and B. Horsthemke, *J. Med. Genet.*, 2005, **42**, 289–291.

CHAPTER 15

Information Content of Ejaculate Spermatozoa and its Potential Utility in Toxicology and Infertility Based Research Programmes

DAVID MILLER,[a] MARTIN BRINKWORTH[b] AND DAVID ILES[c]

[a] Reproduction and Early Development Research Group, Department of Obstetrics and Gynaecology, University of Leeds, Level D, Clarendon Wing, Leeds General Infirmary, Belmont Grove, Leeds LS2 9NS
[b] School of Biomedical Sciences, Richmond Building, University of Bradford, Bradford BD7 1DP
[c] School of Biological Sciences, University of Leeds, LS2 9JT

15.1 Introduction

The paternal contribution to the zygote is carried by the body's smallest cell (the spermatozoon) to the largest cell (the oocyte) and is generally forgotten about once that task has been completed. Indeed, until comparatively recently, it was thought that only the sperm head gained access to the ooplasm, a fundamental misconception that one can still read in some textbooks. The likely reason for this misconception is the still widely held view that the contribution of the male gamete consists solely of the paternal haploid genome and nothing more. However, we now know that in all mammals bar the mouse, the sperm also introduces a centriole, essential for organising the zygote's mitotic spindle and in probably all species, a phospholipase activating factor that signals successful entry of the sperm to the ooplasm and triggers egg activation.[1,2]

We also know that the sperm carries RNA into the egg, consisting mostly of mRNAs with some evidence for non-coding RNAs including transcribed SINE and LINE elements.[3,4] Of some interest is the observation that at least the large ribosomal RNAs are *not* present [Figure 1(A)]. The efforts of several independent laboratories have now persuaded the scientific community that spermatozoal RNA is not an artefact of semen sample processing. There are a number of reasons for this scepticism, based primarily on the observation that sperm are transcriptionally inert.[5] Studies that are almost 30 years old (and recently revisited[6]) showed that this was the case and later studies explained why this must be so.[7–10] In all somatic cells and spermatogonia through to final diploid stages of spermatogenesis, germ cell chromatin is packaged by histones where the machinery of transcription can gain selective access to DNA.[5] Following meiosis, however, there is a gradual substitution of histones first by transition proteins and then by more basic, arginine-rich protamines, the consequence of which is a far greater compaction of the chromatin, required to allow the one metre or so of DNA to be squeezed into a sparse nuclear volume.[11] It also means that the DNA becomes progressively inaccessible to transcription factors and the transcriptional apparatus itself eventually shuts down.[11] To overcome the requirement for nascent protein synthesis beyond this stage, stored mRNAs (*i.e.*, mRNAs transcribed well before this global shutdown occurs) are present long after DNA compaction has essentially been completed. The translated products of these stored mRNAs naturally include the very proteins that participate in the repackaging. Eventually, as the spermatid condenses and loses most of its cytoplasm, so the translational

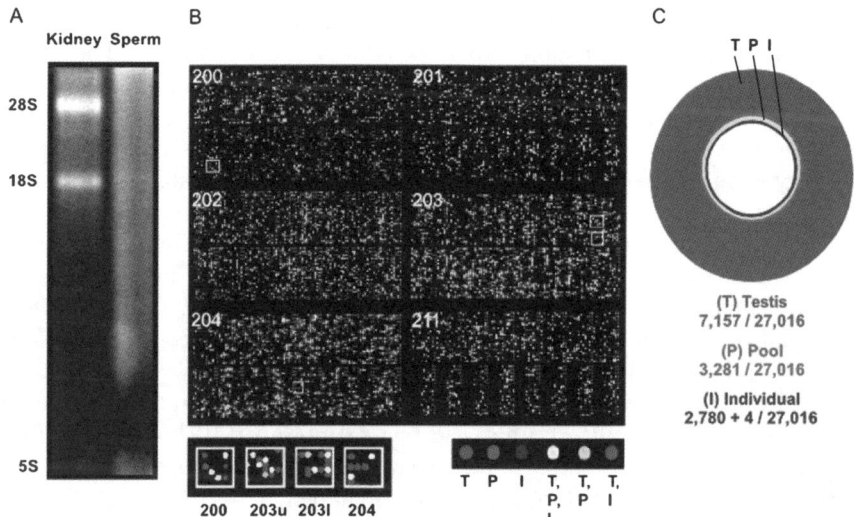

Figure 1 *(A) Agarose gel of total RNA isolated from kidney and spermatozoa. Note the absence of 28S and 18S rRNAs from the latter as well as the smear of RNA that generally indicates mRNA. (B) Representation of the Genefilter probe set hybridised to testis (T), pooled (P) and individual (I) spermatozoal samples. Colour combinations are indicated. (C) Relationship between the same samples illustrated by Venn diagram*

machinery also shuts down and the testicular spermatozoon then matures through epididymal transport into a gamete ready for ejaculation.

15.2 Composition of Spermatozoal RNA

Despite the high degree of compaction of spermatozoal chromatin, early reports on the composition of human and rat sperm RNA indicated the presence of small nuclear species located in and around the nucleus.[12] Subsequently, target-directed RT-PCR identified many mRNA species including c-myc, protamine and transition proteins, β-actin, N-cadherins, calcium channels, integrins, cyclic-nucleotide phosphodiesterases (PDEs), aromatase, heat shock proteins, nitrous oxide synthetases and oestrogen receptors.[13–21] Target-independent strategies such as cloning and sequencing of randomly primed amplicons hinted at a more heterogeneous mix of mRNAs.[3] The overall complexity (in human spermatozoa) was only realised when array-based tests were applied and showed that at least 3500 individual mRNA species persisted in ejaculate spermatozoa.[22] Moreover, evidence to date shows that these mRNAs are fully processed in that they lack intronic sequences.

Curiously, both 28S and 18S rRNAs are not present to any functional level precluding the possibility of mRNA translation *de novo* (Figure 1A,[3,22] although see ref 18). However, the selective retention of this complex population mRNAs during the final condensation of spermatids deserves explanation. Ribsosmal RNAs could conceivably be lost during spermatozoal condensation by their association with cytoplasmic proteins in translation complexes. It is, however, difficult to understand why mRNAs and not rRNAs should be free of these complexes at the time of condensation. Another possibility relates to the poor morphology of human spermatozoa and the presence of high levels of abnormal forms in even the most fertile ejaculate.[23] Indeed, abnormal morphology is no longer considered a priority criterion for judging human sperm quality since unlike motility, it appears to bear no clear relationship to fertility outcomes. The presence of abnormal forms suggests the possibility that sperm mRNAs originate from 'immature' spermatozoa that retain more cytoplasm; however, this fails to account for the selective loss of rRNAs. The presence of an equally complex population of mRNAs in the absence of rRNAs in generally pristine murine spermatozoa (David Dix, personal communication) coupled with the reports of mRNAs in the spermatozoa/pollen of other species including plants[24,25] suggests that selective retention of mRNAs in the final mature gamete is a normal part and consequence of spermatogenesis. Possible reasons for this will be outlined later in this chapter.

Regardless of the mechanism driving its selective retention, we believe that spermatozoal RNA encapsulates a transcriptomic record of spermiogenesis and possibly even spermatogenesis. This definition differs from the standard definition of a transcriptome because at the time of sampling, spermatozoa are incapable of transcription. Nevertheless, spermatozoal mRNA should be a useful (and wholly non-invasive) proxy for the testis because it exists as a subset of testis mRNAs. In a recent study, the RNA profiles of a pooled ejaculate from nine fertile men, one fertile individual and a Stratagene testis cDNA library constituted from 19 trauma victims were compared.[22] Totals of 7157, 3281 and 2780 transcripts were detected in the testis, pooled and individual spermatozoal samples, respectively. One important feature of the sample sets was that they existed as hierarchical subsets, such that the individual sample was a subset of the pooled sample bar four unique mRNA species and that all transcripts in both the pooled and individual sample, were subsets of the testis sample. This relationship is illustrated in Figure 1(C). These data indicate a level of inter-sample variability that has since been confirmed by additional work using 17 individual samples and strongly suggests the existence of a subset of core RNAs (*i.e.* those RNAs common to all samples) possibly demarcating the fertile genotype (David Dix, personal communication). The implicit relationship between the inter-sample variant and core (shared) subsets is illustrated in Figure 2(A).

15.3   Utility of Spermatozoal RNA

It is worthwhile speculating on the informational value of the different subsets of spermatozoal RNAs. In the first place, it is quite possible that the morphological sub-types of human spermatozoa that sediment at different rates in density-gradient media, arise as a result of differential retention of cytoplasm. In essence, those (spermatozoa) that sediment to form the pellet of a standard 45%:90% discontinuous gradient will have minimal retention of cytoplasm, high buoyant density and should contain just the common or core subset of 'fertile' transcripts. In contrast, those spermatozoa that collect at the gradient interfaces will have greater retention of cytoplasm, lower buoyant density and may retain a higher level of 'variant' cytoplasmic transcripts. The average human ejaculate could therefore contain molecular information explaining the different sub-types of spermatozoa based on buoyant density and hence, provide a fresh insight into the different phenotypes observed. Since pelleted spermatozoa are generally the most motile sub-population in the ejaculate,[26] comparison of their transcripts with those from poorly motile or immotile sub-populations will hopefully tell us something about the molecular mechanisms that underpin motility in these cells and conversely, the immobility of cells that do not pellet (or swim-up successfully). The relationship between these intra-sample subsets is illustrated in Figure 2(B).

Some reports have already tested the utility of spermatozoal RNA as a molecular resource for infertility investigation. Motility, for example, appears to correlate with a change in the relative quantity of particular mRNA species. Using a semi-quantitative RT-PCR based approach, Lambard showed that the quantity of PRM1 in the poorly motile sub-population of spermatozoa harvested from the interface of a discontinuous Percoll gradient was higher than in cells obtained from the pellet.[18] The same effect was noted for both endodermal (eNOS) and neuronal (nNOS) nitrous oxide synthase mRNAs. A similar study

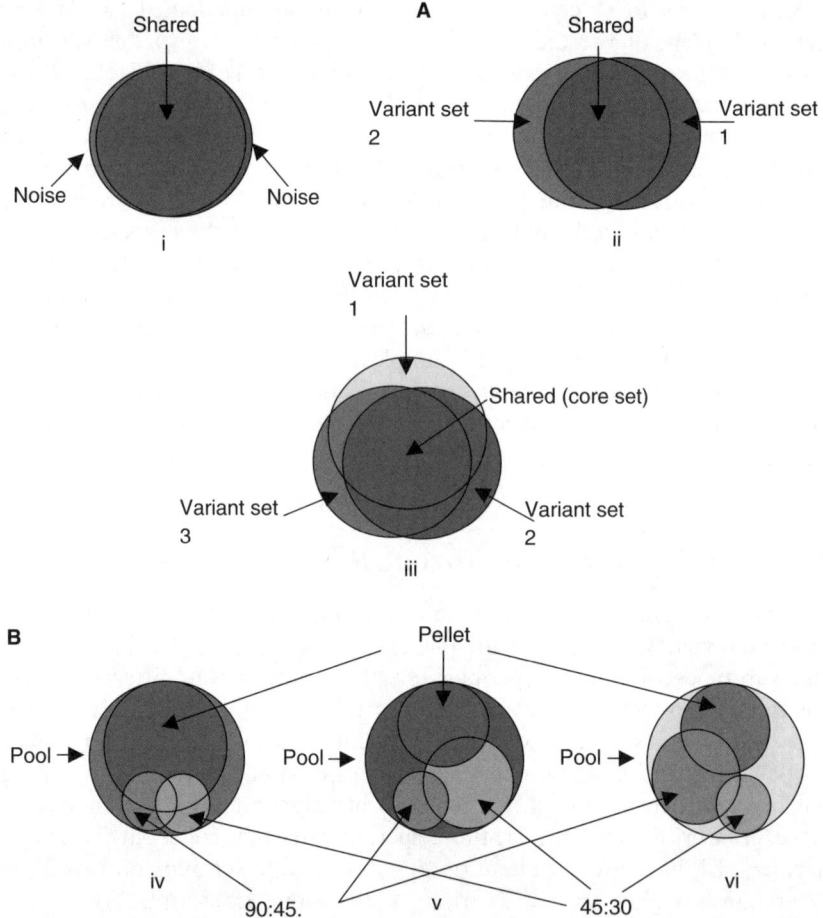

Figure 2 *A(i): Hypothetical inter-sample relationship between RNA populations from two ejaculates obtained from the same individual. While similar, the diagram illustrates the variability (noise) between them. (ii and iii) Hypothetical (inter-sample) relationship between two and three individual samples respectively, showing the shared (core) and individual (variant) mRNA subsets. (B) Hypothetical intra-sample relationship between RNA populations from spermatozoa separated on discontinuous density gradients. (iv) Normozoospermic sample with largest sub-population being derived from the pellet. (v and vi) Two abnormal samples showing reduction in mRNAs isolated from pelleted spermatozoa and expansion in mRNAs isolated from the spermatozoa sedimenting in the different layers*

using microarray-based data coupled with real-time PCR detected quantitative differences in the levels of several spermatozoal mRNAs according to the motility of the sampled population.[27] While it is possible that these quantitative changes truly reflect dynamic changes in the relative presence of spermatozoal mRNAs, a more plausible explanation is that cells failing to pellet in these gradients (or failing to swim-up) are less mature and carry more 'residual' cytoplasm with concomitantly more mRNA template.

A second consideration is that ejaculate spermatozoa are easily obtainable and as indicated earlier, their mRNAs constitute a subset of transcripts expressed in the testis. This subset is likely to be exclusively germ-line in origin making spermatozoa an ideal resource for monitoring the gene expression programme that underpins spermatogenesis in any given individual as well as monitoring for any extraneous influences that might affect its efficiency. Currently, studies aimed at assessing toxicological impact of environmental agents on spermatogenesis are mostly restricted to descriptive studies (effects on sperm count, motility, morphology, *etc.*) because the testis biopsies required for deeper investigations are generally difficult to acquire. Indeed, unless it can be shown to be of some benefit to the patient, it remains ethically questionable to request a testicular biopsy. Sampling ejaculates, however, poses no such restriction since an invasive procedure is unnecessary. Hence toxicological investigations are now open to deeper mechanistic study.[28-30] Information obtained will also be of value in understanding adverse effects on spermatogenesis of unknown origin, as is the case in idiopathic male factor infertility.

Finally, unlike non-obstructive azoospermia and severe oligozoopermia, which often result from the complete loss of one or more genes on the Y-chromosome and which are relatively rare conditions,[31] impaired spermatogenesis is much more likely to be a (a more frequent) consequence of autosomal polygenic effects. The great majority of human male sub-fertility and infertility is probably encompassed by this more nebulous category, which has traditionally been refractory to investigation. Again, unless there are compelling reasons for doing so, the acquisition of testicular biopsy is not normally an option here; but polygenic effects are frequently subtle and except in exceptional circumstances are unlikely to clear the ejaculate of spermatozoa. We would argue that the best resource for investigating the great majority of (idiopathic) male factor infertility are the spermatozoa themselves. Considering the many reports detailing a decline in male fertility over the last 50 years and a corresponding rise in the incidence of testicular cancer, there is an urgent need to assess potential toxicological factors that may influence these possibly related conditions.[32-34] An easily accessible source of germ-line mRNA is surely worth exploiting in this respect.

15.4 Spermatozoal Chromatin

Human spermatozoa are also unusual in that substitution of histones by protamines during spermiogenesis is relatively incomplete (compared with other mammals). Up to 15% of human spermatozoal chromatin continues to be packaged by histones.[35,36] While this level of histone content is high by most mammalian standards, the spermatozoa of one species of dasyurid marsupial is exceptional with up to 25% of the basic protein in its chromatin consisting of histones.[37] Other species appear to be fully protaminated although the presence of smaller quantities of histone-bound DNA in such species cannot be excluded. Considering that there is ample histone-bound DNA to encompass all protein-encoding genes in the human genome, these studies have naturally prompted questions on the functional role of differential DNA packaging in mammalian spermatozoa. One clue is that the histones of mammalian spermatozoa appear to be localised to the nuclear periphery, suggesting a functional ordering of this compartment.[37,38] Exploiting the differential solubility of histones and protamines to salt, it is possible to isolate histone and protamine enriched DNA fractions and to analyse their respective compositions.[39] To date, histone-bound spermatozoal DNA has been shown to be rich in telomeric sequences (the ends of chromosomes) packaged into typical 168 bp nucleosomal octamers comprising H2A, H2B, H3 and H4.[40] Evidence from murine spermatozoa (which typically retain a much lower proportion of nucleohistones estimated at 1–2%) indicates enrichment in short (SINE) and long (LINE) interspersed nuclear elements as well as transcription factor sequences.[41] The evidence for LINE enrichment in murine spermatozoal nucleohistone is interesting, because we have found evidence for a strong representation of one of the LINE open reading frames (ORF2) encoding a reverse transcriptase in the RNA isolated from human spermatozoa.[3] One suggestion for the function of telomeric sequences in human spermatozoal, histone-enriched chromatin is that they help to 'seed' the rapid displacement of protamines by maternal histones following fertilisation. Further analysis of the sequences packaged by histones and protamines in spermatozoa is currently underway in several laboratories. For example, we are currently using arrays designed for comparative genomic hybridisation (CGH) for this purpose. Although very preliminary, initial evidence indicates that the packaging may *not* be random, suggesting a functional requirement.

15.5 DNA Damage Susceptibility

From an environmental perspective, histone-bound DNA, being less condensed than protamine-bound DNA, may be more vulnerable to DNA damage from oxidative radicals generated *de novo* or by external factors that trigger their production. There is very good evidence that the extent of DNA damage in the spermatozoa of infertile men correlates well with spermatozoal dysfunction in terms of motility, abnormal forms, *etc.*[42–44] Moreover, fragmentation of spermatozoal DNA relating with these parameters can also be detected by the single cell gel electrophoresis (comet) assay and the sperm chromatin structure assay (SCSA).[45,46] The integrity of spermatozoal DNA has been shown to correlate with fertility outcomes including both subjective estimates of embryo quality and more objective studies of successful pregnancy and miscarriage rates. It is temping therefore, to speculate whether histone-bound DNA *is* more susceptible to damage and if so, identifying the critically sensitive gene sequences that are packaged by histones may help explain how this damage affects the male genome and in what form (Figure 3). It should also shed light on the ability of

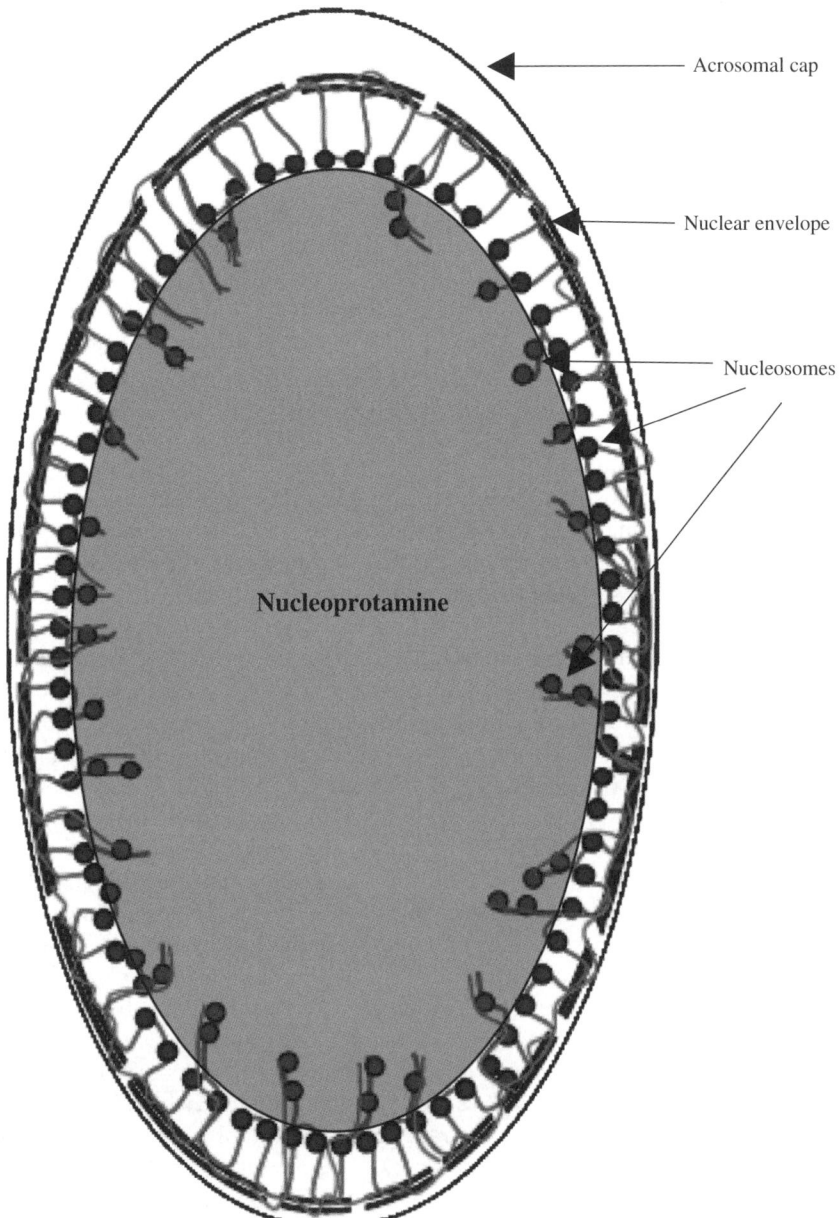

Acrosomal cap

Nuclear envelope

Nucleosomes

Nucleoprotamine

Figure 3 *Possible relationships between protamine (nucleoprotamine) and histone (nucleosomal)-bound DNA and RNA in spermatozoa. In this scenario, the majority of the DNA is packaged by protamines towards the interior of the nucleus with nucleosomal DNA demarcating a zone that lies between the nucleoprotamine and the nuclear envelope. Spermatozoal RNA is associated with the nucleosomal DNA and possibly 'anchors' it to the envelope. The nucleosomal DNA is likely to be more susceptible to damage because of its looser compaction and its proximity to the surface of the spermatozoon*

nascent DNA repair mechanisms in the oocyte to deal with such damage and the fertility outcomes that result from using spermatozoa with damaged DNA.

15.6   Transmission of the Paternal Genome

One clue to the paradox of nucleic-acid complexity in spermatozoa comes from the recent observations that a paternal contribution is not required for mammalian reproduction.[47] The most compelling aspect of this report shows how it is possible to 'fool' an egg into 'thinking' that it has received a spermatozoon and permitting 'normal' development. This feat was achieved by injecting a murine oocyte nucleus carrying a deleted *h19* gene into a normal egg. Syngamy in at least a small number of these gynogenetic constructs led to viable offspring. As *h19* and *Igf2* are reciprocally imprinted in male and female gametes,[48] the elimination of *h19* in the introduced nucleus essentially mimicked the delivery of a spermatozoon (bearing in mind the observation that the mouse is the only mammal that does *not* require a centriole from the spermatozoon). As with somatic cell nuclear transfer, however, the efficiency of the process is very poor, showing that the establishment of an epigenetic balance in the zygote is not readily accomplished. These data suggest that rather than behaving as critical development switches, a more likely role for spermatozoal RNA (including antisense RNA and possibly siRNA[49]) lies in the oocyte's response to fertilisation through reprogramming of the male genome.[50] In this scenario, spermatozoal RNA may act to regulate or even co-ordinate the rapid deprotamination of the spermatozoal nucleus by maternal factors. Indeed, spermatozoal RNA and the differential packaging of spermatozoal chromatin may be linked phenomena. In this respect, the rapid, post-fertilisation de-methylation of the sperm genome (with the exception of imprinted loci and the early transcription reported from the male pronucleus) is a possible outcome of this process.[51–53] Such a mechanism could be mimicked by the sort of experimental manipulation described above, which under normal physiological conditions, would help ensure the requirement for a male contribution to the zygote and hence, the preservation of male-derived genes. Indeed, the general dependence on the paternal centriole for syngamy is also likely to be evolutionarily driven by this imperative. One possible side effect of this 'self preservation' phenomenon might be the rare gestation of molar pregnancies where duplication of the paternal genome occurs. This idea is in agreement with the tendency among paternally imprinted genes to further placental growth, ostensibly to favour the transmission of male-derived genes (irrespective of foetal gender[54]).

15.7   Conclusions

The structural complexity of ejaculated mammalian spermatozoa has evolved over millions of years to optimise the safe delivery of the paternal genome to the egg. However, recent studies have helped to dispel the commonly held notion of the male gamete as a mere purveyor of the male haploid genome. Apart from the contribution of the paternal centriole and an egg-activating factor, there is the selective retention of a complex retinue of mRNAs that are potentially capable of coding for a large number of proteins. Secondly, the differential packaging of DNA into protamine and histone-enriched compartments is likely to have functional relevance. Are these two phenomena mutually exclusive or could they be somehow related? It is perhaps worth noting that to date, no laboratory has reported the successful translation of spermatozoal RNA in cell-free systems that support the translation of mRNAs isolated from other cells and tissues (our own attempts included). This could be due to tenacious inhibitors in the RNA preparations that prevent elongation of peptides; but it could equally be due to the absence of 5' caps on these transcripts. Poly-A tails, however, are present, because the mRNAs from the pooled spermatozoal sample mentioned above[22] were captured using oligo-dT coated magnetic beads.

Whatever their physiological significance, both spermatozoal RNA and the unexplained differential packaging of spermatozoal DNA are worthy of exploration for a number of reasons. The RNA could be very informative of the gene expression programme that underpins the fertile phenotype. Indeed, it is highly likely that spermatozoa retain an RNA record of the structural components of the cell based on the observation that every protein component investigated appears to have a partner transcript. This information will be useful for general studies investigating male infertility and more focused toxicological studies aimed at monitoring environmental, occupational or therapeutic effects on spermatogenesis. For example, we fully anticipate that spermatozoal RNA will help uncover the molecular mechanisms that underpin the infertile phenotype and help assess the role of toxicological factors in any general decline in male sperm production and the co-incident rise in testicular cancer.

Understanding why spermatozoal chromatin is differentially packaged should also help illuminate its significance for DNA damage induced by extraneous factors such as oxidative radicals. If important genes are packaged predominantly by histones, it follows that they are likely to be more susceptible to damage, which may have important effects on fertility outcomes in natural and assisted conception strategies.

References

1. C.M. Saunders, M.G. Larman, J. Parrington, L.J. Cox, J. Royse, L.M. Blayney, K. Swann and F.A. Lai, *Development*, 2002, **129**, 3533.
2. C. Simerly, G.J. Wu, S. Zoran, T. Ord, R. Rawlins, J. Jones, C. Navara, M. Gerrity, J. Rinehart, Z. Binor, R. Asch and G. Schatten, *Nat. Med.*, 1995, **1**, 47.
3. D. Miller, D. Briggs, H. Snowden, J. Hamlington, S. Rollinson, R. Lilford and S.A. Krawetz, *Gene*, 1999, **237**, 385.

4. G.C. Ostermeier, D. Miller, J.D. Huntriss, M.P. Diamond and S.A. Krawetz, *Nature*, 2004, **429**, 154.

5. N.B. Hecht, *Bioessays*, 1998, **20**, 555.

6. S. Grunewald, U. Paasch and H.J. Glander, *Andrologia*, 2005, **37**, 69.

7. P.M. Bhargava, *Nature*, 1957, **179**, 1120.

8. K.A. Abraham and P.M. Bhargava, *Biochem. J.*, 1963, **86**, 298.

9. E.B. Premkumar and P.M. Bhargava, *Nat. New Biol.*, 1972, **240**, 139.

10. J. MacLaughlin and C. Terner, *Biochem. J.*, 1973, **133**, 635.

11. R. Balhorn, L. Brewer, M. Corzett and J. Cosman, *Biol. Reprod.*, 1999, **60**, M34.

12. C.A. Pessot, M. Brito, J. Figueroa, Concha II, A. Yanez and L.O. Burzio, *Biochem. Biophys. Res. Comm.*, 1989, **158**, 272.

13. G. Kumar, D. Patel and R.K. Naz, *Cell Mol. Biol. Res.*, 1993, **39**, 111.

14. M.H. Chiang, N. Steuerwald, H. Lambert, A. Steinleitner and E.K. Main, *J. Immunol.*, 1993, **150**, A283.

15. T.J. Durkee, M. Mueller and M. Zinaman, *Am. J. Obstet. Gynecol.*, 1998, **178**, 1288.

16. L.O. Goodwin, D.S. Karabinus and R.G. Pergolizzi, *Mol. Hum. Reprod.*, 2000, **6**, 487.

17. L.O. Goodwin, D.S. Karabinus, R.G. Pergolizzi and S. Benoff, *Mol. Hum. Reprod.*, 2000, **6**, 127.

18. S. Lambard, I. Galeraud-Denis, G. Martin, R. Levy, A. Chocat and S. Carreau, *Mol. Hum. Reprod.*, 2004, **10**, 535.

19. S. Lambard, I. Galeraud-Denis, P.T. Saunders and S. Carreau, *J. Mol. Endocrinol.*, 2004, **32**, 279.

20. W. Richter, D. Dettmer and H.J. Glander, *Mol. Hum. Reprod.*, 1999, **5**, 732.

21. A. Rohwedder, O. Liedigk, J. Schaller, H. Glander and H. Werchau, *Mol. Hum. Reprod.*, 1996, **2**, 499.

22. G.C. Ostermeier, D.J. Dix, D. Miller, P. Khatri and S.A. Krawetz, *Lancet*, 2002, **360**, 772.

23. W.H.O., "*WHO Laboratory Manual for the Examination of Human Semen and Sperm-Cervical Mucus Interaction*", ed. W.H. Organisation, World Health Organisation, 1999.

24. E. Rejon, C. Bajon, A. Blaize and D. Robert, *Mol. Reprod. Dev.*, 1988, **1**, 49.

25. M.L. Engel, A. Chaboud, C. Dumas and S. McCormick, *Plant. J.*, 2003, **34**, 697.

26. T.F. Kruger and K. Coetzee, *Hum. Reprod. Update*, 1999, **5**, 172.

27. H. Wang, Z. Zhou, M. Xu, J. Li, J. Xiao, Z.Y. Xu and J. Sha, *J. Mol. Med.*, 2004, **85**, 317.

28. J.C. Rockett, J. Christopher Luft, J. Brian Garges, S.A. Krawetz, M.R. Hughes, K. Hee Kirn, A.J. Oudes and D.J. Dix, *Genome Biol.*, 2001, **2**, RESEARCH0014.

29. H. Ren, K.E. Thompson, J.E. Schmid and D.J. Dix, *Toxicol. Sci.*, 2003, **263**, 72(S).

30. J.C. Rockett and D.J. Dix, *Xenobiotica*, 2000, **30**, 155.
31. S.J. Silber, R. Alagappan, L.G. Brown and D.C. Page, *Hum. Reprod.*, 1998, **13**, 3332.
32. S.M. Duty, M.J. Silva, D.B. Barr, J.W. Brock, L. Ryan, Z. Chen, R.F. Herrick, D.C. Christiani and R. Hauser, *Epidemiology*, 2003, **14**, 269.
33. E. Carlsen, A. Giwercman and N.E. Skakkebaek, *I. Med. J.*, 1993, **86**, 85.
34. H. Moller, *Hum. Reprod.*, 2001, **16**, 1007.
35. M. Gardiner-Garden, M. Ballesteros, M. Gordon and P.P. Tam, *Mol. Cell Biol.*, 1998, **18**, 3350.
36. J.M. Gatewood, G.R. Cook, R. Balhorn, C.W. Schmid and E.M. Bradbury, *J. Biol. Chem.*, 1990, **265**, 20662.
37. L.L. Soon, J. Ausio, W.G. Breed, J.H. Power and S. Muller, *J. Exp. Zool.*, 1997, **278**, 322.
38. C. Pittoggi, L. Renzi, G. Zaccagnini, D. Cimini, F. Degrassi, R. Giordano, A.R. Magnano, R. Lorenzini, P. Lavia and C. Spadafora, *J. Cell Sci.*, 1999, **112**(Pt 20), 3537.
39. R. Balhorn, B.L. Gledhill and A.J. Wyrobek, *Biochemistry*, 1977, **16**, 4074.
40. I.A. Zalenskaya, E.M. Bradbury and A.O. Zalensky, *Biochem. Biophys. Res. Comm.*, 2000, **279**, 213.
41. C. Pittoggi, A.R. Magnano, I. Sciamanna, R. Giordano, R. Lorenzini and C. Spadafora, *Mol. Reprod. Dev.*, 2001, **60**, 97.
42. D. Sakkas, F. Urner, D. Bizzaro, G. Manicardi, P.G. Bianchi, Y. Shoukir and A. Campana, *Hum. Reprod.*, 1998, **13**(Suppl 4), 11.
43. C. Cho, H. Jung-Ha, W.D. Willis, E.H. Goulding, P. Stein, Z. Xu, R.M. Schultz, N.B. Hecht and E. M. Eddy, *Biol. Reprod.*, 2003, **69**, 211.
44. D. Anderson, M.M. Dobrzynska, T.W. Yu, L. Gandini, E. Cordelli and M. Spano, *Teratog. Carcinog. Mutagen.*, 1997, **17**, 97.
45. I.D. Morris, S. Ilott, L. Dixon and D.R. Brison, *Hum. Reprod.*, 2002, **17**, 990.
46. M. Tomsu, V. Sharma and D. Miller, *Hum. Reprod.*, 2002, **17**, 1856.
47. T. Kono, Y. Obata, Q. Wu, K. Niwa, Y. Ono, Y. Yamamoto, E.S. Park, J.S. Seo and H. Ogawa, *Nature*, 2004, **428**, 860.
48. M.S. Bartolomei, A.L. Webber, M.E. Brunkow and S.M. Tilghman, *Genes & Development*, 1993, **7**, 1663.
49. G.C. Ostermeier, R.J. Goodrich, J.S. Moldenhauer, M.P. Diamond and S.A. Krawetz, *J. Androl.*, 2005, **26**, 70.
50. D. Miller, G.C. Ostermeier and S.A. Krawetz, *Tr. Mol. Med.*, 2005, **11**, 156.
51. J.M. Trasler, *Semin. Cell Dev. Biol.*, 1998, **9**, 467.
52. P.G. Adenot, Y. Mercier, J.P. Renard and E.M. Thompson, *Development*, 1997, **124**, 4615.
53. F. Aoki, D.M. Worrad and R.M. Schultz, *Dev. Biol.*, 1997, **181**, 296.
54. R. Goshen, V. Tannos, Z. Benrafael, N. Degroot, B. Gonik, A.A. Hochberg and O. Lustig, *Fert. Ster.*, 1994, **62**, 903.

Germline Mutagenesis

CHAPTER 16
Origin of Paternal Mutations

JAMES F. CROW

Genetics Laboratory, University of Wisconsin, 425-G Henry Mall, Madison WI 53705, USA

I would like to dedicate this chapter to R. A. Fisher, who was associated with the city of Bradford in his formative years.[1] For 2 years, starting in 1917 he was on the faculty of Bradford College, teaching physics and mathematics to bored and mischievous students. He was a poor teacher and he hated it, but he managed to do all sorts of other things, such as write dozens of reviews and begin to formulate the statistical ideas that later made him famous. His blockbuster, "The correlation between relatives on the supposition of Mendelian inheritance," was published during this period.[2] This masterpiece laid the foundation for all of quantitative genetics; and it was mainly written while he was a schoolteacher before coming to Bradford.

After a distinguished career as a student at Cambridge, Fisher was at a loose end. He wanted badly to enlist in the British Army, but his poor eyesight led to repeated rejection. So he taught at Rugby, Haileybury, and on a Naval training ship before moving to Bradford College. While at Bradford he also farmed, with considerable help from his wife and her sister. His now-famous paper was completed in 1916, but he had difficulty in publishing it and it was finally printed by the Royal Society of Edinburgh 2 years later.

At the end of the War, Fisher was adrift with no job in sight. Finally, he was offered a position at the Rothamsted Experimental Station. There he began the studies that would soon lead to his being recognized as the greatest statistician of the century. I think it was good for both farming and statistics that Fisher gave up an agricultural career and became a statistician. The city of Bradford should be proud that Fisher once lived here.

My purpose in this chapter is to summarize current information about sex and paternal age effects on human spontaneous mutation. It will serve as background for the more specialized articles to follow. This is mainly based on a longer article[3] to which the reader is referred for more details.

16.1 Historical

As early as 1912, the German physician,Wilhelm Weinberg, noted that children with achondroplasia (short-limbed dwarfism) tended to be late-born in the sibship.[4] This is a dominant trait, and since reproduction is rare, all the affected children came from normal parents. Weinberg made the brilliant suggestion that their being late-born would argue for a mutational origin. Coming at a time when mutation was only vaguely understood, this was remarkably prescient.

Weinberg had no way of distinguishing among paternal age, maternal age, and birth order, and clarification of this waited for more than 40 years. In 1955 Penrose[5] showed that the effect is mainly, if not entirely, paternal age, as Weinberg must have suspected. Shortly before, Haldane[6] had reported a 10-fold higher mutation rate for haemophilia in males than in females.

16.2 Male Excess and Paternal Age Effect for Base Subsitutions

All this suggested that the explanation lies in the greater number of cell divisions between zygote and sperm than between zygote and egg, thereby providing more opportunity for mutation in males, especially in older fathers who have had a larger number of divisions than younger ones.[7]

It is interesting that Weinberg studied achondroplasia, for this is a condition with an extreme sex and age effect; the mutations occur almost entirely in males and there is a steep nonlinear increase with paternal age. As I shall show later, this is exceptional and probably has a different mechanism. It is likely that with a more typical trait Weinberg would have missed discovering the mutational origin of many dominant conditions. It is not the only instance in which a correct conclusion was reached from atypical data.

The ensuing years brought a number of confirmations. In addition to haemophilia, other X-linked traits, Lesch-Nyhan syndrome, ornithine trans-carbamylase deficiency (OTC), and Rett syndrome also showed a greater male mutation rate.[3] Of course, for X-linked traits it is the grandparental rather than the paternal age that is important. Until the discovery of molecular methods, it was not possible to identify the parentage of autosomal mutations.

Risch *et al.*[8] have analyzed paternal age for a number of dominant traits. Usually there is a substantial paternal age effect. There is often a maternal age effect also, but typically nothing more than would be expected from the correlation of maternal and paternal ages.

16.3 Exceptions to the Sex and Paternal Age Effect

So far I have considered only mutations due to base substitutions, but small deletions (<20 bp) will not show such an effect. The frequency, rather than being much higher in males, is roughly equal in the two sexes and there is no age effect in either sex. This suggests that the event may occur only once in the life cycle, perhaps at meiosis.

Duchenne muscular dystrophy and neurofibromatosis I show a smaller sex difference and paternal age effect. The reason is that these are very large genes and a substantial fraction of the mutations are intragenic deletions. The reduced sex and paternal age effects are caused by the base-substitution rate being diluted by these deletions.[3]

Retinoblastoma is particularly good as an illustration. The disease is the result of two mutations, often one germinal and one somatic. The somatic rate is essentially the same in the two sexes, but the germinal rate is higher in males and there is an age effect. All this is expected by the hypothesis that the greater number of cell divisions in the male is the culprit. The overall male/female mutation ratio is 5.7. About 62% of the mutations are base substitutions and 38% are indels (insertions/deletions).[9,10] Using these numbers we estimate the male/female ratio for base substitutions to be 18.1, consistent with the number of cell divisions.[3]

In a recent analysis of the human database of some 21,000 mutations in about 1000 genes,[11] base substitutions account for about 70% and small indels about 23%, so we can expect to find a substantial sex and paternal (or grandpaternal) age effect for dominant-autosomal and X-linked recessive traits.

16.4   Hot Spots

The mutations that I have been discussing are scattered along the chromosome; this is typical. Yet some of the most striking examples of male-excess and paternal age effect are associated with hot-spot mutations, that is, mutations that occur in one or two codons, usually at a single nucleotide. Weinberg's classical example, achondroplasia, is of this type. It is a mutation in the fibroblast growth factor receptor (FGFR3), and mutations are in the same codon. One, a transition (purine → purine, or pyrimidine → pyrimidine) converts GGG into AGG; a second, a transversion (purine → pyrimidine, or vice versa), converts GGG into CGG. As expected the first has a higher rate, and I believe this is the highest codon rate known in humans.[12]

Another example is Apert syndrome, in which the mutations are in FGFR2. The mutations are at two adjacent codons TCG and CCT. The transversions TCG → TGG and CCT → CGT both produce Apert syndrome.[13] Curiously, all the hot-spot mutations so far known occur in three genes: FGFR3, FGFR2, and RET.[3]

16.5   Sperm Analysis

Recently it has become possible to confirm the clinical observations by direct sperm analysis. The most thoroughly analyzed is Apert syndrome.[13,14] The technique depends on finding a restriction enzyme that cuts at the appropriate site and permits identification of mutant sperm. The age distribution of mutant sperm corresponds very well to the clinical data on paternal age effect.

Yet despite the general agreement with the clinical data, there are curious features. For one thing the variance from one donor to another is enormous. Equally striking, many of the donors had a heterozygous single nucleotide polymorphism (SNP) close to the site of interest. It shows, on average, the expected 1:1 segregation among the sperm, but the variance is enormous. How can we account for this?

Wilkie and his group believe that the mutation rate is moderate, but that the mutant spermatogonial cells enjoy a selective advantage. It is easy to show that a small selective advantage of 0.002 is sufficient to mimic the clinical data.[3]

There are several arguments for premeiotic selection.[3,13,14] This is a bold, imaginative suggestion. If this is upheld by future studies, it will open a new direction for human population genetics and will have implications for other long-lived organisms. It seems surprising that a gene that causes a devastating disease in somatic tissue could have a selective advantage in spermatogonia. Yet strange things happen in biology; for example, cancer mutations that confer a selective advantage in certain tissues. This is an exciting subject and we can look forward to most interesting results in the near future.

A possible mechanism is that the mutation causes a change from asymmetrical cell division, leading to linear increase, to symmetrical division leading to exponential increase.[13] This need not happen often to account for both the striking increase with paternal age and the enormous variance. There need not be any increased cell division rate; the simple change from linear to exponential kinetics is more than sufficient. To understand possible mechanisms, it will be interesting to examine other instances in which regulated growth is replaced by unregulated, as in cancers.

Similar sperm studies on achondroplasia have shown quite different results, disagreement with the clinical data and little increase with paternal age.[15] It is possible, however, to explain these results.[14] I shall leave it for future research to explain the discrepancy.

16.6   Premeiotic Selection or Sperm Competition

Superficially, premeiotic selection and sperm competition are similar. Both lead to a greater mutant frequency than would be expected from mutation rates alone. But there are reasons for preferring premeiotic selection. It provides a good explanation of the paternal age effect, whereas there is no obvious reason for sperm competition to do this. Furthermore, the enormous variance is more consistent with premeiotic than postmeiotic mechanisms. Meiotic drive is largely ruled out by the finding that Apert heterozygotes produce sperm in a

1:1 ratio.[16] This does not say, however, whether the two kinds of sperm are equally functional.

Other mechanisms, such as premutational lesions, cannot be ruled out. Yet, at the moment, premeiotic selection, perhaps through conversion from linear to exponential cell division, seems like the best bet.

16.7 Conclusion

Considering all the evidence, there are three main classes of Mendelizing mutations (more than one can occur in the same gene)

(i) Small deletions and duplications (<20 bp)

- No significant age effect
- Roughly equal numbers of paternal and maternal origin

(ii) Base substitutions

- Mainly but not entirely paternal
- Large paternal age effect, which may be nonlinear

(iii) Hot spots

- Essentially all paternal
- Very large, nonlinear paternal age effect
- May be due to selection, rather than high mutation rate

Usually those base substitutions with a moderate sex and age effect occur at many sites. For example, in mild X-linked haemophilia, transitions make up about 62% of the mutations, transversions 23%, indels <50 bp, 8%, and indels >50 bp, 7%. These are scattered along the chromosome.[17] In contrast, mutations with extreme sex and age effects tend to be in hot spots. But again a rigid classification breaks down. A recent study of Noonan's syndrome[18] showed that of 14 mutations in which the parent of origin could be determined, all were paternal and there was a large age effect. Yet, the mutations are not at one hot spot, although a large fraction of them are located at one site. So this system of classification, like all such systems, has exceptions and is fuzzy at the borders. I hope it is useful nevertheless.

I have not included nondisjunction, well known to occur disproportionately in the female and to be strongly age-dependent. There are also a number of other mechanisms, such as various chromosome changes, changes in repeat number of microsatellites, and a terminal inversion causing severe haemophilia, that are strongly sex-dependent.[19]

But the above-mentioned categories include the great bulk of those changes that are inherited in Mendelian fashion.

References

1. J.F. Box, *R. A. Fisher, the Life of a Scientist*, Wiley, NY, 1978.
2. R.A. Fisher, *Trans. Roy, Soc. Edinb.*, 1918, **52**, 399.
3. J.F. Crow, *J. Rad. Res.*, 2006, **47**(Suppl. B), 75.
4. W. Weinberg, *Arch. Rassen u. Gesel. Biolog.*, 1912, **9**, 710.
5. L.R. Penrose, *Lancet*, 1955, **269**, 312.
6. J.B.S. Haldane, *Ann. Eugen.*, 1947, **13**, 262.
7. F. Vogel and A. Motulsky, *Human Genetics*, Springer Verlag, Berlin, 1997.
8. N. Risch, E.W. Reigh, M.W. Wishnick and J.G. McCarthy, *Am. J. Hum. Genet.*, 1987, **41**, 218.
9. T.P. Dryja, J.F. Morrow and J.M. Rapaport, *Hum. Genet.*, 1997, **100**, 446.
10. D.R. Lohman, *Hum. Mutat.*, 1999, **14**, 283.
11. S.E. Antonarakis, M. Krawczak and D.N. Cooper, *Eur. J. Pediatr.*, 2000, **159**, 173.
12. G.A. Bellus, T.W. Hefferon, R.I. Ortiz de Luna, J.T. Hecht, W.A. Horton, M. Machado, I. Kaitila, I. McIntosh and C.A. Francomano, *Am. J. Hum. Genet.*, 1995, **56**, 368.
13. A. Goriely, G.A.T. McVean, M. Rojmyr, B. Ingemarsson and A.O.M. Wilkie, *Science*, 2003, **301**, 643.
14. A.O.M. Wilkie, *Cytokine Growth Factor Rev.*, 2005, **16**, 187.
15. L. Tiemann-Boege, W. Navidi, R. Grewal, D. Cohn, B. Eskenazi, A.J. Wyrobek and N. Arnheim, *Proc. Natl. Acad. Sci. USA*, 2002, **99**, 14952.
16. D. Moloney, *Ann. Rev. Coll. Surg. Engl.*, 2001, **83**, 1.
17. J. Becker, R. Schwaab, A. Moller-Taube, U. Schwaab, W. Schmidt, H.H. Brackmann, T. Grimm, K. Olek and J. Oldenberg, *Am. J. Hum. Genet.*, 1996, **58**, 657.
18. M. Tartagli, V. Cordeddu, H. Chang, A. Shaw, K. Kalidas, A. Crosby, M.A. Patton, M. Sorcini, I. fan der Brgt, S. Jeffery and B.D. Gelb, *Am. J. Hum. Genet.*, 2004, **75**, 492.
19. J.P. Antonarakis, J.P. Rossiter, M. Young, J. Horst, P. de Moerloose, S.S. Sommer, R.P. Ketterling, H.H. Kazazian, C. Negrier and C. Vinciguerra, *Blood*, 1995, **86**, 2206.

CHAPTER 17

Redox Regulation of DNA Damage in the Male Germ Line

R.J. AITKEN, S.D. ROMAN, M.A. BAKER AND
G. DE IULIIS

ARC Centre of Excellence in Biotechnology and Development and Discipline of Biological Sciences, University of Newcastle, Callaghan NSW 2308, Australia

17.1 Introduction

Defective sperm function is the largest single cause of human infertility, which affects approximately 1 in 20 males in developed countries.[1,2] Not only is human semen quality poor but there is some evidence that it might have deteriorated further over the past 50 years.[3–5] While the universality of this effect is hotly disputed, in certain countries the secular trends in semen quality appear to be well established.[6,7] There is also strong evidence that where such trends are observed, it is a birth cohort effect. In other words, the key issue is not so much the age at which you had your semen analysed, as the year in which you were born.[4,6] The possible existence of secular trends in semen quality is also reflected in the incidence of testicular cancer, which has risen dramatically since the second world war in all countries that have been examined, increasing the lifetime risk from 0.3 to 0.8%.[8] In addition, certain studies have revealed a recent increase in other pathologies of the male reproductive tract such as hypospadias and cryptorchidism.[9] It has even been suggested that these changes in semen quality and male reproductive tract development have a common origin during foetal development, which gives rise to a condition referred to as 'the testicular dysgenesis syndrome'. According to this hypothesis exposure of women to xenobiotics during pregnancy has an impact on the differentiation of the male reproductive tract leading to pathologies such as testicular cancer and infertility in the offspring.[10] The discovery of a statistical relationship between the incidences of testicular cancer and maternal smoking during pregnancy in Nordic countries adds weight to this idea.[11] Although other studies have failed to find any relationship between parental smoking or drinking and germ cell tumours in the offspring,[12] there is a great deal of accumulated experimental and epidemiological data to support the general concept that exposure of parental germ cells to xenobiotics can cause morbidity in the offspring. However, the focus of much of this evidence is not so much on the mothers' exposure to germ cell toxicants, but that of the fathers.[7]

17.2 The Central Hypothesis

According to this model, environmental toxicants, including endocrine disruptors, can have a significant impact on the male germ line, causing genetic or epigenetic damage in the spermatozoa. This damage is thought to be induced post-meiotically when the germ cells have differentiated into spermatids and spermatozoa.[7] Premeiotic DNA damage is much less likely since proof reading and repair in the spermatogonial stem cell population are so efficient that these cells have one of the lowest spontaneous mutation rates in the body.[15]

Following their release from the germinal epithelium, spermatozoa are particularly vulnerable to DNA damage because they have lost their capacity to undergo apoptosis and DNA repair is no longer possible. Moreover following spermiation, the gametes can no longer take advantage of the protection afforded to them by the nurse cells of the testes, *i.e.*, the Sertoli cells. As isolated spermatozoa, these cells must spend another 2 weeks travelling through the epididymis, undergoing a period of post-testicular maturation during which they acquire many of the functional attributes that they will need to fertilise the oocyte, including capacities for co-ordinated movement, zona recognition and acrosomal exocytosis. In the specific case of human spermatozoa, ejaculation may be followed by an additional, prolonged, anadromous phase during which these cells must spend up to 6 days swimming around the female reproductive tract waiting for an egg. As a result of this prolonged, isolated existence, human spermatozoa are characterised by relatively high rates of DNA damage as measured by a wide variety of assays.[16–20] Such a phenomenon sets our species apart from all other mammals that have been investigated. Furthermore, DNA damage in human sperm is of considerable clinical importance because it is associated with reduced rates of fertilisation *in vitro*, impaired pre-implantation development of the embryo, increased rates of early pregnancy loss and poor fertility following natural or assisted conception.[21–29] Perhaps most serious of all is the high incidence of morbidity recorded in the offspring of men exhibiting significant DNA damage in their spermatozoa, including dominant genetic disease, infertility and cancer.[7,21]

If this DNA damage in the male germ line is truly post-meiotic, then we would not expect to find mutations in these cells. Instead we should find various kinds of DNA damage (single- and double-stranded breaks, abasic sites, oxidised bases, base adduct formation, *etc.*), the specific nature of the lesion depending on the aetiology of the damage.

This damaged DNA will subsequently be carried into the oocyte by the fertilising spermatozoon and will be the subject of repair as soon as the chromatin decondenses. Aberrant repair of this paternally derived DNA damage within the ooplasm has the potential to create genetic mutations and/or epigenetic defects in the embryo. Moreover, if this aberrant repair precedes the S phase of the cell cycle within the zygote, the genetic/epigenetic damage will be in every cell in the body including the germ line. In this way xenobiotic-mediated DNA damage in spermatozoa has the potential to create mutations or epigenetic defects that may be vertically transmitted to subsequent generations (Figure 1). This mechanism may explain the correlations observed between DNA damage in the spermatozoa of heavy smokers and the increased incidence of childhood cancer in their offspring.[30,31] Such a mechanism may also explain

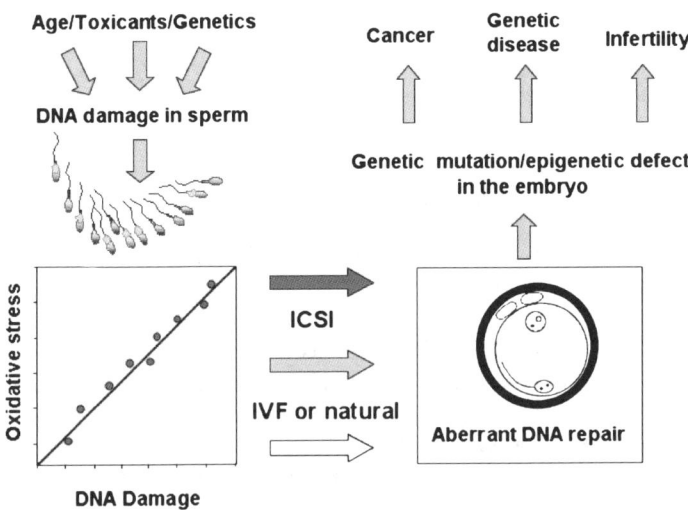

Figure 1 *A hypothesis for the mechanisms by which male-mediated toxicity might be mediated. According to this model, a wide variety of intrinsic and extrinsic factors including attack by xenobiotics, age and genetic defects, lead to a state of oxidative stress in the male germ line that, in turn, leads to the induction of DNA damage. Such stress may involve a loss of antioxidant protection and/or exposure to high levels of ROS. The most vulnerable stage of germ cell development for such stress to occur is during the late stages of haploid germ cell differentiation and maturation (spermatid to spermatozoon) when the spermatozoa have lost their capacity for DNA repair, cannot undergo apoptosis, have ejected their cytoplasm (with its complement of antioxidant enzymes) and, in our species at least, may be committed to two to three weeks of solitary existence before coming into contact with an egg. If the levels of oxidative stress are sufficiently low, the DNA in the sperm nucleus will be damaged but the spermatozoa will have still retained their capacity for fertilisation by IVF or natural means. If oxidative stress is severe and DNA damage is high, fertilisation cannot occur because of the collateral damage that has been done to the lipids of the sperm plasma membrane. However, if ICSI is used as the method of insemination, then such DNA-damaged spermatozoa might successfully fertilize the egg. The damaged DNA that is brought into the egg by the fertilizing spermatozoa must then be repaired within the ooplasm. If a mistake occurs in this repair process prior to the S phase of the zygotic cell cycle, then mutations (or epigenetic defects) will be created that will be present in every cell in the body including the germ line. Mutations induced in this way may then be responsible for a range of paternally mediated developmental defects including abortion, dominant genetic disease, infertility and childhood cancer.*

the relationship between age-related increases in sperm DNA damage and the powerful impact of paternal age on the incidence of dominant genetic disease in children, including conditions such as Apert syndrome or achondroplasia.[7,32,33] The parallel importance of epigenetic effects has also been emphasised by the recent discovery of a trans-generational male infertility phenotype, which is transmitted through the male germ line following exposure of gestating female rats to endocrine disruptors during the phase of gonadal sex determination.[34] Overall, this hypothesis emphasises the ultimate importance of the male germ line in the causation of human disease. It also raises key questions concerning the origins of DNA damage in human spermatozoa, the types of DNA damage induced, the mechanisms of DNA repair in the zygote and the nature of any genetic/epigenetic defects created as a result of deficiencies in this process.

17.3   Causes of DNA Damage in Spermatozoa

To date there are three major hypotheses concerning the origins of DNA damage in the male germ line. The first implicates an abortive apoptotic process. During the early stages of spermatogenesis, germ cells can respond to DNA damage detected at cell cycle checkpoints by initiating an apoptotic cascade that culminates in cell death. However, as germ cell development proceeds, so the ability to undergo apoptosis is progressively lost. Specifically, it has been suggested that germ cells at advanced stages of development may be capable of activating some components of the apoptotic cascade, such as endonuclease activation and phosphatidyl serine exposure, but they are unable to follow the process through to completion. As a result, gametes may be generated that are functionally competent but contain fragmented DNA as a consequence of endogenous endonuclease activation.[35] Although this hypothesis sounds highly plausible, analyses of DNA fragmentation rates in human spermatozoa have failed to find any significant correlations with the expression of apoptotic markers such as annexin V binding or Fas.[36–38]

A second hypothesis is that the DNA damage seen in defective human spermatozoa is the result of defective chromatin packaging during a critical stage of spermiogenesis.[39] This proposal envisages that relief of the torsional stresses associated with chromatin packaging in round and elongating spermatids involves the repeated transient nicking of DNA.[40] These DNA nicks are subsequently repaired by poorly characterised mechanisms possibly involving the participation of transition proteins and histone H2AX phosphorylation.[41,42] Any impediment to this process due to defects in the structure of the chromatin or the activity of the DNA repair system may lead to the generation of functional gametes expressing high levels of DNA fragmentation.

An example of this kind of mechanism in action may possibly be seen in the high proportion of infertile patients exhibiting protamine deficiencies in their spermatozoa.[43,44] Where the protamine-2 gene has been disrupted in mice, a phenotype is observed characterised by poor compaction of the DNA and, importantly, high levels of DNA damage in the gametes. Moreover, the use of such spermatozoa in assisted conception protocols involving ICSI (Intra-Cyto-plasmic Sperm Injection) was associated with poor blastocyst formation rates, which highlights the importance of paternal DNA damage in establishing the viability of the embryo.[45] Across a large number of studies, poor compaction of sperm DNA has been repeatedly associated with DNA damage in the male germ line.[46,47] It is possible that such poorly compacted DNA somehow impedes the assembly and/or operation of the chromatin repair machinery of the testes and, as a result, gametes are generated possessing high levels of DNA fragmentation.

It is also possible that poor compaction of sperm DNA reflects the incom-plete protamination of this material and it is the residual, histone-rich regions of the chromatin that are the sites of DNA damage. If this is the case, and certain regions of the genome are consistently complexed with histones in the male germ line, we might be able to detect DNA damage hot spots in spermatozoa. Such a finding would be of considerable diagnostic value since it would allow the development of probes that specifically target the vulnerable areas of the genome when screening for DNA damage in human spermatozoa. Unfortunately, the literature offers little consensus concerning the identity of histone-rich regions in mammalian spermatozoa. While the obvious candidate genes, the protamines, have been suggested to possess a high histone content on their promoters,[48] other studies have found the histone-rich areas of sperm chromatin to include epsilon and gamma globin genes[49] or retrotransposon-associated LINES and SINES.[50]

A third possibility is that poor DNA compaction may increase the suscep-tibility of sperm DNA to oxidative stress. Thus the ability of protamines to stabilise sperm DNA by creating numerous inter- and intra-molecular disulfide bonds between proximate cysteine residues results in the formation of a highly stabilised chromatin structure that is very resistant to oxidative stress.[51] This protection appears to be particularly dependent on the stability brought about by the cross-linking process, rather than the protamination of the chromatin *per se*. One of the natural models that allow us to draw this conclusion is epididymal maturation. As spermatozoa leave the testes and enter the caput epididymis the sperm chromatin is protaminated, but these proteins have yet to establish the disulfide bridges that will lead to the ultimate stabilisation of this material. The oxidative process that establishes these cross-links is probably ubiquitous among Eutherian mammals and occurs as spermatozoa transit the epididymal tubules, so that mature spermatozoa recovered from the cauda epididymis are highly stabilised, cross-linked cells.[52] Comparison of the sus-ceptibility of spermatozoa from the caput and cauda epididymis to oxidative attack reveals a dramatic difference, with the cross-linked caudal cells being significantly more resistant to this form of stress than their non-cross-linked counterparts recovered from the caput epididymis.[53]

An identical conclusion has been reached by comparing the susceptibility of sperm chromatin from Eutherian and Metatherian mammals to oxidative stress.[54] Thus, while the protamines of all Eutherian mammals possess cysteines that become cross-linked during the process of epididymal maturation, this is not true in marsupials, a majority of which express protamines that do not possess cysteines and, therefore, cannot be cross-linked. Exposure of spermatozoa from the mouse (heavily protaminated and cross-linked), human ($\sim 85\%$ protaminated and cross-linked) and wallaby (fully protaminated and not cross-linked) to oxidative stress revealed highly significant differences in susceptibility of these cells to the genotoxic effects of hydrogen peroxide *in vitro*. The nuclear DNA of wallaby spermatozoa was much more susceptible to peroxide attack than the DNA of either human or murine gametes.[54] Exactly how the stabilisation of sperm chromatin during epididymal transit affords significant protection against a small, highly permeable oxidant such as hydrogen peroxide is unclear at the present time. Limited modelling of protamine–DNA complex formation has been achieved for bull sperm chromatin and the results suggest a highly stable, close-packed, hexagonal lattice of DNA, which is created through an extended network of inter- and intra-molecular disulfide bonds.[55] It is possible that within such highly stabilised structures, the protamine molecules act as sacrificial antioxidants. If this was the case, one would expect the poorly compacted spermatozoa associated with male infertility, which express residual free thiols on their protamines, to offer better protection than the fully oxidised, cross-linked protamines of normal functional cells. However, this is not the case, free thiol expression by human sperm chromatin is positively, not negatively, associated with DNA damage.[56] Thus, while free thiols are excellent radical scavengers, they also possess pro-oxidant properties. In particular, the redox active nature of thiol moieties may allow any reduced, protamine-based cysteine residues to act as redox centres within the sperm chromatin. Such oxido-reductase activity has the potential to induce redox cycling behaviour in the presence of transition metal ions such as Fe(III) and Cu(II), resulting in the production of ROS and the creation of oxidative DNA damage. By contrast, the protective action of fully cross-linked protamine complexes may stem from the ability of these proteins to bind transition metals and quench their redox activity. The close association between sperm DNA and protamines in fully compacted sperm chromatin might therefore serve to isolate the DNA from the damaging, redox-active, metal centres, by acting as a trap for transition metals such as Cu(II).[57]

17.4 Oxidative Stress

The importance of oxidative stress in creating the DNA damage seen in the spermatozoa of male patients is supported by numerous independent lines of evidence (i) the spermatozoa of subfertile patients exhibit high levels of redox activity;[58] (ii) the spermatozoa of such patients also exhibit high levels of DNA damage in a manner that is significantly correlated with the level of spontaneous redox activity exhibited by these cells;[59] (iii) oral antioxidant treatment can reduce the degree of DNA damage exhibited by human spermatozoa;[60] (iv) induction of redox activity in these cells significantly increases the level of DNA damage;[53,61] (v) exposing spermatozoa to exogenously generated reactive oxygen species causes DNA damage in both the nuclear and mitochondrial genomes;[51,62] (vi) infertile patients exhibit significantly elevated levels of oxidised, DNA-base damage (8-hydroxy, 2′-deoxyguanosine)[63] and the 8-OH-dG/dG ratios in such patients correlate with DNA fragmentation as measured by SCSA (Sperm Chromatin Stability Assay).[64] Given the apparent importance of oxidative stress in the aetiology of DNA damage in human spermatozoa, important questions are raised about the origins of the reactive oxygen species (ROS) that cause this damage.

17.5 Origins of Oxidative Stress

Oxidative stress can be created in cells by excessive exposure to ROS or by deficiencies in antioxidant protection. In the case of ROS exposure, two potential sources of these pernicious molecules have been identified: phagocytic leukocytes and the spermatozoa themselves. Every human semen sample is contaminated with leukocytes, the average concentration being around 30×10^3 ml^{-1}.[65] Although this sounds like a low level of contamination, the most common type of leukocyte found in human semen samples, the neutrophil, is 1000-fold more active in generating ROS than a spermatozoon on a cell-for-cell basis. As a result, concentrations of leukocytes well below the threshold for clinical leukocytospermia ($>1 \times 10^6$ ml^{-1}) can generate significant quantities of ROS. Moreover, the neutrophils present in human semen samples appear to be in an activated state because highly significant correlations have been observed between leukocyte concentrations in such material and ROS production, giving 'r' values in the order of 0.8.[65]

Despite the highly significant nature of this correlation, leukocyte contamination is not invariably associated with oxidative stress and impaired sperm function. In a majority of patients, the first time spermatozoa will see leukocytes originating in sites distal to the vas deferens, will be at the moment of ejaculation. At this juncture, the spermatozoa will be protected by the powerful antioxidant properties of seminal plasma.[65,66] Only when the antioxidant protection offered by the male reproductive tract has been overwhelmed will leukocyte contamination have a negative impact on sperm function. Of course, this protection is effectively removed when washed sperm suspensions are prepared for assisted conception therapy. Under these circumstances, free radical-generating leukocytes have ample opportunity to attack the spermatozoa, inducing significant levels of DNA damage and disrupting the fertilizing potential of these cells.[67,68]

Although different publications have variously asserted that the chemiluminescence signals generated by washed human sperm suspensions emanate exclusively from either the spermatozoa or the contaminating leukocytes, the truth is that both sources of ROS are active.[69] In order to resolve the specific contribution made by spermatozoa to oxidative stress in the ejaculate, it is essential that all traces of leukocyte contamination are removed from the sperm suspension. In this context, protocols have been described for both the efficient detection of leukocyte contamination and the selective removal of these cells using paramagnetic particles coated with anti-CD45, the common leukocyte antigen.[70,71]

However, there are very few studies in which these stringent conditions have been met. Where this has been achieved, the results unequivocally identify defective spermatozoa as a second source of redox activity,[69] the intensity of which varies inversely with the functional competence of the gametes.[73]

The cellular origins of the ROS generated by defective human spermatozoa is the subject of intense speculation. Although aberrant production of free radicals by mitochondria is a possible source of oxidative stress in the spermatozoa of infertile men, there is no convincing data to support this contention. Another possibility, for which there is considerable evidence, is that the spermatozoa generating high levels of ROS are the product of a dysfunctional spermiogenetic process that results in the release of defective cells exhibiting morphological defects in the sperm mid-piece, characterised by the retention of excess residual cytoplasm.[74,75] During the terminal stages of spermiogenesis, Sertoli cells actively remove the residual cytoplasm from differentiating spermatozoa, just before these cells are released from the germinal epithelium. In most mammals, a majority of the cytoplasm that remains after spermiogenesis is remodelled into a discrete, spherical, cytoplasmic droplet that is discharged from the sperm tail during epididymal transit. Intriguingly, human spermatozoa have lost the ability to create and shed a cytoplasmic droplet. In these cells, any residual cytoplasm left after spermiation snaps back into the neck region of the cell and remains as testimony to the inadequate spermatogenic process that brought it into being. The presence of such excess residual cytoplasm has been correlated with ROS production by several independent groups.[75–78] One suggested mechanism by which such residual cytoplasm might induce ROS production is through the provision of excess substrate to a putative NADPH oxidase on the sperm surface.[79]

The proposed relationship between cytoplasmic retention and ROS production stems from the elevated presence of several cytoplasmic enzymes in such defective spermatozoa, including lactic acid dehydrogenase, SOD, creatine kinase and glucose-6-phosphate dehydrogenase. Most of these enzymes are simply passive markers of the cytoplasmic space. The pathological entity is glucose-6-phosphate dehydrogenase.[75] This enzyme controls the rate of glucose oxidation through the hexose monophosphate shunt and the latter, in turn, generates the NADPH needed to fuel ROS production by a putative NADPH oxidase enzyme such as Nox 5; a free radical–generating oxidase recently detected in the male germ line.[79] By removing most of the sperm cytoplasm during spermiogenesis, the testes ensure that these cells are only able to generate a limited supply of NADPH; just enough to meet the needs of the protective glutathione cycle and support the ROS-dependent elements of sperm capacitation.[80] However, if excess residual cytoplasm is retained because of mistakes during spermiogenesis, then there is the potential to generate excess NADPH that will, in turn, fuel the production of ROS and thereby damage the functional competence of these cells.

A third possible cause of free radical generation by human sperm populations is the presence of substances that can redox cycle within the sperm cytosol and generate ROS. A variety of phenolic compounds have been found to induce DNA damage in human spermatozoa as a result of redox-cycling mechanisms that are amenable to antioxidant suppression. The thyroid hormones, triiodothyronine (T3) and L-thyroxine sodium salt (T4), and the neurotransmitter noradrenaline (NA) have all been shown to act in this way.[81] Similarly, a variety of phenolic estrogens have been found to induce DNA damage in human spermatozoa *via* oxidative mechanisms involving the generation of hydrogen peroxide.[82] In addition, potentially redox-active metals such as nickel have been shown to redox cycle in male germ cells, inducing high levels of DNA damage and creating dominant lethal effects following mating.[83] Whether exposure to natural compounds, xenobiotics or heavy metals is involved in mediating the oxidative DNA damage seen in the spermatozoa of male patients is a key question that has yet to be resolved.

17.6 Conclusions

DNA damage in the male germ line has been associated with impaired fertilisation, poor embryo quality, low blastocyst formation rates and increased incidences of abortion. The origins of DNA damage in the male germ line are the subject of intense speculation. Although both paternal age and xenobiotic exposure are recognised as important causative factors, the precise nature of the DNA damage detected in human spermatozoa and the mechanisms responsible for its induction are unknown. At present, a majority of the evidence implicates oxidative stress as a major, but not the only, contributor to DNA damage in human spermatozoa. Such oxidative stress may arise as a result of redox-cycling phenomena, involving natural compounds (*e.g.*, catechol estrogens) or xenobiotics (*e.g.*, endocrine disruptors). In addition, defects in spermiogenesis leading to the retention of excess residual cytoplasm by spermatozoa have also been associated with ROS generation by these cells and impaired sperm function. Leukocytic infiltration and defects in the antioxidant protection afforded to spermatozoa during their transit through the male reproductive tract may also contribute to the levels of oxidative stress suffered by the male germ line. Given this background, our current research is addressing the causes of oxidative stress in human spermatozoa with a view to the ultimate development of strategies for its amelioration.

Acknowledgments

We are grateful to Jodi Powell for the organisation of our semen donor panel, Ms Jordana Mulder for technical assistance and the ARC Centre of Excellence in Biotechnology and Development for financial support.

References

1. R.I. McLachlan and D.M. de Kretser, *Med. J. Aust.*, 2001, **174**, 116.
2. M.G.R. Hull, C.M.A. Glazener, N.J. Kelly, D.I. Conway, P.A. Foster, R.A. Hunton, C. Coulson, P.A. Lambert, E.M. Watt and K.M. Desai, *BMJ*, 1985, **291**, 1693.

3. E. Carlsen, A. Giwercman, N. Keiding and N.E. Skakkebaek, *BMJ*, 1992, **305**, 609.

4. D.S. Irvine, E. Cawood, D. Richardson, E. MacDonald and R.J. Aitken, *BMJ*, 1996, **312**, 461.

5. J. Auger, J.M. Kunstmann, F. Czyglik and P. Jouannet, *N. Engl. J. Med.*, 1995, **332**, 281.

6. M. Ulstein, A. Irgens and L.M.S. Irgens, *Acta Obstet. Gynecol. Scand.*, 1999, **78**, 332.

7. R.J. Aitken, *Reprod. Fertil. Dev.*, 2004, **16**, 655.

8. J. Toppari, A.M. Haavisto and M. Alanen, *Cad. Saude Publ.*, 2002, **18**, 413.

9. R.M. Sharpe, *Int. J. Androl.*, 2003, **26**, 2.

10. R.J. Aitken, P. Koopman and S.E. Lewis, *Nature*, 2004, **432**, 48.

11. A. Pettersson, M. Kaijser, L. Richiardi, J. Askling, A. Ekbom and O. Akre, *Int. J. Cancer*, 2004, **109**, 941.

12. Z. Chen, L. Robison, R. Giller, M. Krailo, M. Davis, K. Gardner, S. Davies and X.O. Shu, *Cancer*, 2005, **103**, 1064.

13. J.A. Knight and L.D. Marrett, *J. Occup. Environ. Med.*, 1997, **39**, 333.

14. X.O. Shu, P. Stewart, W.Q. Wen, D. Han, J.D. Potter, J.D. Buckley, E. Heineman and L.L. Robison, *Cancer Epidemiol. Biomarkers Prev.*, 1999, **8**, 783.

15. K.A. Hill, V.L. Buettner, A. Halangoda, M. Kunishige, S.R. Moore, J. Longmate, W.A. Scaringe and S.S. Sommer, *Environ. Mol. Mutagen.*, 2004, **43**, 110.

16. J.G. Sun, A. Juriscova and R.F. Casper, *Biol. Reprod.*, 1997, **56**, 602.

17. D.S. Irvine, J.P. Twigg, E.L. Gordon, N. Fulton, P.A. Milne and R.J. Aitken, *J. Androl.*, 2000, **21**, 33.

18. D.P. Evenson, L.K. Jost, D. Marshall, M.J. Zinaman, E. Clegg, K. Purvis, P. de Angelis and O.P. Claussen, *Hum. Reprod.*, 1999, **14**, 1039.

19. J.L. Fernandez, L. Muriel, M.T. Rivero, V. Goyanes, R. Vazquez and J.G. Alvarez, *J. Androl.*, 2003, **24**, 59.

20. S.E. Lewis, *Hum. Fertil. (Camb).*, 2002, **5**, 102.

21. R.J. Aitken and C.G. Krausz, *Reproduction*, 2001, **122**, 497.

22. D. Sakkas, F. Urner, D. Bizzaro, G. Manicardi, P.G. Bianchi, Y. Shoukir and A. Campana, *Hum. Reprod.*, 1998, **13**(Suppl. 4), 11.

23. E.H. Duran, M. Morshedi, S. Taylor and S. Oehninger, *Hum. Reprod.*, 2002, **17**, 3122.

24. I.D. Morris, S. Ilott, L. Dixon and D.R. Brison, *Hum. Reprod.*, 2002, **17**, 990.

25. D.T. Carrell, A.L. Wilcox, L. Lowy, C.M. Peterson, K.P. Jones, L. Erickson, B. Campbell, D.W. Branch and H.H. Hatasaka, *Obstet. Gynecol.*, 2003, **101**, 1229.

26. S. Loft, T. Kold-Jensen, N.H. Hjollund, A. Giwercman, J. Gyllemborg, E. Ernst, J. Olsen, T. Scheike, H.E. Poulsen and J.P. Bonde, *Hum. Reprod.*, 2003, **18**, 1265.

27. R.A. Saleh, A. Agarwal, E.A. Nada, M.H. El-Tonsy, R.K. Sharma, A. Meyer, D.R. Nelson and A.J. Thomas, *Fertil. Steril.*, 2003, **79**(Suppl. 3), 1597.

28. M. Bungum, P. Humaidan, M. Spano, K. Jepson, L. Bungum and A. Giwercman, *Hum. Reprod.*, 2004, **19**, 1401.

29. M.R. Virro, K.L. Larson-Cook and D.P. Evenson, *Fertil. Steril.*, 2004, **81**, 1289.

30. C.G. Fraga, P.A. Motchnik, A.J. Wyrobek, D.M. Rempel and B.M. Ames, *Mutat. Res.*, 1996, **351**, 199.

31. B.T. Ji, X.O. Shu, M.S. Linet, W. Zheng, S. Wacholder, Y.T. Gao, D.M. Ying and F. Jin, *J. Natl. Cancer. Inst.*, 1997, **89**, 238.

32. N.P. Singh, C.H. Muller and R.E. Berger, *Fertil. Steril.*, 2003, **80**, 1420.

33. J.F. Crow, *Nat. Rev. Genet.*, 2000, **1**, 40.

34. M.D. Anway, A.S. Cupp, M. Uzumcu and M.K. Skinner, *Science*, 2005, **308**, 1466.

35. B. Sotolongo, T.T. Huang, E. Isenberger and W.S. Ward, *J. Androl.*, 2005, **26**, 272.

36. M. Muratori, M. Maggi, S. Spinelli, E. Filimberti, G. Forti and E. Baldi, *J. Androl.*, 2003, **24**, 253.

37. M.H. Moustafa, S.K. Sharma, J. Thornton, E. Mascha, M.A. Abdel-Hafez, A.J. Thomas, Jr. and A. Agarwal, *Hum. Reprod.*, 2004, **19**, 129.

38. C.M. McVicar, N. McClure, K. Williamson, L.H. Dalzell and S.E. Lewis, *Fertil. Steril.*, 2004, **81**(Suppl. 1), 767.

39. D. Sakkas, E. Mariethoz, G. Manicardi, D. Bizzaro, P.G. Bianchi and U. Bianchi, *Rev. Reprod.*, 1999, **4**, 31.

40. L. Marcon and G. Boissonneault, *Biol. Reprod.*, 2004, **70**, 910.

41. G. Boissonneault, *FEBS Lett.*, 2002, **514**, 111.

42. A. Celeste, S. Petersen, P.J. Romanienko, O. Fernandez-Capetillo, H.T. Chen, O.A. Sedelnikova, B. Reina-San-Martin, V. Coppola, E. Meffre, M.J. Difilippantonio, C. Redon, D.R. Pilch, A. Olaru, M. Eckhaus, R.D. Camerini-Otero, L. Tessarollo, F. Livak, K. Manova, W.M. Bonner, M.C. Nussenzweig and A. Nussenzweig, *Science*, 2002, **296**, 922.

43. R. Balhorn, S. Reed and N. Tanphaichitr, *Experientia*, 1988, **44**, 52.

44. L. Mengual, J.L. Ballesca, C. Ascaso and R. Oliva, *J. Androl.*, 2003, **24**, 438.

45. C. Cho, H. Jung-Ha, W.D. Willis, E.H. Goulding, P. Stein, Z. Xu, R.M. Schultz, N.B. Hecht and E.M. Eddy, *Biol. Reprod.*, 2003, **69**, 211.

46. T.E. Schmid, A. Kamischke, H. Bollwein, E. Nieschlag and M.H. Brinkworth, *Hum. Reprod.*, 2003, **18**, 1474.

47. D. Sakkas, G.C. Manicardi, M. Tomlinson, M. Mandrioli, D. Bizzaro, P.G. Bianchi and U. Bianchi, *Hum. Reprod.*, 2000, **15**, 1112.

48. S.M. Wykes and S.A. Krawetz, *J.Biol.Chem.*, 2003, **278**, 29471.

49. M. Gardiner-Garden, M. Ballesteros, M. Gordon and P.P.L. Tam, *Mol. Cell. Biol.*, 1998, **18**, 3350.

50. C. Pittoggi, L. Renzi, G. Zaccagnini, D. Cimini, F. Degrassi, R. Giordano, A.R. Magnano, R. Lorenzini, P. Lavia and C. Spadafora, *J. Cell. Sci.*, 1999, **112**, 3537.

51. D.E. Sawyer, B.G. Mercer, A.M. Wiklendt and R.J. Aitken, *Mutat. Res.*, 2003, **529**, 21.

52. D.P. Evenson, R.K. Baer and L.K. Jost, *Mol. Reprod. Dev.*, 1989, **1**, 283.
53. H. Chen, M.P. Cheung, P.H. Chow, A.L. Cheung, W. Liu and W.S.O. Oiscorrect, *Reproduction*, 2002, **124**, 491.
54. L.E. Bennetts and R.J. Aitken, *Mol. Reprod. Dev.*, 2005, **71**, 77.
55. I.D. Vilfan, C.C. Conwell and N.V. Hud, *J. Biol. Chem.*, 2004, **279**, 20088.
56. A. Zini, K.M. Kamal and D. Phang, *Urology*, 2001, **58**, 80.
57. R. Liang, S. Senturker, X. Shi, W. Bal, M. Dizdaroglu and K.S. Kasprzak, *Carcinogenesis*, 1999, **20**, 893.
58. R.J. Aitken and J.S. Clarkson, *J. Reprod. Fertil.*, 1987, **83**, 459.
59. D.S. Irvine, J.P. Twigg, E.L. Gordon, N. Fulton, P.A. Milne and R. J. Aitken, *J. Androl.*, 2000, **21**, 33.
60. E. Greco, M. Iacobelli, L. Rienzi, F. Ubaldi, S. Ferrero and J. Tesarik, *J. Androl.*, 2005, **26**, 349.
61. J. Twigg, N. Fulton, E. Gomez, D.S. Irvine and R.J. Aitken, *Human Reprod.*, 1998, **13**, 1429.
62. R.J. Aitken, E. Gordon, D. Harkiss, J.P. Twigg, P. Milne, Z. Jennings and D.S. Irvine, *Biol. Reprod.*, 1998, **59**, 1037.
63. H. Kodoma, R. Yamaguchi, J. Fukada, H. Kasai and T. Tanaka, *Fertil. Steril.*, 1997, **68**, 519.
64. I. Oger, C. Da Cruz, G. Panteix and Y. Menezo, *Zygote*, 2003, **11**, 367.
65. R.J. Aitken, K. West and D. Buckingham, *J. Androl.*, 1994, **15**, 343.
66. R. Jones, T. Mann and R.J. Sherins, *Fertil. Steril.*, 1979, **31**, 531.
67. R.J. Aitken, D.W. Buckingham, J. Brindle, E. Gomez, H.W. Baker and D.S. Irvine, *Hum. Reprod.*, 1995a, **10**, 2061.
68. J.P. Twigg, D.S. Irvine, P. Houston, N. Fulton, L. Michael and R.J. Aitken, *Molec. Human Reprod.*, 1998, **4**, 439.
69. R.J. Aitken and K. West, *Int. J. Androl.*, 1990, **13**, 433.
70. R.J. Aitken, D.W. Buckingham, K. West and J. Brindle, *Am. J. Reprod. Immunol.*, 1996, **35**, 541.
71. C. Krausz, K. West, D. Buckingham and R.J. Aitken, *Fertil. Steril.*, 1992, **57**, 1317.
72. R.J. Aitken, D. Buckingham, K. West, F.C. Wu, K. Zikopoulos and D.W. Richardson, *J. Reprod. Fertil.*, 1992, **94**, 451.
73. E. Gomez, D.S. Irvine and R.J. Aitken, *Int. J. Androl.*, 1998, **21**, 81.
74. B. Rao, J.C. Soufir, M. Martin and G. David, *Gamete. Res.*, 1989, **24**, 127.
75. E. Gomez, D.W. Buckingham, J. Brindle, F. Lanzafame, D.S. Irvine and R.J. Aitken, *J. Androl.*, 1996, **17**, 276.
76. E. Gil-Guzman, M. Ollero, M.C. Lopez, R.K. Sharma, J.G. Alvarez, A.J. Thomas, Jr. and A. Agarwal, *Human Reprod.*, 2001, **16**, 1922.
77. M. Ollero, E. Gil-Guzman, M.C. Lopez, R.K. Sharma, A. Agarwal, K. Larson, D. Evenson, A.J. Thomas, Jr. and J.G. Alvarez, *Human Reprod.*, 2001, **16**, 1912.
78. A. Zini, M.K. O'Bryan, L. Israel and P.N. Schlegel, *Urology*, 1998, **51**, 464.
79. B. Banfi, G. Molnar, A. Maturana, K. Steger, B. Hegedus, N. Demaurex and K.H. Krause, *J. Biol. Chem.*, 2001, **276**, 37594.

80. R.J. Aitken, D. Harkiss, W. Knox, M. Paterson and D.S. Irvine, *J. Cell Sci.*, 1998, **111**, 645.
81. M.M. Dobrzynska, A. Baumgartner and D. Anderson, *Mutagenesis*, 2004, **19**, 325.
82. D. Anderson, T.E. Schmid, A. Baumgartner, E. Cemeli-Carratala, M.H. Brinkworth and J.M. Wood, *Mutat. Res.*, 2003, **544**, 173.
83. K. Doreswamy, B. Shrilatha, T. Rajeshkumar and Muralidhara, *J. Androl.*, 2004, **25**, 996.

CHAPTER 18

Advances in the Direct Measurements of Partial Chromosomal Duplication, Deletions and Breaks in Human and Murine Sperm by Sperm FISH

ANDREW J. WYROBEK,[a] THOMAS E. SCHMID,[a,b] JACK BISHOP[c] AND FRANCESCO MARCHETTI[a]

[a] Biosciences Directorate, Lawrence Livermore National Laboratory, PO Box 808, Livermore CA 94550
[b] School of Public Health, University of California Berkeley, Berkeley CA
[c] National Institute of Environmental Health Sciences, Research Triangle Park, NC

18.1　Introduction

Abnormal pregnancy outcomes are relatively common in humans and defects reaching birth can be both traumatic and expensive for family and society.[1,2] Every year in the United States, about 2 million pregnancies are lost before the 20th week of gestation, about 7% of newborns have low birth weight, and 5% of babies are born with a birth defect.[1] More than half of these birth defects are associated with significant health or viability consequences for the affected offspring. Although the types of abnormal reproductive outcomes are well characterised, their relative maternal and paternal contributions and aetiology are poorly understood.[3]

Chromosome abnormalities are the major types of genomic defects associated with abnormal pregnancy outcomes and they are generally detrimental to the viability, development and health of human embryos and offspring. About 1% of newborns carry numerical or structural chromosome defects.[4,5] Furthermore, half of all spontaneous abortions and a major fraction of developmental and morphological birth defects are associated with *de novo* chromosomal abnormalities.[6] Chromosomal defects that arise during gametogenesis of either parent can result in constitutive defects in offspring, while those that arise shortly after fertilisation result in mosaic offspring.

Aneuploidy and structural aberrations are two distinct classes of heritable chromosome abnormality with differing aetiologies and consequences on pregnancy outcomes. Aneuploidy is detected in $\sim 26\%$ of spontaneous abortions and in $\sim 0.3\%$ of newborns.[7] The paternal contribution to aneuploidies depends on the chromosome involved and ranges from about 10 to 100%.[8] Autosomal trisomy appears to be predominantly maternal in origin (*e.g.*, trisomy 21, 18, 16, 13), while sex chromosomal aneuploidies (*e.g.*, 45, X, 47, XXY, 47, XYY, 47, XXX) have a substantial paternal contribution.[9]

Structural chromosomal aberrations are detected in $\sim 6\%$ of spontaneous abortions, and 0.6% of all live births have a major chromosomal abnormality or other serious genetic disorders.[10,11] Conventional cytogenetics at birth will miss smaller genomic defects, as well as any that are lethal before birth. It has been estimated that 80% of structural aberrations detected in foetuses or newborns are paternal in origin,[11,12] and most of the *de novo* gene mutations for ~ 20 autosomal diseases are also predominantly of paternal origin.[13] Chromosomal aberrations in sperm are thought to arise by different mechanisms from those that give rise to aneuploidy, because they have the prerequisite of DNA strand breakage followed by the formation of a fragment or a rearrangement between two chromosomes.[14]

The aetiologies of numerical and chromosomal defects in sperm are generally unknown, and they have been attributed to chance, genetic susceptibility, mosaicism and prior exposures to DNA damaging agents. Decades of testing in rodent breeding assays have identified various toxicants that, when given to males or females before or after mating, can have profoundly deleterious effects on reproduction; these include infertility, lethality during development, malformations and cancer among offspring.[15,16] Of special concern are long-term, low-dose exposures to mutagens such as smoking and air pollution that affect large numbers of individuals, or short exposures of smaller groups to very high doses of mutagens such as cancer chemotherapies.[17,18] Each year, more than 20,000 children or young adults of reproductive age in the United States are treated for cancer with combination chemotherapies, many of which contain agents known to be mutagenic, clastogenic or aneugenic in model systems.[17,19] As cancer treatments become more effective and more patients regain fertility after treatment, there is the growing concern that chemotherapy may induce damage in germ cells of survivors and increase their risk for abnormal reproductive outcomes. However, there are very few human data for combination regimens, and only a few of the component drugs have been individually tested in laboratory animals. In addition to medical exposures, several environmental exposures, lifestyle factors and certain occupational exposures are known to affect sperm quality,[20] raising concern that these exposures may also affect the genetic integrity of the germ cells.

There is increasing evidence that paternally-mediated abnormal reproductive outcomes are a consequence of abnormal reproductive physiology, predisposing genetic factors,[9] past and present male environmental exposures[21] or random errors that occur during spermatogenesis.[22] Elucidating the relative contribution of these factors from epidemiological surveys of affected offspring is difficult because sample sizes of offspring with specific defects are small and there is prenatal selection against certain genomically defective embryos. Considerable progress has been made in developing effective biomarkers to detect genomic damage directly in sperm where greater statistical power can be attained by detecting changes in the frequencies of defective sperm of small numbers of well-characterised men.[18]

The first strategies for detecting chromosomally defective sperm using FISH were developed in the early 1990s for the detection of aneuploid sperm.[18] Aneuploidy in sperm can arise from inherent susceptibilities to malsegregation,[23] errors due to disturbances in recombination,[24] or perturbations of kinetochores and microtubules.[25] Sperm aneuploidy can be induced by cancer chemotherapy[26] as well as by various other factors.[18]

This chapter overviews the recent advances in the field of sperm FISH for detecting structural chromosomal aberrations in human and mouse sperm, and contrasts these hybridisation strategies to the earlier methods for detecting sperm aneuploidy. Methods for detecting chromosome aberrations in sperm share the principle that two or more loci on a single target chromosome are interrogated simultaneously using two or more fluorescent DNA probes. The numbers detected and their spatial relationships indicate whether chromosomal damage is present. We also compare here, baseline frequencies of defective sperm among adult males (humans and mice), and provide early examples of their applications to the study of genetic factors, lifestyle and exposures on the production of sperm with partial chromosomal duplication, deletions and breaks.

18.2   FISH Methods for Detecting Human Sperm Carrying Structural Chromosomal Aberrations

The hamster-egg cytogenetic technique was the first to provide a direct assessment of human sperm cytogenetics including the detection of sperm with chromosomal aberrations.[27–29] However, it was a very difficult and inefficient procedure, rarely yielding more than 20 sperm karyotypes per semen analysis.

Van Hummelen and colleagues[14] in our lab developed the first FISH method for detecting chromosomal aberrations in human sperm, and we used the hamster-egg cytogenetic technique as the reference standard in its development. This sperm FISH method, called the AM-8 assay, uses probes specific for the centromeric and telomeric regions of chromosome 1p plus a centromeric probe for chromosome 8 [Figure 1(A)]. With the AM-8 assay, it is possible to detect sperm carrying terminal duplications and deletions in chromosome 1p,

A. AM-related sperm FISH assays
AM8 , AM16 , and AM18

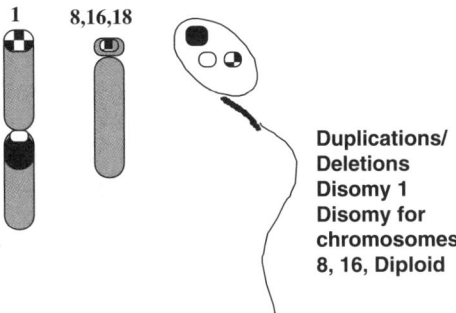

Duplications/
Deletions
Disomy 1
Disomy for
chromosomes
8, 16, Diploid

B. ACM sperm FISH Assay

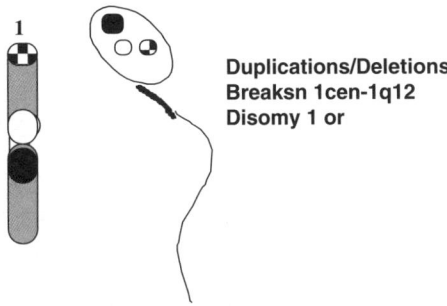

Duplications/Deletions
Breaksn 1cen-1q12
Disomy 1 or

C. 9*9-6-Y sperm FISH Assay

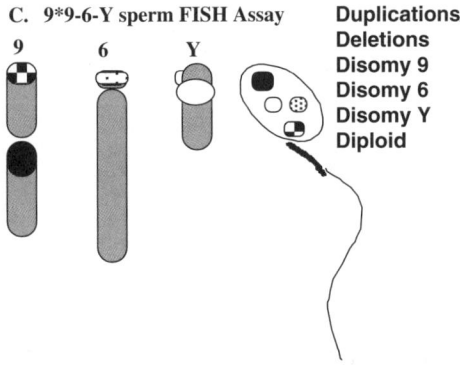

Duplications/
Deletions
Disomy 9
Disomy 6
Disomy Y
Diploid

Figure 1   *Sperm FISH labelling strategies for detecting human sperm that carry chromosomal aberrations (A) AM-related assays (AM-8, AM-16 and AM-18) for detection of deletions and duplications of chromosome 1p, disomies involving chromosomes 1 and others, as well as diploidy in human sperm; (B) ACM assay for detection of deletions and duplications of chromosome 1p, breaks in the classical satellite region of 1q, as well as numerical abnormalities in human sperm; (C) 9/9-6-Y assay for detection of partial duplications and deletions of chromosome 9, disomies involving chromosomes 9, 6 and Y, diploidy in human sperm*

aneuploidy involving chromosome 1 and 8 and sperm diploidy. Initial verification based upon detection of meiotic segregation products by a reciprocal translocation carrier, demonstrated a strong correlation between the frequencies of sperm carrying structural abnormalities identified by FISH *vs.* the frequencies of abnormal metaphases identified by the hamster-egg cytogenetic technique.[30] The numerous types of specific chromosomal aberrations detected by the AM-8 sperm FISH, assay are shown in Table 1.

Baumgartner *et al.*[31] later modified this FISH strategy to detect sperm carrying terminal duplications and deletions in chromosome 1 using chromosome 16 as a reference chromosome (AM-16, Figure 1A). This new FISH method was also substantiated by evaluation of the baseline frequencies of abnormal sperm in the semen of healthy donors who had been previously evaluated by the hamster-egg cytogenetic technique.

Sloter *et al.*[5] made a substantial modification of the AM hybridisation strategy to detect chromosomal breaks within chromosome 1 in addition to duplications and deletions of 1p. This new method [Figure 1(B)] utilised DNA probes specific for three different regions of chromosome 1 to detect abnormal sperm that carry breaks in the alpha satellite (1cen-1q12) region of

Table 1 *Baseline frequencies of the human sperm carrying chromosomal aberrations and numerical defects detected by various sperm FISH aberration assays*

Chromosome target Sperm FISH method Sperm genotype	1p AM-8[a] n=3[e]	1p AM-16[b] n=4[e]	1p and 1q ACM[c] n=4[e]	9q 9/9-6-Y[d] N=3[f]
Structural aberrations				
Duplication, centromeric	2.1 ± 0.9	6.5 ± 2.6	0.9 ± 0.4	4.6 ± 1.3
Deletion, centromeric	0.7 ± 0.8	1.0 ± 1.2	0.8 ± 0.3	9.3 ± 5.3
Duplication, telomeric	3.2 ± 1.0	3.8 ± 1.0	4.5 ± 0.5	8.6 ± 3.5
Deletion, telomeric	2.9 ± 2.1	1.8 ± 1.5	4.1 ± 1.3	6.0 ± 2.2
Numerical abnormalities				
Disomy 8	1.9 ± 1.3			0.1 ± 0.1
Nullisomy 8	4.4 ± 2.8			0
Disomy 16		3.5 ± 1.3		0.3 ± 0.2
Nullisomy 16		3.5 ± 1.3		1.6 ± 0.5
Disomy 1	1.7 ± 1.3	5.0 ± 2.4	8.9 ± 0.7	0.3 ± 0.2
Nullisomy 1	0.4 ± 0.3	1.3 ± 1.5	1.2 ± 0.8	1.6 ± 0.5
Disomy 6				4.0 ± 1.0
Nullisomy 6				
Disomy Y				3.6 ± 1.7
Nullisomy Y				

Notes: Per $10^4 \pm$ SE; no entry means that the endpoint was not tested.
[a] Van Hummelen *et al.*, 1996.
[b] Baumgartner *et al.*, 1999.
[c] Sloter *et al.*, 2000.
[d] Bosch *et al.*, 2003.
[e] Healthy, non-smoking men.
[f] Healthy men, 1 smoker, 2 non-smokers.

chromosome 1, and the usual duplications and deletions of the 1p and 1cen regions as well as aneuploidy of chromosome 1. Again, the corresponding baseline frequencies of sperm with structural defects were substantiated by comparison with data obtained by the hamster-egg cytogenetic technique on semen from the same men.

In 2003, Bosch and colleagues[32] developed a four-colour sperm-FISH assay to detect chromosomal aberrations involving chromosome 9 (9/9-6-Y assay). They utilised DNA probes for the centromeric and subtelomeric regions of chromosome 9q as well as centromeric probes for chromosomes 6 and Y. With this probe combination [Figure 1(C)] they were able to detect duplications and deletions in chromosome 9q, aneuploidies involving chromosomes 6, 9 and Y and sperm diploidies.

Also in 2003, Liu and colleagues[33] reported a minor adaptation of the AM hybridisation strategy developed by Van Hummelen[14] by substituting the chromosome 8 centromeric probe for one for chromosome 18 (AM-18, Figure 1A).

18.3 FISH Methods for Detecting Mouse Sperm Carrying Chromosome Structural Aberrations

To date, only one multi-colour FISH method has been developed for the detection of chromosomal aberrations in an animal model. Hill and colleagues[34] developed a labelling strategy for detection of chromosomal aneuploidy and breakage in murine sperm. The assay employs a combination of DNA probes for the centromeric (C) and telomeric (T) regions of chromosome 2q plus a probe for chromosome 8 to detect three types of damage (1) duplications and deficiencies involving chromosome 2q; (2) aneuploidies involving chromosome 2 and 8; and (3) sperm diploidy. Table 2 lists the baseline frequencies of sperm with structural and numerical chromosomal abnormalities in healthy young adult B6C3F1 male mice. Sperm carrying structural aberrations involving chromosome 2 were more

Table 2 *Baseline frequencies of mouse sperm with chromosomal aberrations and numerical defects using the mouse CT8 assay*

Sperm genotype	Per 10^4 sperm $\pm$ SE
Structural aberrations (chr. 2)	
Duplication, centromeric	0.4 ± 0.2
Deletion, centromeric	0.6 ± 0.3
Duplication, telomeric	0.6 ± 0.3
Deletion, telomeric	0.9 ± 0.2
Numerical abnormalities	
Disomy 2	0.1 ± 0.1
Nullisomy 2	0
Disomy 8	0.3 ± 0.2
Nullisomy 8	1.6 ± 0.5
Diploidy	5.9 ± 1.1
Complex abnormalities	0.1 ± 0.1

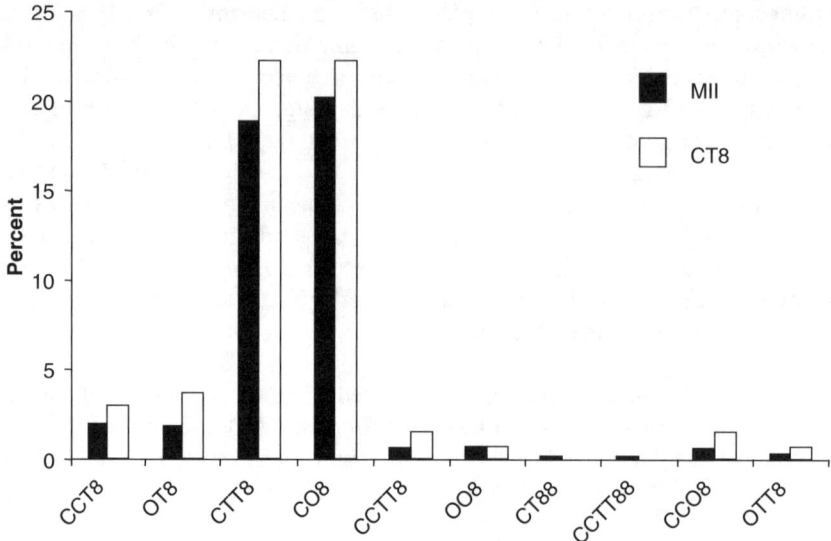

Figure 2 *Meiotic confirmation of percentages of sperm with structural and numerical abnormalities in T(2;14) translocation carrier mice as detected by the mouse CT8 assay and by analysis of metaphase II (MII) spermatocytes. 0=deficiencies of a marker chromosome, C=chromosome 1 centromere, T=chromosome 1 telomere and 8=chromosome 8 (adapted from reference 34).*

common than sperm aneuploidy for chromosomes 2 and 8 combined. Diploid sperm were the most common anomaly found in mouse sperm. Sperm with duplications or deletions of the centromeric and telomeric region of chromosome 2 occurred at similar frequencies. The validity of results from the CT8 assay were corroborated by strong correlations between the frequencies of various chromosome anomalies seen by sperm FISH with those predicted from the analysis of metaphase II (MII) chromosomes in sperm and spermatocytes from T(2;14) translocation carriers, which exhibit elevated frequencies of sperm with chromosomal abnormalities involving chromosome 2 (Figure 2). As shown in Figure 2, there was good agreement between the MII predictions and the observations made by the CT8 sperm assay for all the FISH phenotypes. These results demonstrate that the CT8 assay reliably detects mouse sperm with structural aberrations in target chromosome 2. The mouse CT8 assay provides the first robust rodent screen for potential male germ cell aneugens and clastogens, but additional FISH assays are needed for the sperm of rat and other species.

18.4 Mechanisms of Induction of Chromosomal Aberrations Detectable by Sperm FISH

Numerical and structural chromosomal defects can arise in male germ cells spontaneously (*i.e.*, *de novo*), or after exposure to mutagenic agents. The probability of mutagenic exposures producing defective sperm depends on the susceptibility of the exposed germ-cell stage. The type of chromosomal defect, and thus the optimal assay for its detection, varies with germ-cell stage (Table 3). Stage sensitivity to exposure is typically assessed by controlling the time between exposure and sperm sampling. For example, when assessing the effects of exposure on meiosis, semen samples are obtained more than 40 days after exposure in humans (more than 21 days in the mouse) to represent sperm that were in meiosis or before during the time of exposure.

Sperm aneuploidy arises from segregation errors that occur during either of the two meiotic divisions. Most of the chemicals tested for their effect on sperm aneuploidy using sperm FISH methods are primarily active during meiosis or before meiosis.[17,18] Sperm carrying duplications and deletions may arise before or during meiosis, that is, after at least one cell division has occurred, through either unequal crossing over or breakage. To be detectable by sperm FISH, the breakage/exchange must occur between the two chromosomal regions for which fluorescent probes are being used. Recent data with the CT8 assay[35] showed etoposide-induced chromosomal damage in spermatogonial stem cells that resulted in sperm with duplications and deficiencies. Such sperm are probably products of the meiotic malsegregation of translocations induced in spermatogonia or stem cells (Figure 3).

The post-meiotic period, when maturing sperm undergo major nuclear and morphological changes,[36,37] is perhaps the most vulnerable phase of spermatogenesis for the induction of DNA lesions that can be transmitted by the sperm. This vulnerability occurs, in part, because this is when spermatids become repair deficient.[38,39] Sperm lesions induced in this period may accumulate and be transmitted to the egg where they have the potential to be converted into chromosomal aberrations if improperly repaired.[40] The ACM assay,[5] which is currently the only human sperm-FISH assay that can detect chromosomal breaks (Table 3), has been used to detect damage induced during this critical post-meiotic period.

Table 3 *Germ cell stage sensitivity of chromosomal defects and optimal sperm FISH assays for their detection*

Chromosomal defect	Sensitive spermatogenic stage	Optimal sperm FISH assay
Breaks	Post-meiosis	ACM assay
Aneuploidy	Meiosis	ACM assay, AM assays, aneuploidy assays[18]
Duplications/deletions, products of reciprocal translocations	Stem cells, meiosis	ACM assay, AM assays

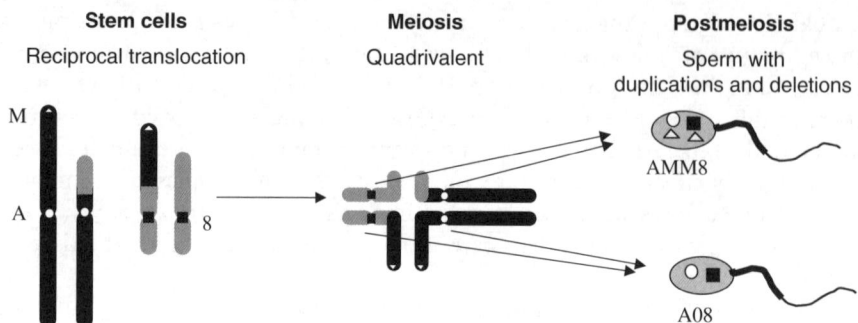

Figure 3   *Sperm products predicted from stem cells carrying reciprocal translocations. Reciprocal translocations in stem cells form quadrivalents in meiosis that undergo meiotic segregation and generate sperm with partial chromosomal duplications and deletions*

18.5   Interspecies Comparison of Baseline Frequencies of Chromosomal Aberrations

The genome-wide baseline frequencies of sperm with duplication and deletions in young adult men and mice (Tables 1 and 2) were compared by using the ratio of the percentage of the genome that was interrogated between the two probes on the target chromosome. For example, chromosome 2 of the mouse represents $\sim 6.4\%$ of the male haploid genome,[41] therefore the number of sperm with duplications and deletions as detected by the CT8 assay were multiplied by 15.6 (*i.e.*, 100/6.4) to obtain the genome-wide estimate. As shown in Figure 4, the genome-wide frequency of human sperm carrying chromosome structural aberrations is estimated to be more than ~ 6-fold higher than that for mice. These results are in agreement with the notion that humans have higher incidences of chromosome imbalances in their germ cells compared with other mammalian and non-mammalian species,[9] but further data on additional target chromosomes are needed for both mouse and human sperm.

18.6   Applications of the Novel Sperm FISH Methods for Detecting Sperm Carrying Chromosomal Aberrations

These FISH methods for detecting human sperm with chromosomal aberrations have been used in pilot investigations to examine the effects of chemotherapy and of age, the frequency of unbalanced meiotic products of reciprocal translocation carriers, and differences between fertile and infertile men. Brief summaries of these studies are presented below.

Van Hummelen and colleagues[42] applied the AM-8 assay to semen from cancer patients to investigate the induction and persistence of defective sperm after treatment. They surveyed three patients before, during and after NOVP

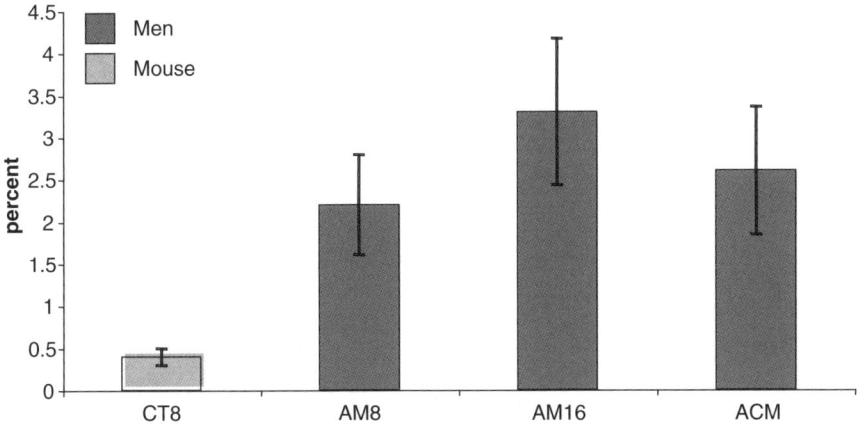

Figure 4 *Estimated genome-wide frequencies of human and mouse sperm carrying chromosome structural aberrations based on sperm FISH analyses*

chemotherapy and four patients 3–25 years after MOPP chemotherapy. Patients who were undergoing chemotherapy had significantly higher frequencies of sperm with 1p telomeric duplications and significantly higher frequencies of disomic and diploid sperm. Semen samples provided many years after chemotherapy by four donors, showed significantly higher levels of diploid and disomic sperm for one donor and a higher frequency of sperm with 1p telomeric loss for another.

Van Hummelen and colleagues[30] also applied a similar AM-10 sperm FISH method to analyse the meiotic products of reciprocal translocation carriers using multi-colour DNA probes that were specific for the chromosomes involved in the reciprocal translocation (a telomeric and a centromeric probe on chromosome 1 plus a centromeric probe on chromosome 10). They reported that frequencies of reciprocal sperm products from adjacent I segregation deviated significantly from the expected 1:1 ratio and the assay allowed them to evaluate recombination events in the interstitial segments after adjacent II segregation.

Sloter and colleagues[43] applied the human ACM method to investigate the effects of age on the frequencies of structural abnormalities in sperm of healthy non-smoking men, aged 22–80 years with no current history of reproductive problems or known exposure to genotoxic agents. Structural defects accounted for ~70% of the chromosomal abnormalities detected in ~255,000 total sperm evaluated by the ACM assay. On average, partial duplications and deletions were ~2-fold in men aged 65–80 years *vs.* 20–29 years. Chromosomal breaks within the 1cen-1q12 region were also more prevalent in sperm from these older men. No age effect was detected for numerical abnormalities involving chromosome 1. These data suggest that healthy older men carry significantly higher frequencies of structural chromosomal abnormalities in their sperm than healthy young men.

Bosch and colleagues[32] evaluated the effects of age on the incidence of structural aberrations and aneuploidy using a four-colour sperm-FISH assay. They found a significant age-related increase in the frequency of sperm with duplications and deletions for the centromeric and subtelomeric regions of chromosome 9, and chromosome 9 disomy as well as diploidy. However, the frequencies of duplications and deletions for chromosome 9 were significantly higher than those for chromosome 1, suggesting there are either chromosomal or technical variations between these two assays that need to be addressed.

Schmid and colleagues[44] applied the ACM method to the examination of sperm from infertile, oligozoospermic patients for structural and numerical chromosomal abnormalities. There was a significant increase in the average frequencies of sperm with duplications and deletions, as well as breaks, in the sperm of infertile patients compared with the healthy concurrent controls. The level of breaks within the 1cen-1q12 region was also significantly elevated. It was concluded that oligozoospermia is associated with structural chromosome abnormalities, suggesting that oligozoospermic men carry a higher burden of transmissible, chromosome damage.

18.7 Implications for Future Studies

These newly developed FISH-assays for detecting structural chromosomal aberrations directly in sperm are promising bio-indicators of paternal risks associated with infertility, spontaneous abortions, aneuploidy syndromes, chromosome rearrangements and certain chromosomal diseases in offspring. These sperm methods also provide direct approaches for assessing damage from exposure to chromosome-breaking agents as well as genetic predisposition to such damage. They are the first practical way to identify physiological and environmental factors that increase the risks of chromosomal rearrangements in spermatogenic stem cells. As demonstrated by our preliminary comparison of human *vs.* mouse baselines of damage, sperm-FISH strategies are equally applicable to any species of interest in biology, genetics, ecology and toxicology, though validation must remain an important consideration in the development of any new sperm FISH method.

The use of sperm FISH methods in human and animal studies has a major advantage over epidemiological surveys of human offspring or animal breeding studies because it provides a higher level of sensitivity and statistical power, and small increases can be detected by analysing sperm from relatively few donors.[45] The major limitation is that sperm FISH analyses have not yet been automated.

These new sperm FISH methods for chromosomal aberrations are joining a growing battery of assays for genomic damage in sperm including sperm aneuploidy assays,[18] DNA fragmentation assays,[46] chromosome breakage assays,[47,48] gene mutation assays in sperm[13,49] and trinucleotide repeat-length variation.[50–54] Additional sperm methods are still needed for other endpoints known to be important in development, such in imprinting profiles of paternal genes, which are known to control important aspects of development.[55]

Acknowledgments

We thank our long standing colleague, Professor Brenda Eskenazi, School of Public Health, University of California, Berkeley because our collaborative project with her laboratory provided the motivation for developing the human sperm-FISH assays for chromosomal aberration detection. We also thank E. Sloter and J. Nath for their invaluable contributions in the development of the ACM assay. This work was performed under the auspices of the U.S. DOE by the University of California, LLNL under contract W-7405-ENG-48 with funding support from NIEHS IAG Y01-ES-8016-5 and from Superfund P42ES04705 from the National Institute of Environmental Health Sciences.

References

1. U.S. Bureau of the Census, *Statistical Abstract of the United States*, 112th edn, U.S. Bureau of the Census, Washington, 1992.
2. U.S. Congress and Office of Technology Assessment, *Technologies for detecting heritable mutations in human beings*, Vol. OTA-H-298, US Government Printing Office, Washington, 1986.
3. E. Sloter, J. Nath, B. Eskenazi and A.J. Wyrobek, *Fertil. Steril.*, 2004, **81**, 925.
4. A.C. Chandley, *J. Med. Genet.*, 1991, **28**, 217.
5. E.D. Sloter, X. Lowe, I.D. Moore, J. Nath and A.J. Wyrobek, *Am. J. Hum. Genet.*, 2000, **67**, 862.
6. T.J. Hassold, *Curr. Top. Dev. Biol.*, 1998, **37**, 383.
7. T. Hassold, P.A. Hunt and S. Sherman, *Curr. Opin. Genet. Dev.*, 1993, **3**, 398.
8. M.A. Abruzzo and T.J. Hassold, *Environ. Mol. Mutagen.*, 1995, **25**(Suppl 26), 38.
9. T. Hassold and P. Hunt, *Nat. Rev. Genet.*, 2001, **2**, 280.
10. E.B. Hook, P.K. Cross and D.M. Schreinemachers, *JAMA*, 1983, **249**, 2034.
11. M.D. Shelby, G.L. Erexson, G.J. Hook and R.R. Tice, *Environ. Mol. Mutagen.*, 1993, **21**, 160.
12. A.M. Estop, C. Marquez, S. Munne, J. Navarro, K. Cieply, V. Van Kirk, M.R. Martorell, J. Benet and C. Templado, *Am. J. Hum. Genet.*, 1995, **56**, 452.
13. R.L. Glaser, K.W. Broman, R.L. Schulman, B. Eskenazi, A.J. Wyrobek and E.W. Jabs, *Am. J. Hum. Genet.*, 2003, **73**, 939.
14. P. Van Hummelen, X.R. Lowe and A.J. Wyrobek, *Hum. Genet.*, 1996, **98**, 608.
15. M.D. Shelby, *Mutat. Res.*, 1996, **325**, 159.
16. K.L. Witt and J.B. Bishop, *Mutat. Res.*, 1996, **355**, 209.

17. A.J. Wyrobek, T.E. Schmid and F. Marchetti, *J. Natl. Cancer Inst. Monogr.*, 2005, **34**, 31.
18. A.J. Wyrobek, T.E. Schmid and F. Marchetti, *Environ. Mol. Mutagen.*, 2005, **45**, 271.
19. J. Byrne, S.A. Rasmussen, S.C. Steinhorn, R.R. Connelly, M.H. Myers, C.F. Lynch, J. Flannery, D.F. Austin, F.F. Holmes, G.E. Holmes, L.C. Strong and J.J. Mulvihill, *Am. J. Hum. Genet.*, 1998, **62**, 45.
20. A.J. Wyrobek, L.A. Gordon, J.G. Burkhart, M.W. Francis, R.W. Kapp, Jr., G. Letz, H.V. Malling, J.C. Topham and M.D. Whorton, *Mutat. Res.*, 1983, **115**, 73.
21. A.F. Olshan and E. van Wijngaarden, *Adv. Exp. Med. Biol.*, 2003, **518**, 147.
22. J.F. Crow, *Nat. Rev. Genet.*, 2001, **1**, 40.
23. D. Warburton and A. Kinney, *Environ. Mol. Mutagen.*, 1996, **28**, 237.
24. I. Lorda-Sanchez, F. Binkert, M. Maechler, W.P. Robinson and A.A. Schinzel, *Hum. Genet.*, 1992, **89**, 524.
25. C.R. Preston, J.A. Sved and W.R. Engels, *Genetics*, 1996, **144**, 1623.
26. W.A. Robbins, J.E. Baulch, D. Moore II, H.U. Weier, D. Blakey and A.J. Wyrobek, *Reprod. Fertil. Dev.*, 1995, **7**, 799.
27. E. Rudak, P.A. Jacobs and R. Yanagimachi, *Nature*, 1978, **274**, 911.
28. R.H. Martin, S. Ernst, A. Rademaker, L. Barclay, E. Ko and N. Summers, *Cytogenet. Cell Genet.*, 1997, **78**, 120.
29. B.F. Brandriff, M.L. Meistrich, L.A. Gordon, A.V. Carrano and J.C. Liang, *Hum. Genet.*, 1994, **93**, 295.
30. P. Van Hummelen, D. Manchester, X. Lowe and A.J. Wyrobek, *Am. J. Hum. Genet.*, 1997, **61**, 651.
31. A. Baumgartner, P. Van Hummelen, X.R. Lowe, I.D. Adler and A.J. Wyrobek, *Environ. Mol. Mutagen.*, 1999, **33**, 49.
32. M. Bosch, O. Rajmil, J. Egozcue and C. Templado, *Eur. J. Hum. Genet.*, 2003, **11**, 754.
33. X.X. Liu, G.H. Tang, Y.X. Yuan, L.X. Deng, Q. Zhang and L.K. Zheng, *Yi Chuan Xue Bao*, 2003, **30**, 1177.
34. F.S. Hill, F. Marchetti, M. Liechty, J. Bishop, J. Hozier and A.J. Wyrobek, *Mol. Reprod. Dev.*, 2003, **66**, 172.
35. F. Marchetti, F.S. Pearson, J.B. Bishop and A.J. Wyrobek, *Hum. Reprod.*, 2006, **21**, 888.
36. M.L. Meistrich, S.P. Chawla, M.F. Da Cunha, S.L. Johnson, C. Plager, N.E. Papadopoulos, L.I. Lipshultz and R.S. Benjamin, *Cancer*, 1989, **63**, 2115.
37. D. Wouters-Tyrou, A. Martinage, P. Chevaillier and P. Sautiere, *Biochimie*, 1998, **80**, 117.
38. G.A. Sega, *Genetics*, 1979, **92**, s49–s58.
39. R.E. Sotomayor and G.A. Sega, *Environ. Mol. Mutagen.*, 2000, **36**, 255.
40. F. Marchetti and A.J. Wyrobek, *Birth Defects Res. C Embryo Today*, 2005, **75**, 112.
41. C.M. Disteche, A.V. Carrano, L.K. Ashworth, K. Burkhart-Schultz and S.A. Latt, *Cytogenet. Cell. Genet.*, 1981, **29**, 189.

42. P. Van Hummelen, X. Lowe, M.L. Meistrich and A.J. Wyrobek, *Am. J. Hum. Genet.*, 1995, **57**, A79.
43. E.D. Sloter, B. Eskenazi, D. Moore II, F.S. Hill, J. Nath and A.J. Wyrobek, *Environ. Mol. Mutagen.*, 2001, **37**, 70.
44. T.E. Schmid, M.H. Brinkworth, F. Hill, E. Sloter, A. Kamischke, F. Marchetti, E. Nieschlag and A.J. Wyrobek, *Hum. Reprod.*, 2004, **19**, 1395.
45. A.J. Wyrobek, T.E. Schmid and F. Marchetti, in *Encyclopedia of Genetics, Genomics, Proteomics and Bioinformatics*, M.J. Dunn, L.B. Jorde, P.F.R. Little and S. Subraniam (eds), Wiley, Chichester, 2005.
46. D. Evenson and L. Jost, *Methods Cell. Sci.*, 2000, **22**, 169.
47. D. Anderson, T.E. Schmid, A. Baumgartner, E. Cemeli-Carratala, M.H. Brinkworth and J.M. Wood, *Mutat. Res.*, 2003, **544**, 173.
48. D. Sakkas, O. Moffatt, G.C. Manicardi, E. Mariethoz, N. Tarozzi and D. Bizzaro, *Biol. Reprod.*, 2002, **66**, 1061.
49. I. Tiemann-Boege, W. Navidi, R. Grewal, D. Cohn, B. Eskenazi, A.J. Wyrobek and N. Arnheim, *PNAS*, 2002, **99**, 14952.
50. Y.E. Dubrova, V.N. Nesterov, N.G. Krouchinsky, V.A. Ostapenko, R. Neumann, D.L. Neil and A.J. Jeffreys, *Nature*, 1996, **380**, 683.
51. Y.E. Dubrova, M. Plumb, B. Gutierrez, E. Boulton and A.J. Jeffreys, *Nature*, 2000, **405**, 37.
52. R. Barber, M.A. Plumb, E. Boulton, I. Roux and Y.E. Dubrova, *PNAS*, 2002, **99**, 6877.
53. C.M. Somers, B.E. McCarry, F. Malek and J.S. Quinn, *Science*, 2004, **304**, 1008.
54. C.M. Somers, C.L. Yauk, P.A. White, C.L. Parfett and J.S. Quinn, *PNAS*, 2002, **99**, 15904.
55. T. Doerksen and J.M. Trasler, *Biol. Reprod.*, 1996, **55**, 1155.

CHAPTER 19

Radiation-induced Transgenerational Instability in Mice

YURI E. DUBROVA

Department of Genetics, University of Leicester, Leicester, UK

19.1 Introduction

Mutation induction in directly exposed cells is currently regarded as the main component of the genetic risk of ionising radiation for humans.[1] However, recent data on the delayed effects of exposure to ionising radiation challenge the existing paradigm. The results of numerous *in vitro* studies show that ionising radiation can not only induce mutations in the directly exposed cells but also lead to delayed effects, with new mutations arising many cell divisions after the initial irradiation damage.[2] Recent data suggest that *in vivo* exposure to ionising radiation can also result in genomic instability manifested over a certain period of time post-irradiation.[3] The results of some publications however show that the ability of cells to exhibit elevated mutation rates cannot be ascribed to the conventional mechanisms of mutator phenotype and is most likely related to epigenetic events.[2–4]

Apart from the studies on mutation rates in somatic cells, considerable progress has been made in the analysis of radiation-induced instability in the mammalian germ line, where the effects of radiation exposure have been investigated among the offspring of irradiated parents (reviewed in refs 5,6). These transgenerational studies were designed to test the hypothesis that radiation-induced instability in the germ line of irradiated parents could be manifested in the offspring, affecting their mutation rates and some other characteristics. The aim of this paper is to review a number of publications addressing transgenerational instability in mice and other laboratory animals.

19.2 Somatic Effects

Results in a number of publications show that parental exposure to ionising radiation produces transgenerational changes in somatic tissues that cannot be ascribed to the conventional mechanisms of mutation induction in the germ line of irradiated parents. So far, the most interesting findings have been made by studying cancer predisposition and somatic mutation rates in the offspring of irradiated parents.

Carcinogenesis is a multistep process in which somatic cells acquire mutations in a specific clonal lineage.[7] How multiple mutations accumulate in the irradiated cells over a clinically relevant time period remains unclear. It was therefore suggested that ongoing genomic instability could result in the accumulation of mutations over a certain period of time after irradiation, which, together with mutations directly induced in the irradiated cells, may significantly enhance radiation carcinogenesis.[8–10] Given the alleged contribution of radiation-induced genomic instability to stepwise tumour progression, it may therefore appear that parental exposure to ionising radiation could predispose their offspring to cancer.

The incidence of cancer in the offspring of irradiated parents has been extensively analysed in a number of publications (early publications are reviewed refs 6,11). A majority of the studies addressing the issue of transgenerational carcinogenesis were initiated by findings showing a clustering of childhood leukaemia in the vicinity of the Sellafield nuclear reprocessing plant[12] and a substantial increase in the incidence of tumours in the non-exposed first-generation offspring (F_1) of male mice exposed to X-rays or urethane.[13] It should be noted that the experimental evidence for elevated incidence of cancer among the offspring of irradiated parents still remains highly controversial, as some studies have so far failed to confirm these results.[14,15] However, it has been shown that the parental exposure to mutagens may also affect the morphology of tumours in their offspring. For example, the analysis of the transgenerational changes among the offspring of male mice exposed to benz(a)pyrene indicated that the incidence of cancer did not exceed those of the control, whereas the mean number of lung adenomas per tumour in the offspring of exposed parents remained persistently elevated over several generations.[16]

Due to the lack of measurable increases in tumour incidence in the non-exposed offspring of irradiated parents, later studies have characterised the incidence of cancer in the offspring of irradiated male mice exposed to recognise carcinogens.[17–20] In contrast to the data obtained on non-treated offspring, these results clearly showed an elevated incidence of cancer among the carcinogen-challenged offspring of irradiated males. For example, Vorobtsova and colleagues[18] demonstrated that the incidence of skin cancer in phorobol 12-myristate 13-acetate-treated F_1 offspring of irradiated male rats significantly exceeded that among the offspring of non-irradiated parents. Similar data were obtained by analysing the incidence of leukaemia in the F_1 offspring of irradiated mice, treated with ethylnitrosourea.[19] Other data also indicate that treatment with carcinogens can modify the pattern of malignancy among the offspring of irradiated parents. The results of one such study showed that treatment shortened the latent period for the leukaemia and resulted in a switch from the predominant thymic lymphoma in the controls to a predominance of leukaemia in the offspring of irradiated males.[19]

The data on transgenerational carcinogenesis indicate that the offspring of irradiated parents can be genetically unstable and show elevated mutation rates in somatic tissues. Being genetically unstable, the offspring of irradiated parents may be at greater risk of acquiring cancer over their lifetime.

To date, somatic mutation rates in the offspring of irradiated male mice have been analysed using a number of endpoints. For example, several publications have demonstrated an elevated frequency of chromosome aberrations in the F_1 offspring of irradiated male mice and rats.[19,21–23] Transgenerational changes in somatic mutation rates were also observed by studying the frequency of micronuclei and *lac*I mutations in the F_1 offspring of irradiated male mice.[24,25]

The most interesting data on somatic instability in the first-generation offspring of irradiated males were obtained from the analysis of somatic reversions of the pink-eyed unstable mutation, p^{un}.[26,27] The mouse p^{un} mutation is caused by the spontaneous disruption of the *pink-eyed dilute* locus on chromosome 7 resulting in a ~ 70 kb head-to-tail DNA sequence duplication. Spontaneous reversion is caused by deletion of one of the duplicated sequences and restores normal pigment production.[28] The p^{un} locus is highly unstable, 1.8% or more of offspring are mosaic revertants.[29] Reversion is generally detected as a somatic event and is generally attributed to homologous recombination.[28] The first, though weak, evidence for transgenerational increases in the frequency of p^{un} reversion was obtained by the analysis of fur spots in the non-exposed F_1 offspring of irradiated male mice.[26] These results were further supported by the study of p^{un} reversion in retinal pigment epithelium,[27] a technique that provides more accurate assessment of reversion at this locus.[30] The authors not only reported an elevated frequency of p^{un} reversion among the F_1 offspring of irradiated males but also provided an important evidence for a genome-wide elevation of mutation rate. Thus, the results of this study showed that irradiation of males led to reversion of the alleles derived from the non-irradiated mothers at a frequency indistinguishable from that at the alleles from the irradiated males.[27] It should be noted that a recent study on the transgenerational effects in the medaka fish has also reported that paternal irradiation can elevate mutations of an un-irradiated maternal allele.[31]

19.3   Germline Effects

The data describing transgenerational changes in the F_1 offspring of irradiated parents raise the possibility that the same effects may also persist in the germ line. The analysis of transgenerational instability in the germ line is more complicated than that for the effects in somatic tissues and requires, at the least, profiling of the second-generation (F_2) offspring of irradiated parents. In these studies, the F_1 offspring are mated and mutation rates in their germ line are estimated by establishing the frequency of mutation in the F_2 offspring.

The first evidence for the transgenerational increases in germ line mutation rates was obtained by Luning *et al.*[32] In this publication, the frequency of dominant lethal mutations in the germ line of directly irradiated male mice and their first-generation offspring was analysed. The authors observed a significantly elevated mutation rate in the germ line of non-exposed F_1 offspring of male mice, injected with the α-emitter Plutonium-239. This assay detects early-pre-implantation and late-post-implantation mortality in mice, both of which were equally elevated in the F_2 offspring of exposed males. Similar data were later obtained from the analysis of the F_1 offspring of male rats treated with cyclophosphamide.[33] Other studies, analysing the proliferation of early embryonic cells and the frequency of malformations in the F_2 offspring of irradiated parents, confirmed these observations.[34–36]

A few years ago, we initiated a large and still ongoing study aimed to analyse mutation rates in the germ line and somatic tissues of non-exposed F_1 and F_2 offspring of irradiated male mice.[37,38] We first approached this problem using a new, sensitive technique for monitoring germ line mutation in mice, previously developed for the analysis of mutation induction in the mouse germ line by ionising radiation and chemical mutagens.[39–43] This technique employs highly unstable expanded simple tandem repeat (ESTR) loci. Unstable ESTR loci consist of homogenous arrays of relatively short repeats (4–6 bp) and show very high spontaneous mutation rates in both germ line and somatic cells.[44–46] The results of our studies of mutation induction in mice have shown that, in sharp contrast to previously used genetic systems, changes in mutation rate can be detected in very small samples.[40–43] Most importantly, an elevated mutation rate at ESTR loci can be robustly detected at doses far lower than had been possible using standard approaches for monitoring germ line mutation in mice.[40–43,47]

Further analysis of ESTR mutation induction in the mouse germ line has revealed that ESTR mutation is most probably attributable to replication slippage, which is greatly enhanced by replication pausing.[48,49] According to our model, the very high mutation rates at ESTR loci could be directly related to their very large size (500–3500 repeats) and, probably, the presence of hairpin structures within the arrays, which together may cause replication fork pausing and subsequently promote polymerase slippage events. Given the striking similarities in ESTR mutation rates and spectra in the germ line and somatic tissues,[50] it appears highly plausible that germ line and somatic ESTR mutations are attributed to the same replication-dependent mechanisms. The alleged involvement of DNA replication in mutation process at ESTR loci makes them a useful tool for the analysis of transgenerational effects in the mouse germ line.

In our first study, ESTR mutation rates were evaluated in the germ line of F_1 offspring of a male mouse exposed to 0.5 Gy of fission neutrons.[37] The results of this work showed a remarkable 4.5-fold increase in ESTR mutation rate in the F_1 germ line, which was in part attributable to increased mutational mosaicism. These data therefore indicated that transgenerational destabilisation should occur either immediately after fertilisation or on the very early stages of the developing F_1 germ line.

In the later study, ESTR mutation rates were analysed in the germ line of first- and second-generation offspring of inbred male CBA/H, C57BL/6 and

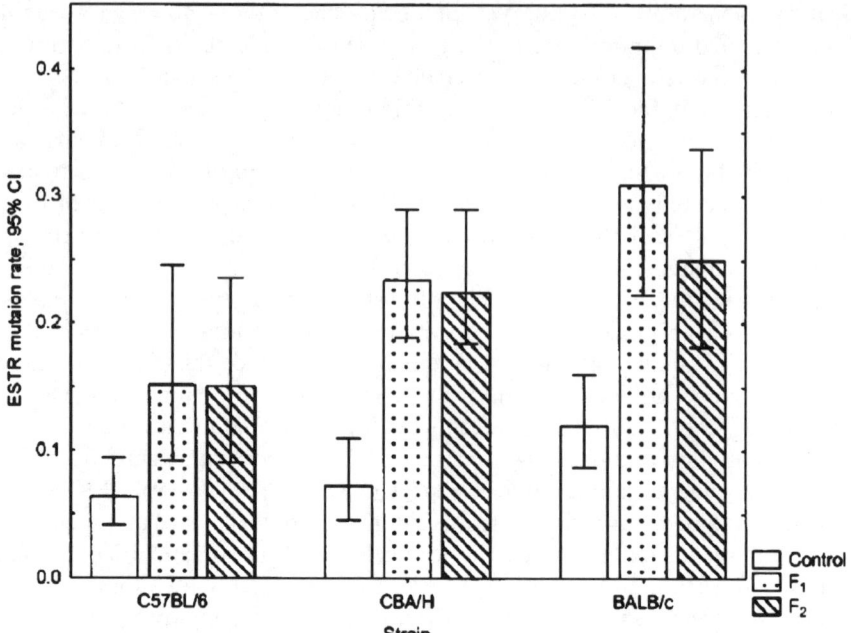

Figure 1   *Transgenerational increases in ESTR mutation rates in the germline of offspring of irradiated male mice. 95% confidence intervals and CI for mutation rate are shown*
(Data taken from ref 38)

BALB/c mice exposed to either high-LET fission neutrons or low-LET X-rays.[38] The main result of this study showed that paternal exposure to ionising radiation results in increased mutation rates in the germ line of two subsequent generations of all inbred strains, demonstrating that transgenerational instability persists at least across two generations (Figure 1). The persistence of transgenerational effects in the mammalian germ line was recently demonstrated by the analysis of the effects of paternal exposure to endocrine disruptors on male fertility across three subsequent generations.[51]

We also compared the transgenerational effects of paternal exposure to high-LET fission neutrons and low-LET X-rays. It is well established that high-LET radiation produces highly complex and localised initial DNA damage, which is different from the sparse damage produced by low-LET radiation, resulting in the unique final biological effects of these different radiation sources.[52] However, it appears that exposure to both types of radiation is capable of inducing genomic instability in somatic cells, though some studies have failed to detect the effects of low-LET exposure (reviewed in ref 53). Our data also demonstrated that paternal exposure to either high-LET fission neutrons or low-LET X-rays results in increased mutation rates in the F_1 and F_2 germ lines.[38]

Despite the fact that ESTR mutation rates were found to be significantly elevated in the germ line of all inbred strains, our data also showed that the extent of transgenerational increase clearly varies with the different strains (BALB/c > CBA/H > C57BL/6). These data are consistent with the results of previous studies showing that BALB/c and CBA mice are significantly more radiosensitive, and display higher levels of radiation-induced genomic instability in somatic cells than C57BL/6 mice.[54-56] As far as the high level of radiation-induced instability detected in the germ line and somatic tissues of BALB/c mice is concerned, it was suggested that this could be attributed to the strain-specific amino-acid substitutions affecting the activity of the 16^{ink4a} cyclin-dependent kinase inhibitor and the catalytic subunit of the DNA-dependent protein kinase.[57,58] Given the wide range of inherited variation in DNA repair capacity,[59] it therefore appears that both exposure and genetics could contribute to the manifestation of radiation-induced genomic instability in humans.

19.4 Mechanisms

The data described here imply that radiation-induced transgenerational instability could substantially enhance the accumulation of mutations and thereby contribute to radiation-induced carcinogenesis. Further analysis of the clinical impact of this phenomenon is currently limited mainly because the mechanisms of underlying radiation-induced genomic instability remain unknown. However, from numerous publications it appears that radiation-induced genomic instability observed *in vitro* and *in vivo*, as well as transgenerational effects may result from a plethora of molecular, biochemical and cellular events, or some each of which could belong to the already characterised pathways of cellular stress response.[2-4] These pathways include the recognition of DNA damage, its repair, cell cycle arrest and apoptosis.[60-62] The results of transgenerational studies described here imply that the germ cells of irradiated parents contain a DNA-signal, which after fertilisation, somehow affects the stability of their offspring and predisposes them to cancer.

As already mentioned, the data on transgenerational instability at the p^{un} locus showed that somatic mutation rates in the F_1 offspring are equally elevated at the alleles derived from the irradiated fathers and non-irradiated mothers.[27] We and others have obtained similar data analysing transgenerational instability at the mouse ESTR loci.[37,38,63] These results show that an increased mutation rate in the offspring of irradiated males results from a genome-wide elevation of mutation rate.

Our data on elevated ESTR mutation rates were obtained on descendants conceived 3 and 6 weeks after the initial paternal exposure to ionising radiation.[38] Given that these stages of the mouse spermatogenesis are transcriptionally active; the results of this study may indicate that instability signal could be induced in transcriptionally proficient cells. However, several recent publications report transgenerational changes in the offspring of male mice irradiated during the late postmeiotic stages of spermatogenesis, where gene expression is practically shut down.[18,21,27,63] Given that pre-mutational lesions in sperm DNA are effectively repaired within a few hours of fertilisation,[64,65] it therefore appears that radiation-induced damage to sperm DNA could later trigger a cascade of events in the fertilised egg. These events are accompanied by a suppression of DNA synthesis in both the irradiated male and non-irradiated female pronuclei,[66] profound changes in the expression of DNA repair genes in the pre-implantation embryo,[67] DNA methylation and histone acetylation.[68] The presence of such dramatic changes in the fertilised egg could result in the delayed effects, similar to those in the progeny of irradiated cells, which may affect the stability of developing embryo. The results showing an unusually high level of mutational mosaicism in the germ line and somatic tissues of F_1 mice[37,63] suggest that the destabilisation could occur at the very earliest stages of development.

The results of the transgenerational studies, together with the data on radiation-induced genomic instability *in vitro*, clearly indicate that these phenomena can be attributed to epigenetic events. For example, our data on the similarly elevated ESTR mutation rates in the F_1 and F_2 offspring of irradiated males rule out the possibility that transgenerational effects are due to radiation-induced mutations at any specific set of genes in the exposed F_0 males.[38] We and others have hypothesised that DNA methylation may be regarded as a strong candidate for such an epigenetic signal resulting in transgenerational mutagenesis.[37,51] DNA methylation and histone modification represent the main mechanisms by which DNA is epigenetically marked.[69,70] Methylation is known to survive the reprogramming of DNA methylation during spermatogenesis and early development[71,72] and can be transmissible through many cell divisions.[73] Alterations in the pattern of DNA methylation might affect genes responsible for maintaining genomic integrity and influence the recognition of DNA damage or its repair. For example, promoter methylation switches off the transcription of the *hMLH1* mismatch repair gene, resulting in colorectal carcinomas and microsatellite instability.[74,75] The transgenerational increases can also be attributed to a change in the expression patterns of genes involved in DNA repair in the offspring of irradiated males. Indeed, recent data have shown persistently altered patterns of expression of some genes in the offspring of irradiated male mice.[76-78]

It should be stressed that the altered expression of DNA repair genes cannot explain the transgenerational increases in mutation rates detected across a number of endpoints, including protein-coding genes, ESTR loci and chromosome aberration. Given that the mechanisms of spontaneous and induced mutation at these systems substantially differ, these observations imply that the efficiency of multiple DNA repair pathways should be simultaneously compromised in the offspring of irradiated parents. The presence of such highly coordinated changes appears to be highly unlikely. The multiplicity of transgenerational changes detected in the offspring of irradiated parents could be attributed to oxidative stress or inflammatory response. The involvement of inflammatory-type processes in the delayed increases in mutation rates in the progeny of irradiated cells has long been suspected.[2-4] As reactive oxygen species are the major source of endogenous DNA damage, including single- and double-strand breaks, abasic sites and a variety of nucleotide modifications,[79] the transgenerational effects could thus be explained by this mechanism. However, transgenerational instability may also be attributed to replication stress. Indeed, the results of recent studies suggest that in human precancerous cells the ATR/ATM-regulated checkpoints are activated through deregulated DNA replication, which leads to a multiplicity of DNA alterations.[80,81] It has also been shown that radiation-induced chromosome instability *in vitro* could be attributed to a long-term delay in chromosome replication.[82] Given that our previous results suggest that ESTR mutation is most probably attributable to replication slippage,[48–50] delayed/stalled replication may therefore provide a plausible explanation for the transgenerational increases in mutation rate at these loci.

19.5 Conclusions

The data reviewed here raise the important issue of the delayed effects of ionising radiation and suggest that persistent transgenerational instability could lead to a significant increase in the mutation load in exposed populations.[5] Given that the results of animal studies have so far provided strong evidence for a variety of transgenerational changes affecting clinically relevant traits such as mortality and cancer predisposition, these may suggest that the genetic risks of ionising radiation for humans is greater than previously thought. However, given the lack of reliable experimental evidence for transgenerational effects in humans, this important issue still awaits its final clarification.

Acknowledgments

I thank the many colleagues and collaborators for their important contribution to this work. This work was supported by grants from the Wellcome Trust.

References

1. UNSCEAR, *Hereditary Effects of Radiation*, United Nations, New York, 2001.
2. W.F. Morgan, *Radiat. Res.*, 2003, **159**, 567.
3. W.F. Morgan, *Radiat. Res.*, 2003, **159**, 581.
4. S.A. Lorimore, P.J. Coates and E.G. Wright, *Oncogene*, 2003, **22**, 7058.
5. Y.E. Dubrova, *Oncogene*, 2003, **22**, 7087.
6. T. Nomura, *Mutat. Res.*, 2003, **544**, 425.
7. L.A. Loeb, K.R. Loeb and J.P. Anderson, *Proc. Nati. Acad. Sci. USA*, 2003, **100**, 776.
8. J.B. Little, *Carcinogenesis*, 2000, **21**, 397.
9. Z. Goldberg, *Oncogene*, 2003, **22**, 7011.
10. L. Huang, A.R. Snyder and W.F. Morgan, *Oncogene*, 2003, **22**, 5848.

11. N.P. Napalkov, J.M. Rice, L.Tomatis and H.Yamasaki, (Eds), *Perinatal and Multigenerational Carcinogenesis*, TARC, Lyon, 1989.
12. M.J. Gardner, M.P. Snee, A.J. Hall, C.A. Powell, S. Downes and J.D. Terrell, *Br. Med. J.*, 1990, **300**, 423.
13. T. Nomura, *Nature*, 1982, **296**, 575.
14. B.M. Cattanach, D. Papworth, G. Patrick, D.T. Goodhead, T. Hacker, L. Cobb and E. Whitehill, *Mutat. Res.*, 1998, **403**, 1.
15. B.M. Cattanach, G. Patrick, D. Papworth, D.T. Goodhead, T. Hacker, L. Cobb and E. Whitehill, *Int. Radiat. Biol.*, 1995, **67**, 607.
16. V.S. Turusov, T.V. Nikonova and Y.D. Parfenov, *Cancer Lett.*, 1990 **55**, 227.
17. T. Nomura, *Mutat. Res.*, 1983, **121**, 59.
18. I.E. Vorobtsova, L.M. Aliyakparova and V.N. Anisimov, *Mutat. Res.*, 1993, **287**, 207.
19. B.I. Lord, L.B. Woolford, L. Wang, V.A. Stones, D. McDonald, S.A. Lorimore, D. Papworth, E.G. Wright and D. Scott, *Br. J. Cancer*, 1998 **78**, 301.
20. K.P. Hoyes, B.I. Lord, C. McCann, J.H. Hendry and I.D. Morris, *Radiat. Res.*, 2001, **156**, 488.
21. I.E. Vorobtsova, *Mutagenesis*, 2000, **15**, 33.
22. K. Kropacova, L. Slovinska and E. Misurova, *J. Radiat. Res.*, 2002, **43**, 125.
23. S. Sanova, S. Balentova, L. Slovinska and E. Misurova, *Neurotoxicol. Teratol.*, 2005, **27**, 145–151.
24. G.A. Luke, A.C. Riches and P.E. Bryant, *Mutagenesis*, 1997, **12**, 147.
25. L.A. Fomenko, G.V. Vasiljeva and V.G. Bezlepkin, *Biol. Bull.*, 2001, **28**, 419.
26. N. Cans and R.H. Schiesti, *Carcinogenesis*, 1999, **20**, 2351.
27. K. Shiraishi, T. Shimura, M. Taga, N. Uematsu, Y. Gondo, M. Ohtaki, R. Kominami and O. Niwa, *Radiat. Res.*, 2002, **157**, 661.
28. Y. Gondo, J.M. Gardner, Y. Nakatsu, D. Durham-Pierre, S.A. Deveau, C. Kuper and M.H. Brilliant, *Proc. Nati. Acad. Sci. USA*, 1993, **90**, 297–301.
29. R.M. Melvold, *Mutat. Res.*, 1971, **12**, 171–174.
30. A.J.R. Bishop, B. Kosaras, N. Carls, R.L. Sidman and R.H. Schiesti, *Carcinogenesis*, 2001, **22**, 641–649.
31. A. Shimada and A. Shima, *Mutat. Res.*, 2004, **552**, 119–124.
32. K.G. Luning, H. Frolen and A. Nilsson, *Mutat. Res.*, 1976, **34**, 539.
33. B.F. Hales, K. Crosman and B. Robaire, *Teratology*, 1992, **45**, 671.
34. M.F. Lyon and R. Renshaw, in *Genetic Toxicology of Environmental Chemicals. Part B: Genetic Effects and Applied Mutagenesis*, C. Ramel, B. Lambert and J. Magnusson, (eds), Alan R. Liss, New York, 1986, 449.
35. L.M. Wiley, J.E. Baulch, O.G. Raabe and T. Straume, *Radiat. Res.*, 1997, **148**, 145.
36. S. Pus, W.-U. Muller and C. Streffer, *Mutat. Res.*, 1999, **429**, 85.
37. Y.E. Dubrova, M. Plumb, B. Gutierrez, E. Boulton and A.J. Jeffreys, *Nature*, 2000, **405**, 37.
38. R. Barber, M.A. Plumb, E. Boulton, I. Roux and Y.E. Dubrova, *Proc. Natl. Acad. Sci. USA*, 2002, **99**, 6877.

39. Y.E. Dubrova, A.J. Jeffreys and A.M. Malashenko, *Nat. Genet.*, 1993, **5**, 92.
40. Y.E. Dubrova, M. Plumb, J. Brown, J. Fennelly, P. Bois, D. Goodhead and A.J. Jeffreys, *Proc. Nati. Acad. Sci. USA*, 1998, **95**, 6251.
41. Y.E. Dubrova, M. Plumb, J. Brown, E. Boulton, D. Goodhead and A.J. Jeffreys, *Mutat. Res.*, 2000, **453**, 17.
42. Y.E. Dubrova and M. Plumb, *Mutat. Res.*, 2002, **499**, 143.
43. C. Vilarino-Guell, A.G. Smith and Y.E. Dubrova, *Mutat. Res.*, 2003, **526**, 63.
44. R. Kelly, G. Bulfield, A. Collick, M. Gibbs and A.J. Jeffreys, *Genomics*, 1989, **5**, 844.
45. M. Gibbs, A. Collick, R.G. Kelly and A.J. Jeffreys, *Genomics*, 1993, **17**, 121.
46. P. Bois, J. Williamson, J. Brown, Y.E. Dubrova and A.J. Jeffreys, *Genomics*, 1998, **49**, 122.
47. R. Barber, M.A. Plumb, A.G. Smith, C.E. Cesar, E. Boulton, A.J. Jeffreys and Y.E. Dubrova, *Mutat. Res.*, 2000, **457**, 79.
48. R.C. Barber, L. Miccoli, P.P.W. van Buul, K.L.-A. Burr, A. van Duyn-Goedhart, J.F. Angulo and Y.E. Dubrova, *Mutat. Res.*, 2004, **554**, 287.
49. Y.E. Dubrova, *Radiat. Res.*, 2005, **163**, 200.
50. C.L. Yauk, Y.E. Dubrova, G.R. Grant and A.J. Jeffreys, *Mutat. Res.*, 2002, **500**, 147.
51. M.D. Anway, A.S. Cupp, M. Uzumcu and M.K. Skinner, *Science*, 2005, **308**, 1466.
52. D.T. Goodhead, *Health Phys.*, 1988, **55**, 231.
53. C.L. Limoli, B. Ponnaiya, J.J. Corcoran, E. Giedzinski, M.I. Kaplan, A. Hartmann and W.F. Morgan, *Adv. Space Res.*, 2000, **25**, 2107.
54. T.H. Roderick, *Radiat. Res.*, 1963, **20**, 631.
55. G.E. Watson, S.A. Lorimore, S.M. Clutton, M.A. Kadhim and E.G. Wright, *Int. J. Radiat. Biol.*, 1997, **71**, 497.
56. B. Ponnaiya, M.N. Comforth and R.L. Ullrich, *Radiat. Res.*, 1997, **147**, 121.
57. S. Zhang, E.S. Ramsay and B.A. Mock, *Proc. Nati. Acad. Sci. USA*, 1998, **95**, 2429.
58. Y. Yu, R. Okayasu, M.M. Weil, A. Silver, M. McCarthy, R. Zabriskie, S. Long, R. Cox and R.L. Ullrich, *Cancer Res.*, 2001, **61**, 1820.
59. H.W. Mohrenweiser, D.M. Wilson and I.M. Jones, *Mutat. Res.*, 2003 **526**, 93.
60. E.C. Friedberg, G.C. Walker and W. Siede, *DNA Repair and Mutagenesis*, ASM Press, Washington, DC, 1995.
61. C.J. Bakkenist and M.B. Kastan, Initiating cellular stress response, *Cell*, 2004, **118**, 9.
62. A. Sancar, L.A. Lindsey-Boltz, K. Unsal-Kacman and S. Linn, *Annu. Rev. Biochem.*, 2004, **73**, 39.
63. S.M. Clutton, K.M. Townsend, C. Walker, J.D. Ansell and E.G. Wright, *Carcinogenesis*, 1996, **17**, 1633.
64. C.L. Limoli, N.I. Kaplan, E. Giedzinski and W.F. Morgan, *Free Radic. Biol. Med.*, 2001, **31**, 10.
65. O. Niwa and R. Kominami, *Proc. Natl. Acad. Sci. USA*, 2001, **98**, 1705.
66. B. Brandriff and R.A. Pedersen, *Science*, 1981, **211**, 1431.

67. W.M. Generoso, K.T. Cain, M. Krishna and S.W. Huff, *Proc. Natl. Acad. Sci. USA*, 1979, **76**, 435.
68. T. Shimura, M. Inoue, M. Taga, K. Shiraishi, N. Uematsu, N. Takei, Z.-M. Yuan, T. Shinohara and O. Niwa, *Mol. Cell. Biol.*, 2002, **22**, 2220.
69. W. Harrouk, A. Codrington, R. Vinson, B. Robaire and B.F. Hales, *Mutat. Res.*, 2000, **461**, 229.
70. T.S. Barton, B. Robaire and B.F. Hales, *Proc. Nati. Acad. Sci. USA*, 2005, **102**, 7865.
71. W. Reik and J. Walter, *Nat. Rev. Genet.*, 2001, **2**, 21.
72. V.K. Rakyan, J. Preis, H.D. Morgan and E. Whitelaw, *Biochem. J.*, 2001, **356**, 1.
73. Roemer, W. Reik, W. Dean and J. Klose, *J. Current Riot.*, 1997, **7**, 277.
74. M. Constancia, B. Pickard, G. Kelsey and W. Reik, *Genome Res.*, 1998, **8**, 881.
75. R. Holliday, *Science*, 1987, **238**, 163.
76. J.M.D. Wheeler, N.E. Beck, H.C. Kim, I.P.M. Tomlinson, N.J.Mc.C. Mortensen and W.F. Bodmer, *Proc. Nail. A cad. Sci. USA*, 1999, **96**, 10296.
77. J.G. Herman, A. Umar, K. Polyak, J.R. Graff, N. Ahuja, J.-P.J. Issa, S. Markowitz, J.K.V. Wilison, S.R. Hamilton, K.W. Kinzler, M.F. Kane, R.D. Kolodner, B. Vogeistein, T.A. Kunkel and S.B. Baylin, *Proc. Nati. Acad. Sci. USA*, 1998, **95**, 6870.
78. A. Daher, M. Varin, Y. Lamontagne and D. Oth, *Carcinogenesis*, 1998, **19**, 1553.
79. E. Baulch, O.G. Raabe and L.M. Wiley, *Mutagenesis*, 2001, **16**, 17.
80. M.M. Vance, J.E. Baulch, O.G. Raabe, L.M. Wiley and J.M. Overstreet, *Int. J. Radiat. Riot.*, 2002, **78**, 513.
81. A.L. Jackson and L.A. Loeb, *Mutat. Res.*, 2001, **477**, 7.
82. J. Bartkova, Z. Horejsi, K. Koed, A. Kramer, F. Tort, K. Zieger, P. Guldberg, M. Sehested, J.M. Nesland, C. Lukas, T. Orntofi, J. Lukas and J. Bartek, *Nature*, 2005, **434**, 864.
83. V.G. Gorgoulis, L.V. Vassiliou, P. Karakaidos, P. Zacharatos, A. Kotsinas, T. Liloglou, M. Venere, R.A. Ditullio, N.G. Kastrinakis, B. Levy, D. Kietsas, A. Yoneta, M. Herlyn, C. Kittas and T.D. Halazonetis, *Nature*, 2005, **434**, 907.
84. K.S. Breger, L. Smith, M.S. Turker and M. Thayer, *Cancer Res.*, 2004, **64**, 8231.

CHAPTER 20
New Genetic Information Generated by Endogenous Reverse Transcription in Sperm Cells

CORRADO SPADAFORA

Istituto Superiore di Sanità, Viale Regina Elena 299, 00161, Rome, Italy

20.1 Introduction

In an article published in 1989,[1] we reported that mouse epididymal spermatozoa take up exogenous DNA spontaneously and that the sperm-bound DNA molecules are subsequently transferred to embryos in IVF assays and further propagated to newborn mice. Despite initial controversy over the reproducibility of the protocol used in those experiments, the peculiar and unique ability of sperm cells to act as vectors not only of their own genome but also of foreign sequences has been widely confirmed. It is now well established that spermatozoa of virtually all species have the spontaneous ability to take up exogenous DNA molecules (see Table 1) and transfer them into oocytes at fertilisation. By exploiting this ability, a variety of experimental protocols of sperm-mediated gene transfer (SMGT), aiming at the generation of transgenic animals, have been developed and applied to numerous species, from echinoids to mammals.[2-4] The SMGT assay schematically represented in Figure 1 refers to the murine system but can be extended to most vertebrate species. A detailed examination of all reported results would be beyond the scope of this review, but some basic conclusions emerging from published data are worth recalling. On one hand, there is a substantial agreement that the new genetic information transferred by sperm cells can be propagated and expressed throughout development and in the adult offspring. On the other, results on the final fate of the exogenous sequences are contradictory and leave the open question as to whether they become integrated into the host genome or are propagated as extrachromosomal structures. Nonintegrated, episomal structures are frequently generated when spermatozoa are directly incubated with foreign

Table 1   *The uptake of exogenous DNA molecules in spermatozoa of various species using different protocols*

Class	Species	Sperm transformation method
Echinoids	Sea urchin	Sperm/DNA incubation
Molluscs	Abalone	Sperm electroporation
Insects	Blow fly	Sperm/DNA incubation
	Honeybee	Sperm/DNA incubation
	Silkworm	Sperm/DNA incubation
	Lucilia cuprina	Sperm/DNA incubation
	Apis mellifera	Sperm/DNA incubation
Fish	Common carp	Sperm/DNA incubation, sperm electroporation
	African catfish	Sperm/DNA incubation
	Tilapia	Sperm electroporation
	Zebrafish	Sperm electroporation
	Salmon	Sperm electroporation
	Loach	Sperm electroporation
Amphibians	*Xenopus laevis*	Sperm/DNA incubation Sperm /REMI[a]
Birds	Rooster	Sperm/DNA incubation, sperm electroporation, sperm lipofection, REM I,[a] LB-SMGT[b]
	Mouse	Sperm DNA incubation, sperm lipofection, ICSI,[c] LB-SMGT[b]
	Rabbit	Sperm/DNA incubation
	Ram	Sperm/DNA incubation
	Goat	Sperm/DNA incubation
	Pig	Sperm/DNA incubation, sperm electroporation
Mammals	Buffalo	Sperm/DNA incubation, sperm electroporation
	Bull	Sperm/DNA incubation
	Human	ICSI[‡]
	Rhesus macaque	

[a] REMI: restriction enzyme-mediated integration.
[b] LB-SMGT: linker-based SMGT.
[c] Intracytoplasmic sperm injection.

sequences in SMGT assays, as reported in a variety of different species,[5–10] while integration is a rare event that has been reported to occur in a single instance.[11] In contrast, integration occurs at high frequency in conditions in which a direct interaction between the exogenous sequences in plasma and the sperm membrane is avoided: this can be achieved by briefly incubating the exogenous DNA with demembranated spermatozoa and then using them in intracytoplasmic sperm injection (ICSI)-mediated trasngenesis,[12] or bypassing the plasma membrane by lipofection,[13,14] or by linker-based mediated trans-genesis.[15] As a whole, these sometimes conflicting reports suggest that SMGT is not a simple and straightforward process and reinforce our intention to investigate the underlying molecular mechanisms thoroughly.

Over the past 15 years, we have concentrated our efforts on clarifying the molecular steps of SMGT, starting with the earliest and fundamental event,

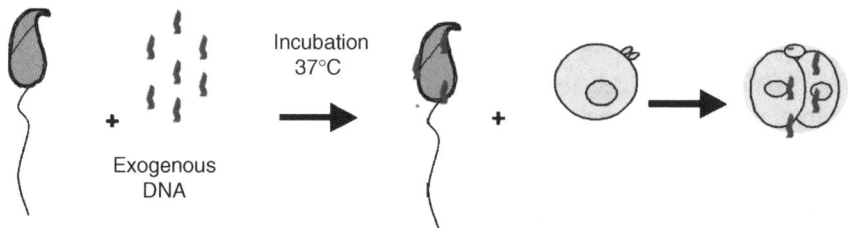

Figure 1 *Schematic representation of SMGT (mouse). Epididymal spermatozoa were obtained surgically from male donors and incubated with plasmid DNA at a concentration ranging from 5 to 50 ng/1.06 spermatozoa for about 30 min at 37°C. Sperm were then used in IVF assays and the foreign DNA sequences were transferred to oocytes and further to two-cell embryos. Embryos can be implanted into foster mothers (not shown) and screening of the offspring is carried out using DNA samples extracted from tail fragments*

that is the interaction of foreign DNA with spermatozoa. These studies have revealed that the binding between sperm cells and exogenous nucleic acid molecules, and their subsequent internalisation in nuclei, do not occur at random, but are well-regulated events mediated by specific factors.[2] Briefly the binding of exogenous DNA always occurs in the subacrosomal segment of the sperm head of epididymal, or ejaculated and thoroughly washed, spermatozoa because it is strongly antagonised in the presence of seminal fluid; a glycoprotein, inhibitory factor-1 (IF-1), abundant in the seminal fluid of mammals and on the surface of spermatozoa in invertebrate species, exerts a powerful inhibitory effect on the interaction of exogenous DNA, thus protecting spermatozoa from undesired intrusions of exogenous nucleic acids in nature; the binding of exogenous DNA to sperm cells is mediated by a class of proteins of 30–35 kDa located on the cell surface that act as DNA-binding substrates; a constant proportion (15–22% in mouse) of the sperm-bound DNA is further internalised in nuclei and this step is strictly CD4-dependent.

Based on these observations, it is clear that the internalisation does not occur as a consequence of passive transfer, but is mediated by a specific regulatory mechanism. Once internalised, the exogenous sequences reach the nuclear scaffold and trigger endogenous nuclease activity in a DNA dose-dependent manner. The activated nucleases cause substantial rearrangements in the exogenous DNA, which becomes fragmented and can undergo recombination events that can eventually yield rare integration events into the sperm genome.[16]

20.2 Mouse Sperm Chromatin Contains Domains with an "Active-Like" Open Conformation and Enriched in RT-Coding Elements

The unexpected finding that foreign DNA sequences can integrate in the sperm genome, albeit rarely,[16] implied that sperm chromatin is not a uniformly compact structure inaccessible to foreign molecules, prompting us to search for "loosely packaged" domains. To this end, we exploited the endogenous nucleases that are activated in spermatozoa in response to long (several hours) incubation in appropriate medium at 37°C.[17] Under these conditions, the nucleases preferentially cleave "accessible" DNA domains in the highly compact sperm chromatin and cause their release from sperm heads in an apoptosis-like process. Using this procedure, we have isolated a nuclease-sensitive fraction from murine sperm chromatin.[18] That fraction contains undermethylated DNA,[19] is organised in nucleosomes and lacks protamines – the major component of the bulk sperm chromatin.[18] These structural features closely resemble those of transcriptionally "active" domains in somatic chromatin. Most importantly, sequence analysis of the DNA from the nuclease-sensitive fraction revealed a high enrichment in sequences of retroposon/ retroviral origin, particularly the LINE1 *ORF2* gene encoding the reverse transcriptase (RT) enzyme.[18] The finding that mouse sperm chromatin contains a nucleohistone fraction is not totally surprising because mammalian spermatozoa had previously been reported to contain core histones; indeed, earlier reports showed that sperm nuclei of rams[20] and humans[21,22] retain variable amounts of chromatin subfractions in which the original histone component is not replaced by protamines. Together, these results show that the chromatin of mature spermatozoa contains RT-coding sequences in a potentially active conformation, thus raising the question if an RT activity is present in the nuclei of mature spermatozoa.

20.3 Exogenous RNA Incubated with Murine Spermatozoa is Reverse Transcribed into cDNA Copies that are Transferred to Embryos at Fertilisation

To assess if a functional RT activity was present in sperm cells, we decided to test RT activity directly, after incubation of mouse epididymal spermatozoa with exogenous RNA molecules. The underlying idea was that if spermatozoa are endowed with a functional RT enzymatic activity, then exogenous RNA molecules taken up by sperm and internalised into nuclei might be used as a substrate and reverse transcribed into cDNA copies. To this end, we used the human poliovirus RNA, a nonretroviral RNA virus whose chromosome replicates through an RNA(−) strand. The choice was made specifically to avoid the synthesis of DNA molecules as intermediates of viral replication, as is the case with retroviruses for example, and avoid any possible artefacts that might have arisen from contamination of our RNA preparation with naturally occurring DNA replication intermediates. We found that poliovirus RNA molecules are efficiently internalised in sperm nuclei and are indeed retrotranscribed into cDNA fragments. Moreover, when RNA-incubated spermatozoa are used in IVF assays, the cDNA molecules are delivered to the oocytes during fertilisation.[23] This surprising finding prompted us to assess the localisation of RT proteins in the sperm nucleus. By immuno-gold electromicroscopy using a specific anti-RT antibody, the endogenous RT molecules were localised on the sperm nuclear scaffold, which we prepared from nuclei by high salt extraction and extensive DNase digestion.[23] These data, in retrospect, are consistent with earlier observations that the male genital tract, including seminal fluid and sperm, is an active site of expression of retroviral genes.[24–26]

Together, these results indicate that a biologically active RT is stored in mature murine spermatozoa, stably associated with nuclear scaffold, and is able to reverse transcribe exogenous RNA sequences taken up by spermatozoa. The finding that these cDNA products are delivered to embryos after fertilisation raised a number of crucial questions (i) whether the cDNA molecules are restricted to early stages of development or persist throughout embryogenesis in adult animals, (ii) whether reverse transcribed sequences behave as biologically active "retro-genes" or do they instead remain in a transcriptionally inert state, and (iii) whether RNA might substitute for DNA in transgenesis strategies.

20.4 cDNA Copies are Mosaic Propagated in Adult F_0 and Mosaic Transmitted to F_1 Progeny

The poliovirus RNA used in our first set of experiments was not suitable to address the questions outlined above, because assessing whether or not the poliovirus cDNA fragments are transcriptionally competent would clearly require much caution and complicate the experiment. We therefore carried out a new set of experiments in which spermatozoa were first preincubated with a RNA population transcribed from a beta-gal gene-containing construct (Figure 2) and then used in IVF assays. PCR-based screenings revealed that beta-gal cDNA molecules that originated in sperm cells were delivered to oocytes at fertilisation, transferred to two- and four-cell embryos and, after

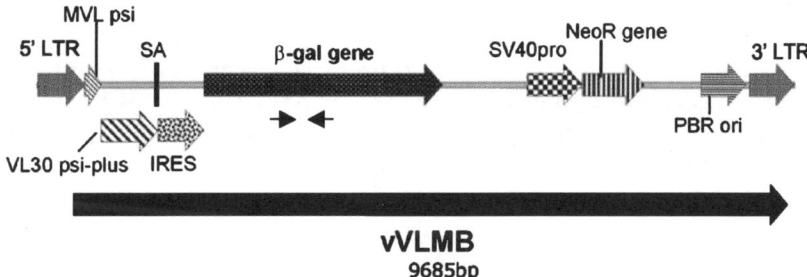

Figure 2   *Linear map of the beta-gal-containing vector vVLMB. Full-length RNA transcribed from this construct was used in incubation experiments with mouse spermatozoa that were then used in IVF assays to generate early embryos, foetuses and adult individuals. The two arrows indicate the positions of the primers used for direct PCR amplification of the beta-gal reporter sequence*

implantation into foster mothers, propagated in different tissues of adult F_0 individuals. Moreover, beta-gal cDNA was sexually transmitted from F_0 animals to the F_1 progeny and, surprisingly, was also mosaically distributed in tissues of the latter.[27] Extensive Southern blot analysis of DNA samples extracted from PCR-positive tissues, from both F_0 and F_1 individuals, failed to detect any clear banding pattern, suggesting therefore that beta-gal cDNA molecules, present in the host nuclei as determined by PCR amplification, are below the resolution power of the method and are highly underrepresented, below one copy per genome. The low abundance, together with the persistent, high degree of mosaicism transmitted from one generation to the next, led to the conclusion that the reverse-transcribed "foreign" cDNA population mostly, if not entirely, constitutes nonintegrated extrachromosomal structures. That conclusion was further strengthened by the impossibility of identifying any integration events using multiple approaches, including the screening of genomic libraries and ligation-mediated PCR analysis. In good agreement with our conclusions, it has been reported that RT-dependent, extrachromosomal structures are frequent guests of eukaryotic nuclei; in particular, reverse transcription of exogenous sequences by endogenous RT is reported to occur spontaneously in transfected[28] and virus-infected[29] cells. In both cases, the integration of cDNAs into the host genome was shown to occur with an extremely low frequency, leading to the suggestion that most of the generated cDNA copies replicate as extrachromosomal structures.[29,30] Moreover, the replication of extrachromosomal structures in eukaryotic nuclei, mainly constituted SINE and LINE-1 repeated elements, and their persistence during embryogenesis and adulthood, are well-described phenomena in a variety of organisms.[31] These structures are associated with genomic instability during development[32] and ageing.[33] Their replication is correlated with endogenous RT activities, of both retroposon[34] and telomeric origin.[35] Interestingly, transgenic sequences can also generate extrachromosomal structures in transgenic animals of a variety of species obtained by SMGT, as briefly mentioned in the introduction[5–8,10] and, in certain instances, also by DNA microinjection.[36–38]

The sporadic integration of foreign sequences are extremely rare events; however, they cannot be categorically excluded. In earlier work, we showed that a specific machinery that is able to mediate the integration of exogenous DNA sequences into the sperm genome is activated when mature spermatozoa are incubated with foreign DNA molecules.[16] In principle, it is possible that this same molecular mechanism is triggered when spermatozoa are exposed to exogenous RNA.

20.5   cDNA Copies are Transcriptionally Competent and are Expressed in Various Tissues

To establish whether the beta-gal-containing reverse transcribed cDNAs are transcriptionallly competent and can give rise to newly acquired genetic traits in the offspring, tissue fragments from PCR-positive organs (spleen, kidney, liver, lung, ovary and testis) were processed for X-gal staining. Several tissue samples (spleen, liver and ovary) were found to be intensely blue stained, indicating that the beta-gal protein was efficiently expressed, whereas other tissues (kidney, lung and testis) were less intensely stained and showed a patchy distribution.[27] Comparable patterns of expression of the beta-gal reporter gene were observed in PCR-positive organs, regardless of their derivation from F_0 or F_1 animals.

As a whole, these data confirm that the reverse transcribed cDNAs are transcriptionally competent, are able to express a new phenotypic trait in founders and are sexually transmitted to the F_1 progeny.

20.6 Risks for Human-Assisted Reproduction

The results summarised thus far show that spermatozoa (i) once depleted of seminal fluid, become extremely reactive towards even very low amounts of foreign nucleic acids; (ii) can capture exogenous RNA and retrotranscribe it into cDNA molecules; and (iii) behave as efficient vectors of both exogenous DNA and newly synthesised RNA-dependent cDNA molecules, which are delivered to early embryos in IVF assays. These intrinsic features of mature spermatozoa may clearly represent a high risk factor for human health when used in assisted reproduction, because exogenous genetic material can be easily internalised (though rarely integrated) in sperm cells depleted of seminal fluid. This risk is highly enhanced when spermatozoa from HIV-positive men, or other sexually transmitted viral diseases, are used for fertilisation. It is known that viral particles, responsible for sexually transmitted diseases such as HIV[39] and herpes,[40] can be taken up by sperm and transferred to oocytes at fertilisation. Under these conditions, spermatozoa may represent a potential threat as vectors of viral transmission from the father directly to the offspring, independent of the role of the mother. Given the ability of the endogenous sperm RT to retrotranscribe cDNA copies from the viral RNA, a critical issue is raised by the possibility that cDNA molecules, either full-length or truncated as a result of partial retrotranscription of the RNA template and not necessarily coding for functional gene products, can be propagated to the offspring. If so, an additional IVF-associated risk, besides the danger of serum-positivity in the offspring, may lead to a disruption of genetic integrity, as the sperm could potentially lead to the creation of an "HIV-transgenic" individual.

20.7 Genesis and Non-Mendelian Inheritance of Extrachromosomal "Retrogenes"

In synthesis, the data summarised in this review indicate that murine sperma-tozoa have the following characteristics: (i) they are endowed with an endogenous RT activity able to generate cDNA copies from exogenous RNA molecules incubated with sperm cells; (ii) cDNA sequences of foreign origin are delivered to oocytes at fertilisation and further transferred to embryos; (iii) these cDNAs are propagated in born founders and transmitted to F_1 progeny in a non-Mendelian fashion; and (iv) the cDNA sequences are present in low copy number (<1 copy per genome), mosaically distributed, are transcriptionally competent and variably expressed in tissues of both F_0 and F_1 individuals. Together, these features strongly suggest that the foreign sequences are not integrated into the host genome but, rather, are maintained as autonomously replicating, extrachromosomal structures. In this conformation, the coding potential of the foreign sequences is maintained intact and the encoded information is expressed throughout development and in adult life.

A new pattern is beginning to emerge, in which new genetic traits can be generated in spermatozoa through a process mediated by endogenous retrotransposon/retroviral machinery and, being efficiently expressed, may cause phenotypic variations in adult individuals. Paradoxically, the new phenotype is not necessarily related to a modified genotype and, more specifically, to an identifiable transgene in the host genome, but to the presence of extrachromosomal sequences in low copy number in positive animals. The structural and functional features of these episomes currently remain elusive. In a speculative scenario, it is possible that foreign sequences taken up by sperm are "captured" by the retrotransposon/retroviral machinery, packaged into defective endogenous retroviral structures and delivered to oocytes at fertilisation. These structures replicate through intermediate DNA sequences – currently detected in positive animals by PCR amplification but not by Southern blotting – throughout embryo development and in tissues of adult animals. The evidence that these sequences are characterised by low copy number and persistent mosaicism suggests that they do not segregate evenly in all cells, but, rather, their distribution may be restricted to a permissive subpopulation of cells. Based on a number of lines of evidence, albeit fragmentary at this stage, these structures are possibly "stored" and allowed to replicate in undifferentiated progenitors and/or "adult" stem cells, but not in terminally differentiated cells and tissues. Although still highly hypothetical, this view is compatible with a body of published data, as described below.

First, retrotransposons and endogenous retroviruses are well known to be highly expressed in poorly differentiated, highly proliferating cells, in both normal cells (*i.e.* in early embryos) and pathological cells (tumours),[41] while being silenced or expressed only at a basal level, in terminally differentiated, nonpathological tissues.[42–44] Similarly, it has recently been shown that long terminal repeat (LTR) enhancers of human endogenous retroviruses are activated in stem/progenitor cells of adult tissues but are inactive in adult somatic cells.[45] Secondly, studies aiming at understanding the formation of processed pseudogenes have revealed that retrotransposon-encoded RT can reverse transcribe cellular mRNAs,[46] generating nonintegrated cDNA copies.[47] As a peculiar aspect of this process, cellular mRNAs can be packaged in defective virions, suggesting that endogenous retroviruses can transfer nonviral RNAs, as well as their own, during cell infection.[48] In this context, it emerges that low-copy, nonintegrated but transcriptionally competent cDNA sequences are generated when the endogenous retroviral machinery is activated.[48,49] Together, these findings indicate the realistic possibility that, within a given tissue, defective endogenous retroviruses may act as potential vectors of foreign RNAs followed by the subsequent production of newly synthesised episomal cDNAs. This may provide a possible explanation for the genesis and expression of new phenotypic traits in positive animals that lack the corresponding transgene. The finding that endogenous RT activities are also present in swine and human spermatozoa (our unpublished results) suggests that this RT-mediated mechanism is not an exclusive feature of mice, but can be generally regarded as an inherent characteristic of male germ cells of mammalian species.

20.8 Conclusions

An increasing body of recently published results challenges the traditional view of spermatozoa as metabolically inert cells and suggests that they may be actively involved in novel, unsuspected functional roles. In this chapter, I have highlighted at least three distinct, though interconnected, novel, functional aspects of these cells. First, the spontaneous ability of spermatozoa, depleted of seminal fluid, to take up, internalise and deliver exogenous DNA or RNA molecules to oocytes at fertilisation, confer on these cells a mutagenic potentiality, reflected in the possibility of generating new genetic traits and inducing phenotypic variations in adult individuals. In this perspective, sperm may be viewed as a potential threat to human health when used in assisted reproduction procedures, particularly in cases in which these cells may be vectors of sexually transmitted viral pathogens and/or causative agents of mutation following the unintentional transmission of exogenous sequences. Second, the expression of genes of retrotransposon/retroviral origin in the germline and the "storage" of a functional RT enzymatic activity in mature spematozoa reveals an unsuspected link between spermatozoa and the retrotransposon/retroviral gene machinery. This association increases the likelihood of spermatozoa being inducers of genome variability because not only exogenous DNA but also RNA sequences are suitable substrates and may represent continuous sources of genome mutation. Finally, the creation of a new genetic trait by reverse transcription and its phenotypic expression in the absence of an identifiable transgene demonstrate the possibile existence of reverse transcription-mediated genetic information, not linked physically to the nuclear genome but biologically active and inheritable in a non-Mendelian way. Interestingly, in a quite different context, the possibility of "a cache of genetic variations beyond that carried in the chromosomes" inherited in a non-Mendelian fashion has been recently reported in *Arabidopsis*.[50] Although still highly hypothetical and requiring experimental confirmation, these results may yield a novel mechanism for the genesis and transmission of genetic information.

Acknowledgments

This work was supported by grants from Istituto Superiore di Sanità (Italian National Institute of Health) (501/1 and 501/2).

References

1. M. Lavitrano, A. Camaioni, V. Fazio, S. Dolci, M.G. Farace and C. Spadafora, *Cell*, 1989, **57**, 717.
2. C. Spadafora, *Bioessays*, 1998, **20**, 955.
3. A.W.S. Chan, C.M. Luetjens and G.P. Schatten, *Curr. Topics Dev. Biol.*, 2000, **50**, 89.
4. K. Smith and C. Spadafora, *Bioessays*, 2005, **27**, 551.
5. H.W. Khoo, K.H. Ang and K.Y. Wong, *Aquaculture*, 1992, **107**, 1.
6. O.J. Rottmann, R. Antes, P. Hoefer and G. Maierhofer, *J. Anim. Breed. Genet.*, 1992, **109**, 64.
7. A.V. Kuznetsov, I.V. Kuznetsova and I.Y.U. Schit, *Mol. Reprod. Dev.*, 2000, **56**, 292.
8. K.O. Robinson, H.J. Ferguson, S. Cobey, H. Vaessin and B.H. Smith, *Insect Mol. Biol.*, 2000, **9**, 625.
9. H.J. Tsai, *Mol. Reprod. Dev.*, 2000, **56**, 281.
10. C. Celebi, P. Auvray, T. Benvegnu, D. Plusquellec, B. Jegou and T. Guillaudeux, *Mol. Reprod. Dev.*, 2002, **62**, 477.
11. M. Lavitrano, M.L. Bacci, M. Forni, D. Lazzereschi, C. Di Stefano, D. Fioretti, P. Giancotti, G. Marfe, L. Pucci, L. Renzi, H. Wang, A. Stoppacciaro, G. Stassi, M. Sargiacomo, P. Sinibaldi, V. Turchi, R. Giovannoni, G. Della Casa, E. Seren and G. Rossi, *Proc. Natl. Acad. Sci. USA*, 2002, **99**, 14230.
12. A.C. Perry, T. Wakayama, H. Kishikawa, T. Kasai, M. Okabe, Y. Toyoda and R. Yanagimachi, *Science*, 1999, **284**, 1180.
13. M. Shemesh, M. Gurevich, E. Harel-Markowitz, L. Benvenisti, L.S. Shore and Y. Stram, *Mol. Reprod. Dev.*, 2000, **56**, 306.
14. H.J. Wang, A.X. Lin, Z.C. Zhang and Y.F. Chen, *Anim. Biotechnol.*, 2001, **12**, 101.
15. K. Chang, J. Qian, M. Jiang, Y.H. Liu, M.C. Wu, C.D. Chen, C.K. Lai, H.L. Lo, C.T. Hsiao, L. Brown, J. Bolen Jr., H.I. Huang, P.Y. Ho, P.Y. Shih, C.W. Yao, W.J. Lin, C.H. Chen, F.Y. Wu, Y.J. Lin, J. Xu and K. Wang, *BMC Biotechnol.*, 2002, **2**, 5.
16. G. Zoraqi and C. Spadafora, *DNA Cell Biol.*, 1997, **16**, 291.
17. B. Maione, C. Pittoggi, L. Achene, R. Lorenzini and C. Spadafora, *DNA Cell Biol.*, 1997, **16**, 1087.
18. C. Pittoggi, L. Renzi, G. Zaccagnini, D. Cimini, F. Degrassi, R. Giordano, A.R. Magnano, R. Lorenzini, P. Lavia and C. Spadafora, *J. Cell Sci.*, 1999, **112**, 3537.
19. C. Pittoggi, G. Zaccagnini, R. Giordano, A.R. Magnano, R. Lorenzini and C. Spadafora, *Mol. Reprod. Dev.*, 2000, **56**, 248.

20. A. Uscheva, Z. Avramova and R. Tsanev, *FEBS Lett.*, 1982, **138**, 50.
21. J.M. Gatewood, G.R. Cook, R. Balhorn, E.M. Bradbury and C.W. Schmid, *Science*, 1987, **236**, 962.
22. J.M. Gatewood, G.R. Cook, R. Balhorn, C.W. Schmid and E.M. Bradbury, *J. Biol. Chem.*, 1990, **265**, 20662.
23. R. Giordano, A.R. Magnano, G. Zaccagnini, C. Pittoggi, N. Moscufo, R. Lorenzini and C. Spadafora, *J. Cell. Biol.*, 2000, **148**, 1107.
24. A.A. Kiessling, *Proc. Soc. Exp. Med.*, 1984, **176**, 175.
25. A.A. Kiessling, R.C. Crowell and R. Connell, *Proc. Natl. Acad. Sci. USA*, 1987, **84**, 8667.
26. A.A. Kiessling, R.C. Crowell and C. Fox, *Proc. Natl. Acad. Sci. USA*, 1989, **86**, 5109.
27. I. Sciamanna, L. Barberi, A. Martire, C. Pittoggi, R. Beraldi, R. Giordano, A.R. Magnano, C. Hogdson and C. Spadafora, *Biochem. Biophys. Res. Commun.*, 2003, **312**, 1039.
28. J. Maestre, T. Tchenio, O. Dhellin and T. Heidmann, *EMBO J.*, 1995, **14**, 6333.
29. P. Klenerman, H. Hengartner and R.M. Zinkernagel, *Nature*, 1997, **390**, 298.
30. E. Segal-Bendirdjian and T. Heidmann, *Biochem. Biophys. Res. Commun.*, 1991, **181**, 863.
31. J.W. Gaubatz, *Mutat. Res.*, 1990, **237**, 271.
32. S. Cohen, S. Menut and M. Méchali, *Mol. Cell. Biol.*, 1999, **19**, 6682.
33. J.W. Gaubatz and S.C. Flores, *Mutat. Res.*, 1990, **237**, 29.
34. J.J. Krolewski and M.G. Rush, *J. Mol. Biol.*, 1984, **174**, 31.
35. S. Cohen and M. Méchali, *EMBO Rep.*, 2002, **3**, 1168.
36. U. Kiessling, K. Becker, M. Strauss, J. Schoeneich and E. Geissler, *Mol. Gen. Genet.*, 1986, **204**, 328.
37. C.C. Mello, J.M. Krame, D. Stinchomb and V. Ambros, *EMBO J.*, 1991, **10**, 3959.
38. A.I. Nikolaev, T.T. Tchkoni, C.A. Kafiani-Eristavi and V. Tarantul, *Mol. Gen. Genet.*, 1993, **236**, 326.
39. B. Baccetti, A. Benedetto, A.G. Burrini, G. Collodel, E. Costantino Ceccarini, N. Cris, A. Di Caro, A.R. Garbuglia, A. Massacesi, P. Piomboni and D. Sollazzo, *J. Cell Biol.*, 1994, **127**, 903.
40. L. Bobroski, A.U. Bagasra, D. Patel, P. Saikumari, M. Memoli, M.V. Abbey, C. Wood, C. Sosa and O. Bagasra, *J. Reprod. Immunol.*, 1998, **41**, 149.
41. C. Spadafora, *Cytogent. Genome Res.*, 2004, **105**, 346.
42. A.A. Kiessling and M. Goulian, *Cancer Res.*, 1979, **39**, 2062.
43. R.I. Salganick, V.P. Tomsons, G.B. Pyrinova, N.P. Korokhov, E.V. Kiseleva and N.B. Khristolyubova, *Biochem. Biophys. Res. Commun.*, 1985, **131**, 492.
44. P. Medstrand and J. Blomberg, *J. Virol.*, 1993, **67**, 6778.
45. W. Pi W, Z. Yang, J. Wang, L. Ruan, X. Yu, J. Ling, S. Krantz, C. Isales, S.J. Conway, S. Lin and D. Tuan, *Proc. Natl. Acad. Sci. USA*, 2004 **101**, 805.

46. C. Esnault, J. Maestre and T. Heidmann, *Nat. Genet.*, 2000, **24**, 363.
47. O. Dhellin, J. Maestre and T. Heidmann, *EMBO J.*, 1997, **16**, 6590.
48. R. Lum and M.L. Linial, *J. Virol.*, 1998, **72**, 4057.
49. A. Tanaka, H. Hara, H.T. Park, J. Wolfert, M. Fujihara, R. Izutani and A. Kaji, *J. Gen. Virol.*, 1992, **73**, 1781.
50. S.J. Lolle, J.L. Victor, J.M. Young and R.E. Pruitt, *Nature*, 2005, **434** 505.

CHAPTER 21

Sperm Abnormalities in Exposed Humans

RADIM J. SRAM[a] AND JIRI RUBES[b]

[a] Institute of Experimental Medicine AS CR and Health Institute of Central Bohemia, 142 20 Prague, Czech Republic
[b] Veterinary Research Institute, 621 32 Brno, Czech Republic

21.1 Introduction

The impact of sperm abnormalities for humans became a hot topic due to its possible effect on male fertility. Also the term "sperm abnormality" was originally related to sperm morphology and sperm head morphology. With new methods used for the human monitoring of genetic damage such as the detection of chromosomal aberrations in sperm by fluorescence *in situ* hybridization (FISH) and the sperm chromatin structure assay (SCSA) results were obtained important to public health and for clinical practice.

Exposure to environmental pollution may contribute to a perceived decline in sperm counts world wide.[1-4] Epidemiological studies in human populations suggest that reproductive impairment can be associated with environmental contamination by endocrine-disruptive chemicals such as polychlorinated biphenyls (PCBs), chlorinated pesticides and 2,3,7,8-tetrahydrochlodibenzo-*p*-dioxin (TCDD).[3,5,6]

What do we know about a possible effect of air pollution due to occupational exposure, life style or environmental exposure?

21.1.1 Occupational Exposure

Data on the impact of pollutants in the chemical industry to induce sperm abnormalities seem to be very limited, if only studies with information about exposure are included (Table 1). Higher levels of Cd and Pb in blood have affected sperm motility and sperm morphology.[7,8] Acrylonitrile is the only exposure inducing DNA strand breakage determined by Comet assay, and sex chromosome disomy.[9] Benzene, ethylbenzene, toluene and xylene have affected sperm count, sperm motility and sperm morphology,[10] benzene also sex chromosome disomy.[11]

Table 1   *The effect of occupational exposure to induce sperm abnormalities*

Pollutant	Exposure	Effect	References
Cadmium	$>10\ \mu g\ L^{-1}$	Sperm motility, sperm morphology	7
Lead	$>200\ \mu g\ L^{-1}$, $400\ \mu g\ L^{-1}$	Sperm count, sperm motility, sperm head morphology	7,8
Acrylonitrile	$>0.8\ mg\ m^{-3}$	Disomy	9
Benzene	$32{-}48$, $84\ mg\ m^{-3}$	Sperm count, sperm motility, sperm morphology, disomy	10
Ethylbenzene	$221{-}234\ mg\ m^{-3}$	Sperm count, sperm motility, sperm morphology	11
Toluene	$190{-}212\ mg\ m^{-3}$	Sperm count, sperm motility, sperm morphology	10
Xylene	$47{-}57\ mg\ m^{-3}$	Sperm count, sperm motility, sperm morphology	10

21.1.2   Life Style

Since many compounds in tobacco smoke are mutagens, it is expected that smoking may affect male reproduction. The recent review of 21 studies by Marinelli *et al.*[12] shows a limited effect of smoking on conventional sperm parameters. Only six studies were performed among healthy men from the general population, three of them showed a decrease of total sperm count, sperm motility and sperm volume.

Using an immunofluorescence method, PAH-DNA adducts were shown to be associated with sperm head morphology, but not related to smoking.[13] A previous study suggested that PAH-DNA adducts in sperm were increased by smoking.[14] The formation of DNA adducts in spermatozoa may be understood to be a potential source of transmissible prezygotic DNA damage. Zenzes *et al.*[15] confirmed the paternal transmission of modified DNA by detection of DNA adducts in spermatozoa of a smoking father and his embryo. Using the ^{32}P-postlabelling method, the bulky DNA adducts were significantly higher between current smokers and never smokers among healthy individuals.[16]

The data on alcohol are more sparse and show a protective effect of moderate alcohol drinking on sperm parameters, especially sperm motility and viability.[12] This effect may be due to the antioxidant effect of some alcoholic beverages, as for example in red wine.

Eskenazi *et al.*[17] studied the association between dietary and supplement intake of micronutrients (zinc and folate) and antioxidants (vitamins C, E and β-carotene) and semen quality in nonsmokers. Higher antioxidant intake was associated with higher sperm numbers and motility.

21.1.3   Environmental Pollution

Studies on the impact of exposure to air pollution on human sperm, using conventional sperm parameters as well as determining DNA damage, have only been done in the Czech Republic.[18–20]

21.2 Biomarkers of Sperm Injury

Biomarkers and their outcomes have been usually associated with fertility status.

21.2.1 Conventional Sperm Parameters

Routine semen analysis includes semen volume, sperm concentration, total number of sperm per sample, percentage of motile sperm, percentage of sperm with normal morphology and percentage of sperm with normal head morphology.[21]

21.2.2 Computer-Aided Sperm Analysis (CASA)

Its outcomes include indicators of sperm progression and sperm vigour, as for example the total number of motile sperm per sample and the total number of progressive sperm per sample.[22,23]

21.2.3 Sperm Chromatin Structure Assay (SCSA)

SCSA detects DNA breakage in sperm cells, using acid-induced danaturation and acridine orange to detect susceptibility of human sperm to *in situ* DNA denaturation. DNA fragmentation is viewed as the molecular precursor to later gross chromosome damage observed under the light microscope.[24,25]

21.2.4 Comet Assay

Comet assay (single-cell gel electrophoresis assay) has been adapted to sperm cells and used to detect DNA single- and double-strand breaks.[26,27]

21.2.5 Sperm FISH Aneuploidy

Sperm FISH aneuploidy uses multi-colour FISH to detect the sex chromosomes and/or selected autosomes in sperm nuclei. Fluorescent chromosome-specific probes label a portion of each chromosome with a unique colour resulting in a fluorescent spot or domain on the sperm nucleus when viewed by fluorescence microscopy. Duplicate fluorescent spots for a single chromosome are indicative of hyperhaploidy (disomy) while duplicate spots for all chromosomes are indicative of diploidy.[28]

21.2.6 DNA Adducts

The ^{32}P-postlabelling method seems to be sensitive to smoking,[16] but no studies with specified air pollution were found.

21.3   Teplice Programme

The Northern Bohemia brown coal basin was perceived as one of the worst environmentally polluted regions in Europe. Conifers in Krušné Hory (Ore Mountains) forming the northern border of this region have been essentially destroyed. This process started more than 20 years ago. First consequences of environmental pollution on known health were a remarkable increase in allergies, immunodeficiencies and respiratory diseases in children. The unfavourable effect of the environment on pregnant women was understood to be the reason for the increase of birth defects and higher numbers of children with low birth weight. An exploratory analysis of data collected prior to 1999 suggested a higher incidence of cancer and reproductive and behavioural effects in this region.[29,30]

The Teplice Programme was initiated by the Czech Ministry of the Environment in 1990 to provide scientifically valid information needed to assess environmental health problems in the Northern Bohemia basin area. In collaboration with the US Environmental Protection Agency (US EPA), the programme started in 1991. Simultaneously, this programme was incorporated into PHARE II, as the EC/HEA/18-CZ project "Impact of Environmental Pollution on the Health of Population (Teplice Programme)." This programme has succeeded in bringing together many different research organizations and government laboratories in the Czech Republic, United States and EC countries to accomplish the multidisciplinary programme.

The hypothesis in the Teplice Programme has been that the air pollution in the Teplice district adversely affects the health of population. The principal objective of this programme was to assess human exposure to toxic air pollutants, to relate ambient concentrations of pollutants to health risks.[18]

21.3.1   Semen Quality

21.3.1.1   First Study

This study examined whether exposure to high levels of air pollution over the entire process of spermatogenesis (about 90 days) was associated with abnormal semen parameters.[19,31] Certain components of air pollution, for example polycyclic aromatic hydrocarbons (PAHs), have altered male reproductive function in test species providing additional rationale for its undertaking.[32,33]

Increased concentrations of SO_2 (sulfur dioxide) in the second month before conception also affected fecundability.[34] In the overall study, 325 young men from both communities, Teplice and Prachatice were examined. Surveys in the fall and late winter included interviews, physical examinations and collection of semen samples. Semen quality measures included concentration, volume, motility and morphology, computer-aided motion analysis and SCSA. Men were classified into exposure groups using relative levels of ambient sulfur dioxide.

Semen was collected, measures included sperm concentrations and volume, percentage of motile sperm, percentage of sperm with normal morphology and percentage with normal head morphology. For the morphology evaluation, 200–300 sperm per sample were assessed from air dried preparations and classified into WHO categories.[20] To estimate the relevant exposure scenarios SO_2 data were grouped by district, sampling time, early fall or late winter, year and exposure window. An examination of these exposure windows showed that mean SO_2 values were similar regardless of the exposure window used. There were no significant relationships observed between measures of sperm production (semen volumes, sperm concentrations or total sperm count) and air pollution in the logistic regression model. The most notable findings of this report were the significant relationships observed between air pollution and sperm morphology with the lower air pollution levels in the late winter in Prachatice with ORs of 0.2 (0.1–0.7) and 0.5 (0.2–2.0) and increasing with medium levels to an OR of 4.1 (1.2–13.9) and high levels to an OR of 10.1 (2.8–36.0). The results were consistent for the linear regression: sperm morphology (very low $b = 3.31$ (0.12, 6.52), low $b = 5.90$ (1.88, 9.93), medium $b = 5.80$ (-8.84, -2.75), $p = 0.0002$ and high $b = -7.25$ (-10.62, -3.89), $p = 0.0001$. These analyses suggest a strong relationship between sperm morphology and air pollution (Table 2) with a possible seasonal effect in the opposite direction. Alteration of sperm head shape may be a significant component of the sperm morphology effect, but does not account for all of it.

SCSA was used for the first time in an epidemiology study on the effect of air pollution. Results show that DNA fragmentation increased in the period of a high pollution.

The main effects associated with air pollution appeared to be postmeiotic effects on sperm motility and morphology. Effects on sperm morphology

Table 2 *Semen outcomes by exposure*

Outcome	N	Low SO₂ = 29.5, PM10 = 35.5, mean ± SD	N	Medium SO₂ = 79.8, PM10 = 61.3, mean ± SD	N	High SO₂ = 164.0, PM10 = 164.7, mean ± SD
Sperm concentration (milmL)	162	59.9 ± 64.3	63	65.4 ± 61.6	47	60.1 ± 46.7
Total count (mil/sample)	162	113.5 ± 130.7	63	100.9 ± 97.6	47	129.1 ± 103.1
Motile sperm[a] (%)	156	36.2 ± 17.1	63	27.9 ± 18.1	37	41.6 ± 40.4
Normal sperm morphology[a] (%)	154	19.8 ± 8.5	62	15.9 ± 5.5	46	13.2 ± 6.5
Normal sperm head morphology[a] (%)	154	39.3 ± 10.9	62	15.9 ± 5.5	46	13.2 ± 6.5
SCSA-DFI[a] (%)	158	19.2 ± 12.2	61	16.2 ± 9.3	47	28.8 ± 20.4

adapted from ref 29.
[a] Different by Kruskal–Wallis test, $p < 0.05$.

suggest probably occurred during spermiogenesis when the normal spermatids were transformed into differentiated sperm cells. Severe alterations in motility and morphology can be associated with infertility. In general, these data suggest that exposure to air pollution for one spermatogenic cycle may increase the risk of altered semen quality. This appears to be reversible since the young men evaluated 6 months after high pollution episodes have improved semen quality.

In the cohort of 25 subjects from the Teplice district aneuploid human sperm by FISH were analysed by FISH.[35] Semen samples were provided by ten men who reported smoking 20 cigarettes/day, and 15 nonsmokers who reported no more than minimal exposure to passive smoking. Sperm FISH aneuploidy was determined with the use of three probes directly labelled for chromosome X, Y and 8 (Table 3). This method allows for an accurate distinction between disomic and diploid sperm nuclei, nullisomic and nonhybridising spermatozoa and meiosis I and meiosis II errors in sex-chromosome aneuploidy and diploidy.[19,35]

There were no statistically significant differences between smokers and non-smokers with regard to occupational exposure, health status, passive smoking, age at first seminal emission and other activities. Smokers produced significantly elevated aggregate frequencies of sperm, aneuploidy for chromosomes X, Y and 8 ($p < 0.01$). The frequencies of sperm with Y disomy were significantly elevated among smokers in comparison with nonsmokers ($p < 0.001$). There have been no previous reports relating an exogenous exposure to increased frequencies of Y disomy in sperm. The finding of increased frequencies of Y disomy in sperm may have implications for the risk of having an aneuploid child. The proportion of affected children may be higher in men who smoke more.

The relationship between air pollution in the district of Teplice, other covariates of interest and sperm aneuploidy was described using Poisson and linear regression modelling.[19] YY8 aneuploidy was significantly associated with

Table 3   *Aneuploidy in human sperm by FISH*

Aneuploidy type	Teplice Spring 1994[35]		Teplice[19]	
	NS	S	Spring 1993	Autumn 1993
	$N = 15$	$N = 10$		
Disomy				
X	5.8 ± 0.6	5.9 ± 0.8	3.7 ± 2.4	2.8 ± 2.9
Y	2.2 ± 0.3	4.5 ± 0.6^b	3.5 ± 2.3	0.6 ± 0.9^a
X – Y	14.0 ± 1.4	20.4 ± 4.4	4.4 ± 4.2	5.8 ± 3.2
8	4.7 ± 0.6	6.5 ± 1.0	2.5 ± 2.1	2.4 ± 3.5
Diploidy				
X – X – 8 – 8	2.3 ± 0.6	4.7 ± 1.4	3.0 ± 3.2	5.0 ± 5.2
Y – Y – 8 – 8	2.6 ± 0.6	2.5 ± 1.1	2.8 ± 2.6	4.2 ± 3.1
X – Y – 8 – 8	19.0 ± 5.3	23.9 ± 5.7	15.0 ± 24.0	17.0 ± 13.0

Note: NS, nonsmokers; S, smokers.
[a] $p < 0.01$,
[b] $p < 0.001$.

the season of heaviest air pollution (OR = 5.25, 95% CI 2.5–11.0; linear modelling of normally transformed YY8, coefficient 1.44, p < 0.0001). No other cytogenetic endpoints were significantly associated with seasonal air pollution. These findings are suggestive of an influence of seasonal air pollution on YY8 disomy. J. Rubes put forward an idea for further study if disomy of chromosome Y in sperm could be used as a marker of sperm injury by chronic exposure to environmental pollutants.

21.3.1.2 Second Study

Later Rubes *et al.*[36] put forward a longitudinal study. Thirty-six men from the Teplice District, recruited among the volunteers participating in the first study, were surveyed on seven occasions over 2 years time (Table 4). All methods corresponded to the first study. Semen outcomes were analysed for changes associated with high levels of air pollution (PM10 particulate matter <10 μm), PAHs, carc-PAHs, SO_2, NO_x). Descriptive statistics for semen outcomes are given in Table 5. No significant associations were found between exposure and sperm concentration, percent motile sperm or percent normal sperm heads. DNA fragmentation index (DFI), obtained using SCSA, showed significant (p < 0.05) positive associations with air pollution categorised as high *vs.* low, and with SO_2 levels.

This study found a significant association between exposure to air pollution and the percentage of sperm with fragmented DNA, indicating that exposure to high levels of air pollution may have damaging effects on sperm DNA.

The sperm aneuploidy assay was conducted for subsets ($N = 15$) of men from the longitudinal study,[36,37] evaluated by the X–Y–8 multicolour sperm FISH method (Table 5). There was no significant effect of season (high or low pollution) on aneuploidy or diploidy for any of the chromosomes studied.

Comparing the first and second studies, the DNA fragmentation in sperm by SCSA was the only biomarker observed to be related to the air pollution exposure. No significant associations with sperm morphology, motility, CASA parameters or sperm aneuploidy. This lack of consistency between studies may be related to a substantial decline of air pollution due to the remedial actions by the Czech government.[38]

With specific reference to the two semen studies described above, mean SO_2 levels for comparable 90 day intervals (late December to late March) were notably higher in 1993[20] (164.0 μg m^{-3}) compared with 1996 (78.5 μg m^{-3}). The same was true for PM10 where the comparable 1993 mean[20] was 184.7 μg m^{-3} compared with 67.8 μg m^{-3} for 1996.

In a subsequent study conducted in 1998, another group of 50 18-year old men were similarly examined (Rubes, unpublished data). Comparison of mean values for the first group of men sampled in spring 1993 to this group sampled in 1998 showed that sperm concentration and percentage of motile spermatozoa were significantly higher (60.1 *vs.* 102.3 mil mL^{-1} and 32.5 *vs.* 62% motile, respectively) in the 1998 group. Taken together, the all these studies provide evidence that exposure to high levels of air pollution is associated with

Table 4 *Air pollution levels for 90 days preceding collection of semen samples*[36]

Date Cycle	Sept. 1995 1	Jan. 1996 2	Feb. 1996 3	March 1996 4	Sept. 1996 5	Feb. 1997 6	Sept. 1997 7
PM10 (ng/mg)[a]	41.4 ± 15.6	51.9 ± 35.0	60.7 ± 38.0	67.8 ± 40.1	26.0 ± 18.3	62.7 ± 44.8	28.7 ± 15.3
PAH total (ng/mg)[a]	39.0 ± 17.2	144.1 ± 91.6	164.1 ± 90.9	140.9 ± 76.0	29.3 ± 11.0	191.0 ± 121.9	31.4 ± 15.4
carc-PAH (ng/mg)[a]	4.8 ± 6.7	34.1 ± 26.4	40.6 ± 28.2	36.5 ± 24.8	1.7 ± 1.6	46.6 ± 33.9	3.2 ± 4.8
SO_2 ($\mu g/m^3$)[a]	27.7 ± 18.4	74.5 ± 83.4	87.5 ± 81.3	78.5 ± 75.3	15.0 ± 13.7	91.0 ± 118.8	19.1 ± 14.5
NO_x ($\mu g/m^3$)[a]	44.3 ± 16.4	79.0 ± 54.7	80.2 ± 55.6	77.5 ± 50.3	39.3 ± 17.9	107.7 ± 67.8	51.8 ± 27.7

Note: PAHs, polycyclic aromatic hydrocarbons extracted from particulate fraction; carc-PAH, carcinogenic fraction of PAH.
[a] Mean ± SD.

Table 5 *Descriptive statistics for semen outcomes obtained by repeated sampling of 36 men seven times over two years*[a]

Semen endpoints Sample	Sept. 1995 1 LOW	Jan. 1996 2 HIGH	Feb. 1996 3 HIGH	March 1996 4 HIGH	Sept. 1996 5 LOW	Feb. 1997 6 HIGH	Sept. 1997 7 LOW
Sperm concentration (mil/mL)	98.6 (74.9–122.4)	78.5 (57.1–99.9)	79.9 (60.9–99.0)	103.1 (76.6–129.6)	92.1 (66.3–117.9)	81.6 (67.9–95.4)	103.6 (65.4–141.7)
Motile sperm (%)	58.5 (52.0–65.0)	55.0 (47.2–62.9)	59.1 (52.0–66.3)	66.3 (61.8–70.7)	62.7 (55.7–69.8)	68.3 (64.4–72.3)	56.2 (50.9–61.6)
Normal sperm morphology (%)	17.5 (14.9–20.1)	14.8 (13.1–16.6)	15.8 (11.8–19.7)	12.7 (10.7–14.6)	11.3 (9.3–13.3)	8.4 (7.5–9.2)	7.9 (6.8–8.9)
Normal sperm head morphology[c] (%)	29.5 (25.8–33.1)	29.0 (26.6–31.3)	26.4 (23.7–29.1)	26.8 (23.6–29.9)	26.0 (23.4–28.6)	24.5 (23.1–26.0)	27.8 (26.4–29.2)
SCSA-DFI[c] (%)	15.1 (12.4–17.8)	20.3 (16.0–24.6)	15.8 (11.88–19.7)	17.4 (13.0–21.7)	13.5 (10.3–21.7)	15.4 (11.3–19.5)	12.2 (9.5–14.8)
Total aneuploidy[b] (#/10,000 sperm)	21.2 (16.0–26.6)	24.2 (17.3–31.0)	22.6 (13.7–31.6)	20.1 (13.0–27.2)	21.3 (13.8–28.7)	18.5 (14.0–23.0)	24.0 (15.6–32.5)

adapted from ref 36.
[a] Values are mean (95% confidence interval) for each group of samples. Pollution is designated as "LOW" or "HIGH" for the 3 months preceding each sample.
[b] Values for total aneuploidy represent the sum of total disomy and total diploidy and are based on 15 men who contributed seven samples each.
[c] "LOW" vs all "HIGH" periods, $p < 0.05$

decreases in semen quality, and that these appear to be reversible once the pollution is lowered.

21.3.2 Dioxin-Like and Estrogen Activity of Xenobiotics

Machala *et al.*[39] studied concentrations of PCB congeners and organochlorine contaminants in ejaculate and whole blood samples from 25 young men from the Teplice district. Their levels in semen and blood samples were similar, and no specific increase of any xenobiotic was observed (Table 6). Using transgenic cell lines, the sample extracts were screened for estrogenic and dioxin-like activities *in vitro*. Relatively high estrogenic activity was detected in several samples of semen and blood. It may be speculated that the effect of higher concentrations of carcinogenic PAHs are adsorbed on respirable particles.

21.4 Conclusions

The results of the studies from the Czech Republic are the only research on the effect of air pollution to human sperm. The impact of air pollution depends on the concentration of pollutants. Probably the most important are respirable particles (PM10) and carcinogenic PAHs (the originally used SO_2 was a substitute and the only pollutant regularly monitored in the previous period).

As sensitive biomarkers the following method may be recommended: sperm morphology, DNA fragmentation by SCSA and sperm aneuploidy. For the other biomarkers there is still the lack of data, which could prove their sensitivity for epidemiological studies.

Observed decreases in the percentage of normal sperm morphology and an increase of DNA fragmentation by SCSA and sperm aneuploidy may have adverse effects on reproductive outcomes such as infertility, delayed conception, spontaneous abortions or adverse developmental outcomes.

This new knowledge should be used for risk assessment. Environmental air pollution can affect the quality of sperm so transmitting genetic damage to embryos and also affecting pregnancy outcome. These effects may be very significant for developing countries, as industrial development is usually followed by air pollution.

Acknowledgments

This review was supported by the Ministry of Environment of the Czech Republic (grants no. VaV/740/5/03), MZE 00027 16201, Academy of Sciences of the Czech Republic (IQS 500 390506) and the Commission of the European Communities (grant no. QLK4-CT-2002-02198 ChildrenGenoNetwork). The authors thank B. Binkova, A. Milcova, D. Zudova, Z. Zudova, M. Vozdova, M. Machala, I. Benes. F. Kotesovec, J. Nozicka, D.P. Evenson, S.D. Perrault, W.A. Robbins and S.G. Selevan for their continual support and use of their data.

Table 6 *Xenobiotics in human ejaculates and blood* (μg/mL)

PCB congener no.	28	52	101	118	138	153	180	Σ
Ejaculate[a]	0.49 ± 0.40	0.60 ± 0.50	0.31 ± 0.21	0.21 ± 0.15	0.61 ± 0.24	0.94 ± 0.67	0.22 ± 0.18	3.35 ± 1.69
Blood[a]	0.74 ± 0.23	0.58 ± 0.18	0.55 ± 0.24	0.29 ± 0.16	1.52 ± 0.61	1.62 ± 0.58	1.08 ± 0.39	6.37 ± 2.05
Organochlorine compounds	*α-HCH*		*β-HCH*		*δ-HCH*		*Lindane*	*HCB*
Ejaculate[a]	0.45 ± 0.36	0.48 ± 0.73	0.14 ± 0.15	5.00 ± 4.06	0.90 ± 0.96			
Blood[a]	0.17 ± 0.14	0.26 ± 0.17	3.95 ± 0.51	0.07 ± 0.04	0.42 ± 0.19			
	o,p'-DDE	*p,p'-DDE*	*o,p'-DDD*	*p,p'-DDD*	*o,p'-DDT*	*p,p'-DDT*		
Ejaculate[a]	1.34 ± 1.06	4.84 ± 7.60	0.26 ± 0.32	0.25 ± 0.30	2.20 ± 3.64	1.25 ± 2.61		
Blood[a]	0.14 ± 0.08	1.52 ± 0.70	0.13 ± 0.07	1.11 ± 0.06	0.30 ± 0.15	0.50 ± 0.32		

adapted from ref 39.
[a] Mean ± SD.

References

1. E. Carlsen, A. Giwercman, N. Keiding and N. E. Skakkebaek, *Br. Med. J.*, 1992, **305**, 609.
2. J. Auger, J.M. Kunstmann, F. Czyglik and P. Jouannet, *N. Engl. J. Med.*, 1995, **332**, 281.
3. J. Toppari, J.C. Larsen, P. Christiansen, A. Giwercman, P. Grandjean, L.J. Guillette Jr., B. Jégou, T.K. Jensen, P. Jouannet, N. Keiding, H. Leffers, J.A. McLachlan, O. Meyer, J. Muller, E. Rajpert-De Meyts, T. Scheike, R. Sharpe, J. Sumpter and N.E. Skakkebaek, *Environ. Health Perspect.*, 1996, **104**(Suppl. 4), 741.
4. A.G. Andersen, T.K. Jensen, E. Carlsen, N. Jorgensen, A.M. Andersson, T. Drarup, N. Keiding and N.E. Skakkebaek, *Hum. Reprod.*, 2000, **15**, 366.
5. R.J. Kavlock, G.P. Daston, C. De Rosa, P. Fenner-Crisp, L.E. Gray, S. Kaatari, G. Lucier, M. Luster, M.J. Mac, C. Maczka, R. Miller, J. Moore, R. Rolland, G. Scott, D.M. Sheehan, T. Sinks and H.A. Tilson, *Environ. Health Perspect.*, 1996, **104**(Suppl. 4), 715.
6. R. Hauser, Z. Chen, L. Pothier, L. Ryan and L. Altshul, *Environ. Health Perspect.*, 2003, **111**, 1505.
7. S. Telisman, P. Cvitkovic, J. Jurasovic, A. Pinzent, M. Gavella and B. Rocic, *Environ. Health Perspect.*, 2000, **108**, 45.
8. M. De Rosa, S. Zarrilli, L. Paesano, U. Carbone, B. Boggia, M. Petretta, A. Maisto, F. Cimmino, G. Puca, A. Colao and G. Lombardi, *Hum. Reprod.*, 2003, **18**, 1055.
9. D.X. Xu, Q.X. Zhu, L.K. Zheng, Q.N. Wang, H.M. Shen, L.X. Deng and Ch.N. Ong, *Mutat. Res.*, 2003, **537**, 93.
10. R. De Cells, A. Feria-Velasco, M. Gonzales-Unzaga, J. Torres-Calleja and N. Pedron-Nuevo, *Fertil Steril.*, 2000, **73**, 221.
11. X. Liu, L.K. Deng and Q. Zhang, *Yi. Chuan. Xue.Bao*, 2001, **28**, 589.
12. D. Marinelli, L. Gaspari, P. Pedotti and E. Taioli, *Int. J. Environ. Health*, 2004, **207**, 185.
13. L. Gaspari, Seong-Sil Chang, R.M. Santella, S. Garte, P. Pedotti and E. Taioli, *Mutat. Res.*, 2003, **535**, 155.
14. M.T. Zenzes, R. Bielecki and T.E. Reed, *Fertil. Steril.*, 1999, **72**, 330.
15. M.T. Zenzes, L.A. Puy, R. Bielecki and T.E. Reed, *Mol. Hum. Reprod.*, 1999, **5**, 125.
16. S. Horak, J. Polanska and P. Widlak, *Mutat. Res.*, 2003, **537**, 53.
17. B. Eskenazi, S.A. Kidd, A.R. Marks, E. Sloter, G. Block and A.J. Wyrobek, *Hum. Reprod.*, 2005, **20**, 1006.
18. R.J. Sram, I. Benes, B. Binkova, J. Dejmek, D. Horstman, F. Kotesovec, D. Otto, S.D. Perreault, J. Rubeš, S.G. Selevan, I. Skalik, R.K. Stevens and J. Lewtas, *Environ. Health Perspect.*, 1996, **104**(Suppl. 4), 699.
19. W.A. Robbins, J. Rubes, S.G. Selevan and S.D. Perreault, *Environ. Epidemiol. Toxicol.*, 1999, **1**, 125.
20. S.G. Selevan, L. Borkovec, V.L. Slott, Z. Zudova, J. Rubes, D.P. Evenson and S.D. Perreault, *Environ. Health Perspect.*, 2000, **108**, 887.

21. WHO, *WHO Laboratory Manual for the Examination of Human Semen and Semen-Cervical Mucus Interactions*, 3rd edn., Cambridge University Press, Cambridge, UK, 1992.
22. S.M. Schrader, R.E. Chapin, E.D. Clegg, R.O. Davis, J.L. Fourcroy, D.F. Katz, S.A. Rothmann, G. Toth, T.W. Turner and M. Zinaman, *Reprod. Toxicol.*, 1992, **6**, 275.
23. S.P. Boyers, R.O. Davis and D.F. Katz, *Curr. Probl. Obstet. Gynecol. Fertil.*, 1989, **12**, 167.
24. D.P. Evenson, L.K. Jost, M.J. Zinaman, E. Clegg, K. Pulvis, P. De Angelis and O.P. Clause, *Hum. Reprod.*, 1999, **14**, 1039.
25. D.P. Evenson and R. Wixon, *Toxicol. Appl. Pharmacol.*, 2005, **207**, 532.
26. C.M. Hughes, S.E.M. Lewis, V.J. McKelvey-Martin and W.A. Thompson, *Mol. Hum. Reprod.*, 1996, **2**, 613.
27. D. Anderson, N. Basaran and M.M. Dobrzynska, *Teratogen. Carcinogen. Mutagen.*, 1997, **17**, 29.
28. S.D. Perreault, J. Rubes, W.A. Robbins, D.P. Evenson and S.G. Selevan, *Andrologia*, 2000, **32**, 247.
29. R.J. Sram, I. Roznickova, V. Albrecht, A. Berankova and E. Machovska, in *Mechanisms of Environmental Mutagenesis-Carcinogenesis*, A. Kappas (ed), Plenum Press, New York, 1990, 255.
30. R.J. Sram, in *Ethical Issues of Molecular Genetics in Psychiatry*, R.J. Sram, V. Bulyshenko, L. Prilipko and Y. Christen (eds), Springer, Berlin, 1990, 94.
31. R.J. Sram, B. Binkova, P. Rossner, J. Rubes, J. Topinka and J. Dejmek, *Mutat. Res.*, 1999, **428**, 203.
32. K.M. MacKenzie and D.M. Angevine, *Biol. Reprod.*, 1981, **24**, 183.
33. E. Ford and C. Huggins, *J. Exp. Med.*, 1963, **118**, 27.
34. J. Dejmek, R. Jelinek, I. Solansky, I. Benes and R.J. Sram, *Environ. Health Perspect.*, 2000, **108**, 647.
35. J. Rubes, X. Lowe, D. Moore, S. Perreault, V. Slott, D. Evenson, S.G. Selevan and A.J. Wyrobek, *Fertil. Steril.*, 1998, **70**, 715.
36. J. Rubes, S.G. Selevan, D.P. Evenson, D. Zudova, M. Vozdova, Z. Zudova, W.A. Robbins and S.D. Perreault, *Hum. Reprod.*, 2005, **20**, 1.
37. J. Rubes, M. Vozdova, W.A. Robbins, O. Rezacova, S.D. Perreault and A.J. Wyrobek, *Am. J. Hum. Genet.*, 2002, **70**, 1507.
38. R.J. Sram (ed), *Teplice Program – Impact of Air Pollution on Human Health*, Academia, Prague, 2001.
39. M. Machala, R. Ulrich, J. Vondracek, J. Rubes and R.J. Sram, *Environ. Epidemiol. Toxicol.*, 2000, **2**, 24.

CHAPTER 22

Oestrogenic Compounds and Oxidative Stress

DIANA ANDERSON,[a] EDUARDO CEMELI,[a] THOMAS E. SCHMID,[a,b,c] ADOLF BAUMGARTNER,[a,d] MARTIN H. BRINKWORTH[a] AND JOHN M. WOOD[a]

[a] Department of Biomedical Sciences, University of Bradford, BD7 1DP, UK

[b] Biology and Biotechnology Research Program, Lawrence Livermore National Laboratory, Livermore, CA, USA

[c] School of Public Health, University of California in Berkeley, Berkeley, CA, USA

[d] Ob/Gyn & Reproductive Sciences, University of California, San Francisco, 533 Parnassus, Box 0720, Rm. 255, San Francisco, CA 94143-0720, USA

22.1 Introduction

22.1.1 Oestrogens and Reactive Oxygen Species

There is concern that oestrogens can affect both human health and the environment.[1] The role of oestrogens as hormones and enhancers of transcription factors for many genes has been thoroughly investigated.[2–5] However, their role in the generation of reactive oxygen species (ROS) leading to oxidative stress and DNA damage has received limited attention in cellular systems. Here, we provide evidence that oestrogens appear to produce ROS, especially hydrogen peroxide, at levels that disrupt DNA structure significantly in spermatozoa and in peripheral blood lymphocytes. Severe oxidative stress and DNA damage requires the production of 10^{-3}M of hydrogen peroxide in cells.[6] DNA damage from hydrogen peroxide is most likely caused by the localised production of highly reactive hydroxyl radicals by the Fenton and the Haber–Weiss reactions [$Fe^{2+} + H_2O_2 \rightarrow Fe^3 + OH^- + OH^{\cdot}$; $O_2^- + H_2O_2 \rightarrow O_2 + OH^- + OH^{\cdot}$].[7] These reactions are not really separate entities. In the Haber–Weiss reaction, superoxide is used as the reductant that allows the FeII-catalysed Fenton chemistry to proceed. The production of hydrogen peroxide by oestrogens and the ensuing oxidative stress can be reversed by the addition of catalase and, to a lesser extent, by superoxide dismutase suggesting that hydrogen peroxide is produced directly from molecular interactions of the oestrogens. An important issue is how the hydrogen peroxide is derived at levels that produce damage. A potential source of hydrogen peroxide is from (a) cytochrome P-450-catalysed reactions and (b) the metabolism of oestrogens containing an aromatic phenol ring to catechols followed by oxidation to semiquinone radical intermediates that produce orthoquinones as the final products.[8] Quinoids, quinoid radicals and phenoxyl radicals have also been implicated.[9] Cytochrome P450 is highly expressed in cells and tissues and is concentrated in the endoplasmic reticulum. In the case of spermatozoa, which do not possess an endoplasmic reticulum, their capacity for P-450-mediated hydroxylations is extremely limited, although aromatisation of androgens to oestrogens, which are the terminal ligands in steroid biosynthesis by the loss of the C-19 methyl group and the formation of an aromatic ring, has been described.[10,11] Aromatase is required for the conversion. It is also possible that there may be some additional cells in the sperm ejaculate preparations (immature germ cells or possibly neutrophils) that are capable of such P-450-mediated chemical conversions. It is known that lymphocytes are capable of P-450 metabolism.[12] Quinoids, quinoid radicals and phenoxyl radicals have also been implicated.[9] P-450 enzymes are very important in the biosynthesis and degradation of oestrogens. These mixed function oxygenases react with a large number of different substrates and form a superfamily with more than 50 members. Some of the most powerful carcinogens are produced by cytochrome P-450.[13–21]

Competitive inhibition of P-450 by oestrogen analogues short circuits the enzyme cycle producing hydrogen peroxide from the Fe^{3+}-O-O_2- catalytic intermediate. By contrast, the P-450-mediated oxidation of the aromatic phenolic group of oestrogens finally to orthoquinones occurs by two single electron transfer reactions generating two superoxide anion radicals from two oxygen molecules.[20] Disproportionation of the two superoxide anions to hydrogen peroxide and oxygen by the superoxide dimutases lowers the localised level of oxidative stress, although quinones and semiquinones can also damage DNA directly.[8,9] The two, one-electron oxidations of oestrogen to generate the corresponding orthoquinone and two molecules of superoxide require that the starting material is a 2-, or 4-catechol oestrogen. If catechol oestrogens are supplied directly to the spermatozoa then superoxide production is definitely activated *via* this mechanism. However, if a simple phenol is used as the starting point, such as 17 β-oestradiol, an initial P-450 hydroxylation must occur, probably interacting with the endoplasmic reticulum in the case of lymphocytes, and probably utilising cells of the the raw ejaculate in the case of sperm. Benzene-related compounds that react through oxygen radical mechanisms such as benzoquinone and hydroquinone have been shown to produce DNA damage in the Comet assay.[12] Despite the fact that the free-radical chemistry associated with oestrogenic compounds is being documented,[9] surprisingly little work has focused on the effects it can cause in cells. We have investigated, in the Comet assay in human sperm and blood, the effects of various oestrogenic compounds, such as diethylstilboestrol, β-oestradiol and nonylphenyl, also including phytoestrogens (daidzein, genistein and equol) and their modulation with CAT, SOD and Vit C.

Beta-oestradiol

Daidzein

Nonylphenyl

Equol

Diethylstilboestrol

Genistein

22.1.2 Flavonoids

In addition, we have also investigated the effect of the flavonoids, quercetin and kaempferol, in modulating the responses of the oestrogens. Flavonoids are low-molecular-weight compounds present in seeds, citrus fruits, olive oil, tea and red wine, with possible anti-oxidant activities *in vivo*. However, only in some cases are their anti-oxidant properties known *in vitro*.[22] Kaempferol and quercetin have been reported to act in a pro-oxidant or anti-oxidant manner depending on their concentration.[23–27] Flavonoids are known to interact with cellular signal pathways controlling the cell cycle, differentiation and apoptosis.[28] These features and their pro-oxidant or anti-oxidant properties seem to depend on the bioavailability of quercetin and kaempferol inside the cell.

There are published reports of the pro-oxidant effects of kaempferol and quercetin on mammalian cells in the Comet assay, indicating that these flavonoids have a concentration-dependent pro-oxidant effect on their own and, in the case of quercetin, anti-oxidant effects once combined with food mutagens.[23] Some other studies support the anti-oxidant capability of quercetin in cells treated with H_2O_2[24,25] or H_2O_2-induced DNA damage, for instance, by NF-kappa B DNA-binding activity.[26] This last study also reported that quercetin inhibits DNA damage when induced *via* γ-radiation rather than by the Fenton reaction. By contrast, kaempferol was not effective or less effective than quercetin[27] in modulating H_2O_2-DNA-induced damage. Kaempferol and, to a lesser degree, quercetin like genistein and daidzein have been shown to have affinity for the oestrogen receptor and to activate transcription *via* the oestrogen receptor[29,30] as is the case for estrogens in general.[1] However, at the doses used, the mechanism of action through oxidative stress is more relevant. Therefore, while the ROS properties of the oestrogen-like compounds (DES is a synthetic hormone, β-oestradiol is an endogenous hormone and genistein and daidzein are phytoestrogens) and the anti-oxidant properties of flavonoids (quercetin and kaempferol) are known, their effects in combination are not. Since flavonoids are present in the diet, it would be useful to determine if damage produced by oestrogens, by whatever mechanism it occurs *in vivo*, could possibly be counteracted by these dietary constituents.

22.2   Methodology

The chemicals were obtained from Sigma-Aldrich. RPMI 1640 and normal-melting-point and low-melting-point agarose were obtained from Gibco (Paisley, UK). Tris-hydrochloride and superfrost slides were obtained from BDH (Poole, UK). Lymphoprep was purchased from Axis-shield (Oslo, Norway); foetal calf serum (FBS) from Nalgene (Rochester, NY, USA); phosphate buffered saline (PBS) from Oxoid (Basingstoke, UK); proteinase K from Roche (Basel, Switzerland); and sodium chloride from Riedle de Haën (Seelze, Germany).

Heparinised blood samples (10 ml) were obtained by venepuncture from healthy 25-year old and 31-year old male donors. Lymphocytes were isolated from whole blood [using lymphoprep (Axis Sheild, Oslo Norway] and aliquots stored in liquid nitrogen; the cells were resuspended in RPMI-1640 for analysis $(2 \times 10^5$ cells $ml^{-1})$ and were treated for 30 min with the oestrogenic compounds.[31] Semen was obtained from a 34-year-old male and analysed according to WHO criteria.[32] The attributes of semen quality were sperm number: 62×10^6/ml; motility: 64%; sperm morphology: abnormal forms: 25%. 10 μl sperm from aliquots frozen at $-80°C$[33] were made up to 1 ml RPMI 1640 medium in a microcentrifuge tube. Each was treated for 1 h with the compounds at $37°C$.[23,32–34] The procedure was then carried out as above for the lymphocytes; however, just before the slides were placed in lysis solution 0.05 mg ml^{-1} proteinase K (Roche Diagnostics, Germany) was added and the slides were left in the incubator overnight at $37°C$. The cells were checked for their viability before the start and after the completion of the experiment using trypan blue dye.[35] Preparation of slides for the Comet assay and subsequent electrophoresis, and staining were carried as previously described in the alkaline assay[31] for the lymphocytes and the sperm,[23,32–34] using a modification of the methods of Singh *et al.*[36,37] Slides were scored using an image analysis system (Kinetic Imaging, Liverpool, UK) attached to a fluorescence microscope

(Leica, Germany) equipped with appropriate filters. The microscope was connected to a computer through a charge coupled device (CCD) camera to transport images to appropriate software (Komet 4.0, Kinetic Imaging Ltd, Liverpool) for analysis. The final magnification was 200 (20 objective × 10 lens). The parameter taken for the lymphocytes was tail moment (arbitrary units). The image analysis software automatically generated the tail moment, which is the tail length multiplied by the intensity. The parameter taken for the sperm was percentage head DNA and is the percentage of total intensity remaining in the head. Because the background damage level of 20% is higher in sperm as opposed to 1.5% in lymphocytes,[23] with such a high background the parameter percentage head DNA allows for more appropriate statistical analysis. Images from 50 cells (25 from each replicate slide) were analysed.

Two independent studies were carried out to provide reproducibility of data both for sperm and for lymphocytes with the oestrogenic chemicals and H_2O_2. CAT, SOD and Vit C, quercetin and kaempferol were each examined at two doses (see Tables 1 and 2 and Figures 1 and 2 for details and dose ranges). Statistical analysis was performed in each study on median values, and the mean ±S.E. values of results are shown. The tail moment data for the lymphocytes and the percentage head DNA for the sperm violated the normality test and equal variance test required for the parametric analysis. Pairwise comparisons of all treatment groups *vs.* the control were performed using Kruskal Wallis ANOVA followed by the Mann–Whitney *U*-test.

22.2.1 Responses of Sperm and Lymphocytes to Oestrogens and Anti-oxidants

The viability of sperm and lymphocytes exceeded 80% in all Comet assays after treatment and this excluded artifactual results due to the toxic effects of the chemicals.

A typical experiment for equol is shown in Table 1 for sperm. Equol induced DNA damage by decreasing the percentage of DNA in the sperm head, and the responses were increased towards control levels by treatment with equol in

Table 1 *Effect of equol (EQ) and catalase (CAT) on DNA damage in the Comet assay in human sperm (% Head DNA)*

		Study 1		Study 2	
		Mean ± SE	*Median*	*Mean ± SE*	*Median*
(a)	Negative control	76.34 ± 1.77	78.56	76.75 ± 1.34	75.04
(b)	250 μM EQ	47.01 ± 1.61	49.13	47.48 ± 1.34	46.02
(c)	250 μM EQ + 100 U ml^{-1} CAT	55.06 ± 1.53	53.31	60.02 ± 1.35	58.53
(d)	250 μM EQ + 500 U ml^{-1} CAT	54.16 ± 1.31	53.43	63.44 ± 1.61	62.43

Notes: Mann-Whitney test results:
A–b: $p < 0.001$ (Study 1); b–c: $p < 0.01$ (Study 1); b–d: $p < 0.01$ (Study 1).
A–b: $p < 0.001$ (Study 2); b–c: $p < 0.001$ (Study 2); b–d: $p < 0.001$ (Study 2).
Study 1 and 2 represent two samples from a single donor.

Table 2   *Effect of daidzein (DA) and catalase (CAT) on DNA damage in the Comet assay in human lymphocytes (Tail moments)*

		Study 1		Study 2	
		Mean ± SE	Median	Mean ± SE	Median
(a)	Negative control	1.52 ± 0.38	0.73	0.65 ± 0.15	0.25
(b)	250 μM DA	4.37 ± 0.66	2.59	5.97 ± 0.94	4.1
(c)	250 μM DA + 100 U ml^{-1} CAT	1.18 ± 0.24	0.55	2.36 ± 0.37	1.38
(d)	250 μM DA + 500 U ml^{-1} CAT	2.19 ± 0.39	1.31	2.33 ± 0.39	1.42

Notes: Mann–Whitney test results
a–b: $p < 0.001$ (Study 1); b–c: $p < 0.001$ (Study 1); b–d: $p < 0.01$ (Study 1).
a–b: $p < 0.001$ (Study 2); b–c: $p < 0.001$ (Study 2); b–d: $p < 0.01$ (Study 2).
Study 1 and 2 represent two samples from a single donor.

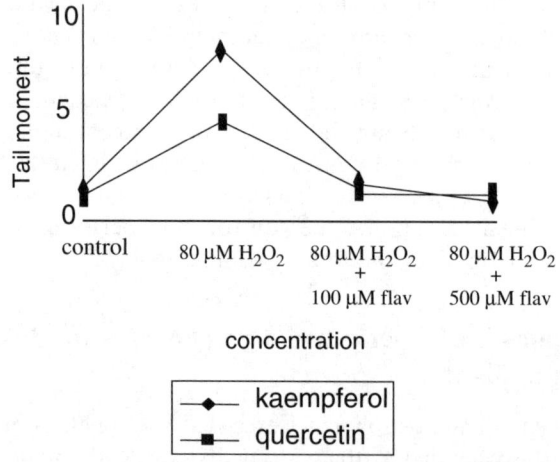

Figure 1   *H_2O_2 -induced DNA damage in lymphocytes modulated by flavonoids (flav)*

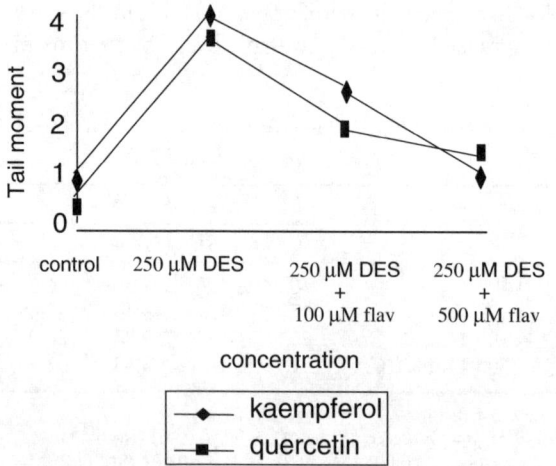

Figure 2   *DES -induced DNA damage in lymphocytes modulated by flavonoids (flav)*

combination with CAT. The data from the two studies are supportive of each other. In Table 2, daidzein induced DNA damage by increasing the tail moments in lymphocytes and decreased the responses towards control levels after treatment with daidzein in combination with CAT. In Table 2, the data are also supportive of each other, in that the profile of response is similar in both studies and give the same statistically significant values. Table 3 shows the results from DNA damage measured in the Comet assay in summary from 84 experiments for all the compounds as well as for hydrogen peroxide. All oestrogenic compounds produced statistically significant positive responses in the Comet assay in both raw sperm ejaculates and lymphocytes. Catalase statistically significantly reduced responses for nearly all agents with some few exceptions as shown in Table 3. Superoxide dismutase and Vit C reduced responses to a lesser extent, in both sperm and lymphocytes.

22.2.2 Responses of Sperm and Lymphocytes to Oestrogens and Flavonoids

Results obtained from 30 experiments on DNA damage measured by the Comet assay in human sperm and human lymphocytes with quercetin and kaempferol are shown in summary in Table 4. H_2O_2, as well as all the oestrogen-like compounds tested, induced significant increases in tail moments in the lymphocytes and significant decreases in %head DNA in the human sperm. The higher concentration of flavonoids (500 μM of quercetin or kaempferol) reduced the DNA damage significantly in all the experiments performed. However, the lower concentration of flavonoids (100 μM of quercetin or kaempferol) did not show a consistent pattern. Figure 1 shows the DNA damage induced by H_2O_2 in lymphocytes and its modulation by kaempferol and quercetin. Figure 2 shows the DNA damage induced by DES in lymphocytes and its modulation by kaempferol and quercetin. Both Figure 1 and Figure 2 are shown in lymphocytes as examples of the similarity of the patterns induced by the compounds.

22.2.3 How Anti-oxidants Affect Responses to Oestrogens in Sperm and Lymphocytes

In humans, CAT and SOD can be endogenous anti-oxidants in cells while Vit C is used as an exogenous nutrient anti-oxidant supplement, and their effects on hydrogen peroxide in both human sperm and lymphocytes in the Comet assay were as anticipated. There is thought to be sufficient P-450 enzyme activity in the lymphocytes and sperm ejaculate for the oxidation of the aromatic phenolic group of oestrogens finally to orthoquinones by two single electron transfer reactions generating two superoxide anion radicals from two oxygen molecules.[20] Disproportionation of the two superoxide anions to hydrogen peroxide and oxygen by the superoxide dismutases lowers the localised level of oxidative stress, although quinones and semiquinones can also damage DNA directly.[8,9,12,21] We have found lymphocytes and sperm on previous occasions

Table 3 *The effect on DNA damge in the COMET assay of various oestrogens in human sperm (SP) and lymphocytes (LY) and their response to treatment with oestrogens in combination with catalase (CAT), superoxide dismutase (SOD) and vitamin C (VIT C)*

Compound/dose (μM)	Expt	Effect/result[b] SP	LY	DOSE	Catalase[a] SP	LY	Superoxide dismutase[a] SP	LY	SP	Vitamin C[a] LY
Equol/250	CAT	+/+	+/+	1	-/-	-/-	-/-	-/-	-/-	-/-
	SOD	+/+	+/+	2	-/-	0/-	0/-	0/-	-/0	-/-
	VIT C	+/+	+/+							
Genistein/250	CAT	+/+	+/+	1	0/0	-/-	-/0	0/-	0/-	-/-
	SOD	+/+	+/+	2	-/-	-/-	-/-	-/0	0/0	0/-
	VIT C	+/+	+/+							
Daidzein/250	CAT	+/+	+/+	1	0/-	-/-	-/-	-/-	0/0	0/0
	SOD	+/+	+/+	2	-/-	-/-	-/-	-/-	0/0	0/0
	VIT C	+/+	+/+							
Diethylstilboestrol/100	CAT	+/+	+/0	1	-/-	-/-	0/0	0/0	-/-	-/0
	SOD	+/+	+/+	2	-/-	0/0	0/0	0/0	0/-	-/0
	VIT C	+/+	+/+							
β-oestradiol/50	CAT	+/+	+/+	1	-/-	0/-	0/-	0/0	-/-	0/0
	SOD	+/+	+/+	2	-/-	-/-	-/-	0/0	-/-	0/0
	VIT C	+/+	+/+							
Nonylphenyl/50	CAT	+/+	+/+	1	0/0	-/-	0/0	0/0	0/0	0/0
	SOD	+/+	+/+	2	0/0	0/-	0/0	0/0	0/0	0/0
	VIT C	+/+	+/+							
Hydrogen peroxide/80	CAT	+/+	+/+	1	-/-	-/-	-/-	0/0	-/-	0/0
	SOD	+/+	+/+	2	-/-	-/-	-/-	0/0	-/-	0/0
	VIT C	+/+	+/+							

Notes: + = positive response with oestrogen; − = reduced response with oestrogen; 0 = no significant response; 1 = lower dose (100 units/ml CAT; 50 units/ml SOD; 0.5 mM Vit C); 2 = higher dose (500 units/ml CAT; 150 units/ml SOD; 1.00 mM Vit C).
[a] = by comparison with oestrogen response ($p < 0.05$).
[b] = by comparison with negative control ($p < 0.05$).

Table 4 *The effect on DNA damage in the Comet assay in human sperm and human lymphocytes treated with oestrogen-like compounds in combination with kaempferol and quercetin*

		Kaempferol		Quercetin
		Sperm[a]	Lymphocytes[a]	Lymphocytes[a]
H_2O_2	80 μM H_2O_2	+/+	+/+	+/+
	+ 100 M flavonoids	−/O	−/−	−/−
	+ 500 μM flavonoids	−/−	−/−	O/−
Diethylstilboestrol	250 μM diethylstilbestrol	+/+	+/+	+/+
	+ 100 μM flavonoids	O/O	O/O	−/−
	+ 500 μM flavonoids	−/−	−/−	−/−
β-oestradiol	70 μM β-oestradiol	+/+	+/+	+/+
	+ 100 μM flavonoids	O/O	−/O	−/O
	+ 500 μM flavonoids	−/−	−/−	−/−
Genestein	250 μM genestein	+/+	+/+	+/+
	+ 100 μM flavonoids	O/O	O/O	O/O
	+ 500 μM flavonoids	−/−	−/−	−/−
Daidzein	250 μM daidzein	+/+	+/+	+/+
	+ 100 μM flavonoids	O/O	−/−	−/−
	+ 500 μM flavonoids	−/−	−/−	−/−

Notes: + = $p > 0.01$ compared with the negative control; O = no significant difference compared with the oestrogen-like compound on its own; − = $p > 0.01$ compared with the oestrogen-like compound on its own.
[a] Results are shown for two comparative studies.

to be sensitive to oxygen radical damage,[23,34] see also refs 38–40 later for sperm). We have also found, on many occasions, lymphocytes to be metabolically active by metabolising exogenous compounds (*e.g.* see ref 12) and also sperm *e.g.* see refs 23,34. Sperm were shown to respond in the same way as lymphocytes after treatment with food mutagens and flavonoids in the Comet assay with or without metabolic activation.[23] Catalase significantly reduced the response of hydrogen peroxide as did SOD in the sperm but not in the lymphocytes. This may be because the rate of production of superoxide radicals in cells may be slower than their conversion to further hydrogen peroxide. Vit C also reduced the response in the sperm but not in the lymphocytes, where it exacerbated responses by acting in a pro-oxidant manner. Anderson *et al.*[31] have previously shown that VitC can be pro-oxidant in the Comet assay. Nonylphenyl appeared to be the oestrogenic compound where responses were least affected by the anti-oxidants.

Thus, with all the compounds, CAT generally reduced the responses, whereas SOD and Vit C did not, probably for the same reasons as outlined for hydrogen peroxide above. This would also suggest that hydrogen peroxide is involved in oxidative reactions of oestrogenic compounds in sperm and lymphocytes *in vitro*, and similar responses might take place *in vivo*, which might ultimately lead to heritable effects and cancer induction, respectively. It cannot be ruled out that *in vivo* inflammation after toxic insult, releasing oxygen radicals from neutrophils and tissue injury, will also contribute to DNA damage invoking the 'innocent bystander' effect. It is also being debated whether human sperm make ROS and if the rate of production is sufficient to account for their physiological effects.[38,39] Hydrogen peroxide concentrations in the range 10–100 μM can have a positive effect on capacitation, but metabolism may decrease hydrogen peroxide concentrations very rapidly and spontaneous capacitation is inhibited by catalase, suggesting that even lower peroxide concentrations may have a positive effect.[39] Given the presence of mitochondria and a plasma membrane NADPH oxidase[41] in these cells, some level of ROS production by spermatozoa is inevitable. This assumption is supported by data including electron spin resonance measurements[42] and biochemical analyses of lipid peroxidation.[43,44] That enzymes such as superoxide dismutase and catalase have been shown to suppress sperm function also suggests that the production of ROS at low levels is biologically important to these cells.[45] Spermatozoa appear to be redox-active cells, and excessive redox activity is associated with defective sperm function.[46,47] However, our results show a property of oestrogenic compounds in sperm and lymphocytes as mediators of oxidative stress and DNA damage, which can be reduced by endogenous and sometimes exogenous nutrient anti-oxidants. They have also shown that comparable levels of H_2O_2 in the present study *in vitro* produce DNA damage in sperm and lymphocytes, and this was also the case for lymphocytes in earlier studies.[12,31] Thus, free radical production at levels comparable to those that may induce capacitation *in vivo* can also give DNA damage. Increasing oxidative stress by oestrogenic compounds over and above the levels of ROS, which may already be present in cells, will probably have adverse effects in terms of capacitation and defective sperm and lymphocyte function. A comparison of the levels of hydrogen peroxide needed to provoke sperm capacitation and induce DNA damage has been published[48] and also, the effects of DNA damage in the male germ line.[49]

22.2.4   How Flavonoids Affect Responses to Oestrogens in Sperm and Lymphocytes

Although plant phenols ingested in the diet are generally thought to be poorly absorbed from the gut by the animal, and largely excreted faecally, there is growing evidence that some are absorbed.[50] A review suggests that significant quantities of quercetin and possibly myrecitin and kaempferol are absorbed by the gut.[28] Another factor relevant to bioavailability is the cell membrane, which may have an important role in limiting the access of these "double-edged" anti-oxidants to cells.[51] Nevertheless, there is evidence that flavonoids accumulate in the cell.[52] This is corroborated by the results of this study, in which quercetin and kaempferol were able to modulate the DNA-damaging effects induced by the oestrogenic compounds that were studied.

The results of the study also confirm the anti-oxidant effect of quercetin in lymphocytes at both concentrations. A high concentration of kaempferol reduced the DNA damage induced by the four oestrogenic compounds and H_2O_2 in sperm and lymphocytes. However, the results of the lower doses of kaempferol were not consistent but were more consistent for quercetin, which was more effective in reducing responses at the lowest concentration. At the higher concentration both were effective. There are several possible biological explanations for the stronger anti-oxidant efficiency of quercetin *vs.* kaempferol. One could be that the absolute concentration and the presence of complementary or synergistic intracellular anti-oxidants may affect the extent of activity of each putative anti-oxidant.[51] For this to occur, it is necessary that the anti-oxidants can permeate the cell membrane, which could limit their bioavailability and, thereby, the anti-oxidant balance. The difference in their reduction potentials, 0.33 V for quercetin and 0.75 V for kaempferol, also could affect their kinetics of eliminating free radicals.[53] Some authors consider that the anti-oxidant activity of free flavonoids is related to the number and position of the hydroxyl groups,[27] which varies between the two flavonoids. Another reason related to the physiology of the donor could be that diet, time of the day, sampling, *etc.*, modulated the anti-oxidant defences in the cell and may have played a role in cases where the effects in the two repeat experiments differed. This has been reported previously for studies with lymphocytes.[54]

The doses of compounds used in the present study generally tend to be higher than those measured *in vivo*. Nevertheless, H_2O_2 in the sperm can approach these levels,[55] and when the oxidative mechanisms of the oestrogens swamp the endocrine-disrupting mechanisms then perhaps more DNA damage can be produced. It is possible at these levels that prostate and breast cancer and reproductive abnormalities could be initiated *in vivo*. The prostate is also known to produce significant levels of catechol oestrogens in this context.[8]

On previous occasions we have shown that DNA damage can be produced by oestrogen-like compounds in lymphocytes and sperm[34,55] as well as the modulating effects of flavonoids on these compounds. The treatment conditions in this study were the same as those that established DNA damage *via* ROS. It is well recognized that these compounds are notorious for acting differentially under different conditions and doses.[31] It is believed that the flavonoids are operating through anti-oxidant effects, but it cannot be excluded that they are also operating through another mechanism such as an antagonistic binding between the compounds, blocking of the endoplasmic reticulum in lymphocytes but not in sperm, inhibition of enzymatic activity, stimulation of DNA repair and all the other mechanisms for anti-genotoxicity. However, it has been shown that with food mutagens an anti-oxidant effect appears to be involved.[34] Plants contain oestrogenic compounds called phytoestrogens. Phytoestrogens, which are natural constituents of our diet, have been investigated at epidemiological, clinical and molecular levels.[56–58] Several authors report an influence on the incidence of hormone-dependent cancers,[59,60] while other authors report beneficial features of phytoestrogens, such as their function as anti-oxidants that protect against oxidatively induced DNA damage,[61] or their use in natural therapies as chemopreventive agents in adults.[62] Therefore, oestrogens and oestrogen-like compounds continue to be surrounded by controversy, in terms of their preventive or initiating responses. Nevertheless, it is clear that oestrogens, whether acting as anti-oxidants or as carcinogens, carry out their functions both *via* receptor mechanisms and *via* the phenolic groups they possess,[8,63] and as a consequence, ROS can be easily generated. This was confirmed by using anti-oxidants such as catalase, SOD (superoxide dismutase) and vitamin C, which showed very clear patterns associated with anti-oxidant modulation of ROS.[55] The flavonoids, also in combination with the oestrogens, were behaving in a manner very similar to that when they were combined with H_2O_2.[64]

22.3   Conclusions

In conclusion, the similar activity of the oestrogenic compounds and H_2O_2 to produce DNA damage and the similar reducing effects of the anti-oxidants and the flavonoids on this damage support our hypothesis that the oestrogens generate ROS that target DNA. Furthermore, it indicates that the flavonoids also possess clear anti-oxidant activity at the higher dose under these conditions. It is believed that the observed activities were not generated by cell-free cell culture conditions because increased responses were observed over and above control values when the compounds were added. Also increasing dose–response relationships have been found after treatment with such oestrogenic compounds in previously reported studies in human sperm ejaculates and lymphocytes, despite the lack of endoplasmic reticulum in sperm.[34] Such findings could have implications for the risk assessment process.

Acknowledgments

At the time, Dr. T.E. Schmid was a Wellcome Trust Travelling Research Fellow, grant reference number: 062288. Dr. E. Cemeli received a Leonardo da Vinci scholarship from CAEB of the Balearic Islands, Spain.

References

1. IEH, Enviromental Oestrogens: Consequences to Human Health and Wildlife, Institute for Environmental Health, Leicester, 1995.
2. R.M. Evans, *Science*, 1988, **240**, 889–895.
3. K.R. Yamamoto, *Annu. Rev. Genet.*, 1985, **19**, 209–252.
4. D.M. Tanenbaum, Y. Wang, S.P. Williams and P.B. Sigler, *Proc. Natl. Acad. Sci. USA*, 1988, **95**, 5998–6003.
5. J.W. Schwabe, L. Chapman, J.T. Finch and D. Rhodes, *Cell*, 1993, **75**, 567–578.
6. D.J. Tobin, N.N. Swanson, M.R. Pittelkow, E.M. Peters and K.U. Schallreuter, *J. Pathol.*, 2000, **191**, 407–416.
7. J.M.C. Gutteridge and B. Halliwell, *Free Radicals in Biology and Medicine*. Oxford University Press, Oxford, 1999.
8. E.L. Cavalieri, K.M. Li, N. Balu, M. Saeed, P. Devanesan, S. Higginbotham, J. Zhao, M.L. Gross and E.G. Rogan, *Carcinogenesis*, 2002, **23**, 1071–1077.

9. J.L. Bolton, *Toxicology*, 2002, **177**(1), 55–65.
10. S.C. Stillman, B.A. Evans and I.A. Hughes, *J.Endocrinol.*, 1990, **127**, 177–183.
11. S.C. Stillman, B.A. Evans and I.A. Hughes, *Clin. Endocrinol.*, 1991, **35**, 533–538.
12. D. Anderson, T.-W. Yu and P. Schmezer, *Environ. & Mol. Mutagenesis*, 1995, **26**, 305–314.
13. M. Ingleman-Sundberg, M. Oscarson and R.A. McLellan, *Trends Pharmacol. Sci.*, 1999, **20**, 342–349.
14. D.R. Nelson, *Arch. Biochem. Biophys.*, 1999, **369**, 1–10.
15. L.L. Wong, *Curr. Opin. Chem. Biol.*, 1998, **2**, 263–268.
16. M.S. Denison and J.P. Whitlock, *J. Biol. Chem.*, 1995, **270**, 18175–18178.
17. T.L. Poulos, *Curr. Opin. Struct. Biol.*, 1995, **5**, 767–774.
18. A.D. Vaz and M.J. Coon, *Biochemistry*, 1994, **33**, 6442–6449.
19. F.J. Gonzalez and D.W. Nebert, *Trends Genet.*, 1990, **6**, 182–186.
20. P.J. O'Brien, *Chem. Biol. Interact.*, 1991, **80**, 1.
21. L. Rossi, G.A. Moore, S. Orrenius and P.J. O'Brien, *Arch. Biochem. Biophys.*, 1991, **251**, 1–15.
22. C. Rice-Evans, *Curr. Med. Chem.*, 2001, **8**, 797–807.
23. D. Anderson, N. Basaran, M. Dobrynska, A. Basaran and T. Wei-Yu, *Teratogen, Carcinogen and Mutagen.*, 1997, **17**, 45–58.
24. S.J. Duthie and V.L. Dobson, *Eur. J. Nutr.*, 1999, **38**, 28–34.
25. S.J. Duthie, A. Ma, M.A. Ross and A.R. Collins, *Cancer Res.*, 1996, **56**, 1291–1295.
26. C.A. Musonda and J.K. Chipman, *Carcinogenesis*, 1998, **19**, 1583–1589.
27. M. Noroozi, W.J. Angerson and M.E. Lean, *Am. J. Clin. Nutr.*, 1998, **67**, 1210–1218.
28. J.M. Gee and I.T. Johnson, *Curr. Med. Chem.*, 2001, **8**, 1245–1255.
29. G.G. Kuiper, J.G. Lemman, B. Carlsson, J.C. Corton, S.H. Safe, P.T. Van Der Saag and J.A. Guswtafsson, *Endocrinology*, 1988, **139**, 4252–4263.
30. S.M. Oh and K.H. Chung, 2004, 74, 1325–1335.
31. D. Anderson, T.-W. Yu, B.J. Phillips and P. Schmezer, *Mutat. Res.*, 1994, **307**, 261–271.
32. WHO, World Health Organization Laboratory Manual for the Examination of Human Semen and Sperm-cervical Mucus Interaction, Cambridge University Press, Cambridge, UK, 2000.
33. D. Anderson, M.M. Dobrzynska, T.-W. Yu, L. Gandini, E. Cordelli and M. Spano, *Teratogen, Carcinogen and Mutagen.*, 1997, **17**, 97.
34. D. Anderson, M.M. Dobrzynska and N. Basaran, *Teratogenesis, Carcinogenesis and Mutagenesis*, 1997, **17**(1), 29–43.
35. B.L. Pool-Zobel, Klein, U.M. Leigebel, F. Kuchenmeister, S. Weber and P. Schmezer, *Clin. Invest.*, 1992, **70**, 299–306.
36. N.P. Singh, M.T. McCoy, R.R. Tice and E.L. Schneider, *Exp. Cell Res.*, 1988, **175**, 184–191.
37. N.P. Singh, D.B. Danner, R.R. Tice, M.T. Mc Coy, G.D. Collins and E.L. Schneider, *Exp. Cell Res.*, 1989, **184**, 461–470.

38. R.J. Aitken and M.A. Baker, *Int. J. Androl.*, 2002, **25**, 191–194.
39. W.C.L. Ford, *Int. J. Androl.*, 2003, **26**, 126.
40. J. Twigg, N. Fulton, E. Gomez, D.S. Irvine and R.J. Aitken, *Humn. Reprod.*, 1998, **13**, 1429–1437.
41. B. Banfi, G. Molnar, A. Maturana, K. Steger, B. Hegedus, N. Demaurex and K.H. Krause, *J. Biol. Chem.*, 2001, **276**, 37594–37601.
42. P.G. Kumar, M. Laloraya and M.M. Laloraya, *Andrologia*, 1991, **23**, 171–175.
43. R. Jones, T. Mann and R. Sherrins, *Fertil. Steril.*, 1979, **31**, 531–537.
44. E. Gomez, D.S. Irvine and R.J. Aitken, *Int. J. of Androl.*, 1998, **21**, 81–94.
45. R.J. Aitken, *Int. J. Androl.*, 2003, **26**, 127.
46. E. Gil-Guzman, M. Ollero, M.C. Lopez, R.K. Sharma, J.G. Alvarez, A.J. Thomas Jr and A. Agarwal, *Hum. Reprod.*, 2001, **16**, 1922–1930.
47. M. Ollero, E. Gil-Guzman, M.C. Lopez, R.K. Sharma, A. Agarwal, K. Larson, D. Evenson, A.J. Thomas Jr and J.G. Alvarez, *Hum. Reprod.*, 2001, **16**, 1912–1921.
48. R.J. Aitken, E. Gordon, D. Harkiss, J.P. Twigg, P. Milne, Z. Jennings and D.S. Irvine, *Biol. Reprod.*, 1998, **59**, 1037–1046.
49. R.J. Aitken, The amoroso lecture, *J. Reprod. Fertil.*, 1999, **115**, 1–7.
50. P.C. Hollman, J.H. de Vries, S.D. van Leewen, M.J. Mengelers and M.B. Katan, *Am. J. Clin. Nutr.*, 1995, **62**, 1276–1282.
51. Szeto, A.R. Collins and I.F.F. Benzie, *Mutat. Res.*, 2002, 500, 31–38.
52. S.M. Kuo, *Adv. Exp. Med. Biol.*, 2002, **505**, 191–200.
53. S.V. Jovanovic and M.G. Simic, *Ann. NY Acad. Sci.*, 2000, **899**, 326–334.
54. G. Oliveri and A. Bosi, in *Chromosomal Aberrations*, G. Obe and A.T. Natarajan (eds), New York, Springer Verlag, 1992, 130–139.
55. D. Anderson, T.E. Schmid, A. Baumgartner, E. Cemeli, M.H. Brinkworth and J. Wood, *Mutat. Res.*, 2003, **544**(2–3), 173–178.
56. A.L. Catapano, *Angiology*, 1997, **48**, 39–44.
57. M. López-Lázaro, *Curr. Med. Chem. Anti-Canc. Agents*, 2002, **2**, 674–691.
58. P.G. Pietta, *J. Nat. Prod.*, 2000, **63**, 1035–1042.
59. M.L. de Lemos, *Ann. Pharmacother.*, **2001 35**, 1118–1121.
60. C.J. Kirk, R.M. Harris, D.M. Wood, R.H. Waring and P.J. Hughes, *Biochem. Soc. Trans.*, 2001, **29**, 209–216.
61. J. Sierens, J.A. Hartley, M.J. Campbell, A.J. Leathem and J.V. Woodside, *Mutat. Res.*, 2001, **485**, 169–176.
62. A. Stark and Z. Madar, *J. Pediatr. Endocrinol. Metab.*, 2002, **15**, 561–572.
63. J.G. Liehr, *Mutat. Res.*, 1990, **238**, 269–276.
64. E. Cemeli, T.E. Schmid and D. Anderson, *Environ. Mol. Mutagen.*, 2004, **44**, 420–426.

CHAPTER 23

DNA Repair Capacities in Testicular Cells of Rodents and Man

GUNNAR BRUNBORG, NUR DUALE, JULIE TESDAL
HAALAND, CHRISTINE BJØRGE, ERIK SØDERLUND,
ERIK DYBING, RICHARD WIGER AND ANN-KARIN
OLSEN

*Norwegian Institute of Public Health, P.O.Box 4404 Nydalen, Oslo N-0403,
Norway*

23.1 Introduction

DNA repair processes of various types are well preserved through evolution reflecting their importance both for the individual and for the species. All cell types in the human body repair their DNA, although the efficiency may vary. There is little reason to assume that germ cells represent an exception, especially because the integrity of germ cell DNA is essential for the offspring and hence also for coming generations. In the male, spermatogenesis is active throughout adult life, during which a population of stem cells is constantly under attack from DNA-damaging agents produced by endogenous processes and exogenous agents; in the order of 100,000 DNA lesions are induced per cell per day.[1] These lesions accumulate and may result in cell death if not repaired, and in mutations transmitted to the offspring if not repaired by the time of replication or repaired incorrectly. If transmitted to the zygote, unrepaired lesions in male germ cells may lead to fetal death (pre- or post-implantation loss). New mutations in the paternal genome will not be eliminated in the fertilised egg unless they affect an essential gene product.

Although many aspects of somatic cell DNA repair functions have been deciphered in detail, there is still limited information on DNA repair in male germ cells. Sperm DNA is haploid, and DNA is packed very differently from somatic cells; these are two of several reasons why DNA repair in male germ cells may be different. Our interest in DNA repair in germ cells was triggered when we studied the testicular toxicity of 1,2-dibromo-3-chloropropane (DBCP).

This compound was first demonstrated to be a human testicular toxicant, due to the finding that production workers showed signs of sterility or reduced fertility, and this was associated with structural changes in the testes in the form of azoospermia.[2] Using rats, it had been reported already in 1961 that DBCP exposure could result in severe organ damage (liver, lung, intestinal mucosa, kidney and testis) after inhalation of DBCP at 10–40 ppm.[3] When studying the relationship between geno- and organ toxicity, we observed a striking correlation between DNA strand break formation and testicular atrophy/necrosis, in four rodent species (Table 1). For both these endpoints, rats were very sensitive whereas mice were almost completely resistant after intraperitoneal administration of DBCP. For hamster and guinea pig there was some but not complete correlation between testicular toxicity and DNA damage. It was later reported that dominant lethal mutations were induced by DBCP in rats but not in mice (however, somatic point mutations are induced[5]). Our studies with primary testicular cells exposed *in vitro*[4] demonstrated similar species-specific patterns of DNA lesion formation as found *in vivo*, suggesting that toxicokinetic differences were not of decisive importance.

From these data, along with other lines of evidence, it was concluded that DNA damage is an initiating event in DBCP-induced testicular toxicity. Furthermore, this was also a strong indication that DNA damage could play an important role in male fertility and reproduction, a notion that is now strongly supported by increased levels of oxidative DNA lesions among sub-fertile men.[6] Our working hypothesis was that the species differences observed with DBCP reflected differences in the formation of active metabolites (most probably episulfonium ions formed *via* glutathione transferase[7]), and/or differences in the repair of DNA damage.[8] Fifteen known male reproductive toxicants were tested using short-term exposures of primary testicular cells from rats or humans (prepared from normal donor testicular biopsies), which were then assayed for induced DNA damage (single-strand breaks (SSBs) and

Table 1   *DBCP-induced testicular organ toxicity, and cellular DNA damage as measured with alkaline elution*

Species	Mean testicular injury 10 days after 170 µmol kg⁻¹ DBCP i.p. Scale 0–4	DNA damage (single strand breaks/alkali labile sites) in testicular cells isolated from animals 1 h after 170 µmol kg⁻¹ DBCP i.p. Arbitrary scale	Relative DNA damage (single strand breaks/ alkali labile sites) in testicular cells. 1 h exposure in vitro. Arbitrary scale	
			25 µM	*100 µM*
Rat	1.4 ± 1.5	7.3	10.9	
Mouse	0.0 ± 0.0	0.2	0.2	0.3
Hamster	0.4 ± 0.9	0.6	0.2	0.7
Guinea pig	2.8 ± 1.3	7.9	0.3	2.7

Source: Adapted from ref 4.

alkali-labile sites) with alkaline elution or single cell gel electrophoresis (comet assay) (Table 2).

Quantitative and qualitative differences between the two species were observed. For six compounds there was no significant increase in DNA damage in either species, four compounds were active in both species, whereas five were active in one species only. Among the latter, four were active in rats only, whereas one compound (acrylamide) was active in humans but not in rats. With respect to DBCP, there was a remarkably low induction of DNA lesions in human testicular cells compared with rats (Table 2), in spite of the fact that this compound is a potent human male reproductive toxicant. These investigations suggested that primary cultures of testicular cells may be used for testing male reproductive toxicants, but extrapolation from rodent to human cells should be done with care.

In these experiments, only immediate DNA damage was assayed; cells were not then incubated without test compound to measure subsequent DNA repair. Could the observed species-specific response be related to differences in repair? Alkaline elution and the comet assay provide little information on the specific nature of the DNA lesions formed; "silent" lesions may exist in the form of alkali-resistant DNA adducts. During repair, such lesions are incised by glycosylases and apurinic/apyrimidinic(AP)-lyases/endonucleases or during nucleotide excision repair (NER). In the latter case, the repair intermediates in the form of strand breaks may be detected by inhibiting resynthesis of the large stretches (about 30 bp) that are inserted during NER. Repair inhibitors were used together with many of the test chemicals in Table 2, but qualitative

Table 2 *DNA damage in primary cultures of testicular cells from rats and humans. Incubation for 30 min at 32°C. Minus (−), not significantly different from unexposed controls; plus (+), significantly increased compared to unexposed controls*

Chemical	Type/use	Concentration range (μM)	Rats	Humans
Methoxychlor	Pesticide	10–100	−	−
Benomyl	Pesticide	10–100	−	−
Thiotepa	Cytostatic	10–1000	−	−
Cisplatin	Cytostatic	30–1000	−	−
Cd^{++}	Industrial chemical	30–1000	−	−
Acrylonitrile	Industrial chemical	30–1000	−	−
Styrene oxide	Industrial chemical	10–300	+	+
1,2-Dibromoethane	Pesticide	30–1000	+	+
Thiram	Pesticide	10–300	+	+
Chlordecone	Pesticide	10–1000	+	+
DBCP	Pesticide	10–300	+	−
Acrylamide	Industrial chemical	30–1000	−	+
1,3-Dinitrobenzene	Industrial chemical	10–300	+	−
Cr^{6+}	Industrial chemical	30–1000	+	−
Aflatoxin	Fungicide	10–300	+	−

Source: Adapted from ref 9.

changes in their overall genotoxicities were not observed. However, in light of our more recent experiments this was indeed to be expected because NER is of low efficiency in testicular cells from both rodents and humans.[10]

The testes represent a challenge with respect to the isolation and cultivation of the different male germ cell types. Thanks to the pioneering studies by M.L. Meistrich and G.P. van der Schans, it has been known for many years that DNA strand breaks induced by ionizing radiation in spermatogenic cells from mice, hamsters and rats are repaired efficiently, that is as efficiently as in somatic cells, and furthermore that elongated spermatids exhibit no repair.[11–13] Our laboratory has later made similar observations with male germ cells from humans, mice and rats, using various methodologies including cells exposed within intact tubules, that is cells in their normal environment. Different zones of rodent tubules contain spermatogenic cells of specific composition; such cells may be removed and analysed by the comet method. The rejoining of DNA SSBs induced by X-rays was shown to be ploidy specific, that is in both primary rat spermatocytes and early spermatids half of the induced breaks were repaired within 8–10 min, whereas in late spermatids this took 15.2 ± 1.5 min (unpublished observations). We may hence conclude that frank DNA SSBs are repaired similarly in male germ cells in all the studied species. What about other types of DNA lesions?

23.2 Limited Nucleotide Excision Repair in Male Germ Cells of Rodents and Man

NER is a major DNA repair pathway acting on a variety of helix-distorting DNA lesions (bulky adducts), such as those caused by UV-light (cyclobutane pyrimidine dimer (CPD) and 6-4 photoproduct (6-4PP)), a wide range of exogenous chemicals including environmental agents such as benzo(a)pyrene (B[a]P) and aflatoxin B1 and by chemotherapeutic agents such as cisplatin. A defect in one of the repair proteins results in recessive syndromes such as xeroderma pigmentosum (XP), associated with very high (1000-fold increased) cancer risk.[14]

NER involves recognition of the DNA lesion, incision of the DNA strand containing the lesion followed by DNA synthesis and ligation to replace an excised oligonucleotide (there are several excellent reviews[15–17]). NER comprises two subpathways: global genomic repair (GGR) which repairs DNA lesions in the entire genome, and transcription-coupled repair (TCR), which removes DNA lesions blocking RNA synthesis, that is, in transcribed genes.[18]

In total, more than 30 proteins are directly involved in NER. The efficiency of repair of different kinds of lesions varies greatly; for example, repair of UV-induced DNA damage *via* GGR is much more rapid for 6-4PPs than for CPDs in Chinese hamster ovary cells (~ 1 h compared with ~ 6 h, respectively[19]).

About 10 years ago, we observed indications of very low DNA repair in primary cultures of rat testicular cells exposed to UV-C.[20] When cells were incubated post-UV in the presence of repair inhibitors as described above, there was no accumulation of DNA SSBs as measured with alkaline elution. We subsequently confirmed this low incision of UV-induced lesions in spermatocytes and round spermatids, using the alkaline comet assay (Figure 1).

This apparent inability to remove bulky DNA lesions would be expected to have important implications for the testes and for the offspring. NER functions were hence studied in more detail in male germ cells of rodents and man. Removal of CPDs was very poor in rat testicular cells in suspension, as well as in cells within intact seminiferous tubules, at low doses of UV-C indicating that the GGR subpathway was inefficient in these cells.[10] The TCR subpathway was

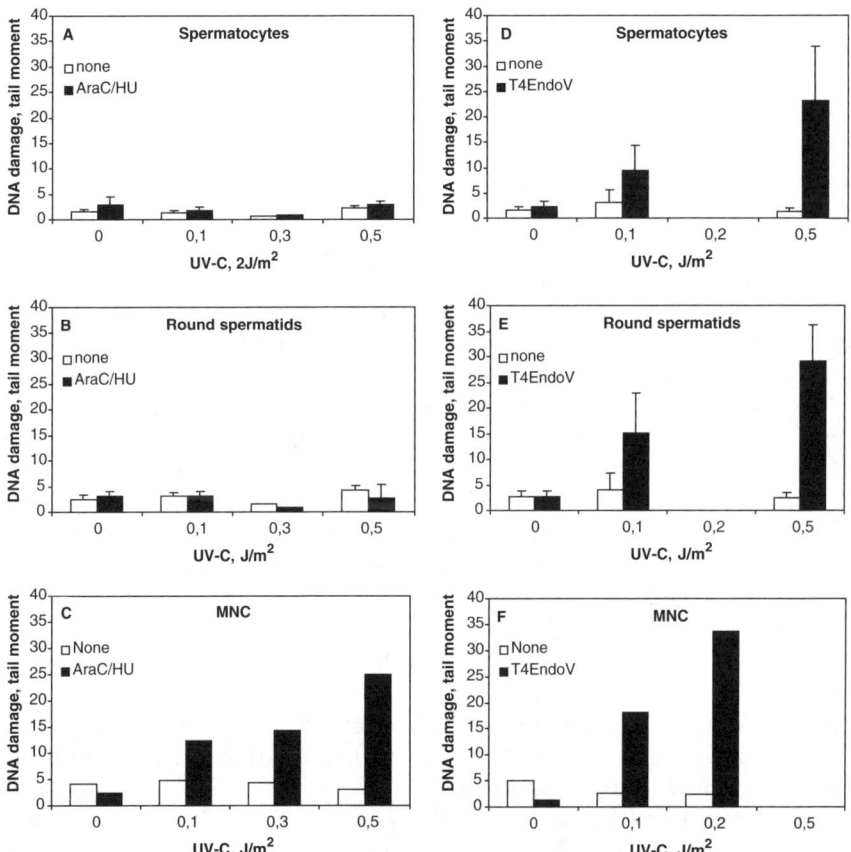

Figure 1 *Accumulation of incised photolesions in rat spermatocytes, rat round spermatids and human lymphocytes (MNC), measured with the alkaline comet assay. (A–C) Cells with or without repair inhibitors as indicated (hydroxyurea 2 mM and AraC 0.1 mM; 1 h at 32°C (spermatogenic cells) or 37°C (lymphocytes)). (D–F) Analysis with the alkaline comet assay with or without T4endoV as indicated. In A, B, D and E, comets were sorted according to their DNA content to identify cell-type (ploidy)-specific DNA damage*
(Adapted from Jansen and co-workers. See ref 10)

also inefficient, according to experiments in which enriched cell samples of rat spermatocytes of zygotene/leptotene, mid-pachytene and diplotene stages were exposed to 30 J m^{-2} UV-C followed by assessment of the degree of repair in the transcribed strand of the meiosis-specific gene *SCP1*.[10] *In vitro* incision activities of protein cell extracts were measured using oligonucleotides containing defined DNA lesions; excision was high with early- and mid-pachytene spermatocytes as opposed to extracts from other stages (diplotene spermatocytes and round spermatids). It is not unexpected that protein extracts exhibit proficient repair activities (as has recently been shown with various rat tissues[21]), while in the live cell repair may still be absent because the distribution, spatial translocation and association of repair proteins are essential for their function. We hypothesised that NER proteins in male germ cells are sequestered by mispaired regions in DNA involved in synapsis and recombination, thus explaining the highly inefficient NER in premeiotic cells.[10] These findings in the rat prompted us to conduct equivalent experiments with human testicular cells, and indeed quantitatively similar traits have been observed (unpublished), indicating that human and rodent male germ cells have similar poor removal of CPDs *via* the NER pathway. This picture may however be more complex, as P.C. Hanawalt and co-workers[22] recently reported efficient repair of 6-4PP in spermatogenic cells, whereas the repair of these lesions was suppressed in postmeiotic cells from aged mice.

Human male germ cells were also studied with respect to expression of a number of NER-associated proteins, using lymphoblastoid cell cultures with or without specific repair deficiencies as positive and negative controls (unpublished data). Protein extracts were prepared from primary testicular cells from human testis biopsies. Low amounts of several NER-associated proteins (DDB2, XPA and XPC) were indeed observed in testicular cells, whereas other proteins (RPA) were present at similar levels as in somatic cells. (A description of the NER proteins and their specific functions is found in ref 23.) One testis biopsy from an individual showed no sperm production and the so-called Sertoli-only characteristics; this particular extract contained significant amounts of some of the proteins (XPA) that were present in diminishing amounts in male germ cells from men with normal spermatogenesis. Similar studies will be continued with rodents.

In general, at the level of mRNA, transcripts of most NER-related enzymes are measured in high amounts.[24–27] Different from somatic cells, in testicular cells transcripts are stored as ribonucleoprotein particles in a translationally repressed state for several days.[28,29] It has been demonstrated that, in haploid spermatids, essentially every mRNA exhibits evidence of translational repression. An apparent high mRNA expression in testicular cells is therefore not necessarily associated with high functional activity.

Low NER is in compliance with observations recently reviewed by Sotomayor and co-workers.[30] In studies of unscheduled DNA synthesis (UDS) in male germ cells exposed to different agents, as many as 59 chemicals plus UV and X-rays were tested in spermatogenic cells from humans, rabbits, rats and mice. Although these aspects were not discussed in the chapter, in general, agents inducing DNA lesions that are removable by NER did not show UDS (2-AAF, aflatoxin B1, B[a]P, N-OH-AAF). Furthermore, low NER may explain the lack of genotoxicity of several known reproductive toxicants shown in Table 2, with cisplatin as a prominent example because this compound produces intra- and interstrand crosslinks normally repaired by NER. Cisplatin-based treatment is particularly efficient with tumours of testicular origin, and we and others have suggested that this may be related to low NER activity.[31,32]

23.3 Base Excision Repair

As opposed to NER, base excision repair (BER) is initiated with the release of altered bases by DNA glycosylases, recognising and excising aberrant bases that cause minor structural changes in DNA. Each DNA glycosylase recognises a specific set of DNA-substrates. The excision of the base generates AP-sites followed by endonuclease cleavage, resynthesis and ligation. BER is subdivided into short-patch repair (SPR) in which only one nucleotide is replaced, and long-patch repair (LPR) in which up to ten nucleotides are replaced.

23.4 Many Base Excision Repair Functions are Equally Efficient in Male Testicular Cells as in Somatic Cells

Many altered bases are repaired by BER; we have studied some of them and their associated DNA glycosylases and downstream activities. In cases of diets low in folic acid, uracil is incorporated into the genome. Uracil residues may lead to mutations and occur in the genome either as a result of erroneous incorporation *via* replication, or by spontaneous deamination of thymine. By measuring enzymatic activities and by immunochemical detection we found that the most important uracil-DNA glycosylase (UDG) was present in human and rat testicular cells in amounts similar to those in somatic tissues.[32] The data suggest that uracil is removed from the male germ cell genome and this is corroborated by Intano and co-workers,[33] reporting UDG also to be found in mouse testicular cells. Furthermore, Grippo and co-workers[34] detected human uracil-DNA glycosylase (UNG) activity in DNA-synthesising male germ cells.

DNA alkylations constitute another class of prevalent DNA lesions; several of these types are repaired *via* the methylpurine DNA-glycosylase (MPG). MPG mRNA is abundant in mice testes.[34–36] Other proteins that repair alkylations are the direct acting "suicide protein" O^6-Methylguanine-DNA-glycosylase (MGMT) and the recently identified AlkB (alkylation repair) human homologue (ABH) proteins,[37,38] each with their specific DNA-substrates. We found that MGMT was expressed in normal amounts in human and rat male germ cells (unpublished), similar to findings in mice.[39] In our studies,

MPG is present in greater amounts in human and rat male germ cells compared to somatic cells.[32] Minor differences were observed between different cellular stages of rat spermatogenesis. DNA lesions induced by exposure to the mutagen methyl-methane sulfonate (MMS) were 5-fold higher in rat compared to human male germ cells, indicating major differences in sensitivity between the two species. Repair of methylated DNA studied at the cellular level was efficient in both human and rat male germ cells, in primary spermatocytes as well as round spermatids, compared to primary somatic cell types. This proficient repair is consistent with the lack of increased mutation rates in male germ cells of Big Blue transgenic mice treated with MMS;[40] large insertions and deletions are however not scored in the latter assay. Hence, male germ cells of both rodents and humans seem well protected against DNA methylations such as those inflicted by MMS. Other mouse germ cell alkylating mutagens, such as ethyl nitrosourea (ENU) and isopropyl methanesulfonate (iPMS), attack other base positions and do induce mutations.[40] Alkylating agents also attack other components of a cell besides DNA bases, and it is now believed that some of these disturb chromatin packaging during spermiogenesis.

23.5   Oxidised Lesions are Repaired Differently in Rodent and Human Spermatogenic Cells

Oxidised lesions and their repair in male germ cells are of special interest. One prevalent and important lesion is 8-oxoG arising from oxidised guanine or misincorporated oxidised dGTP. 8-oxoG is mutagenic and is an inevitable consequence of oxidative metabolism, but is also induced by many environmental mutagens, including B[a]P originating from cigarette smoke and combustion exhaust.[41] Other important lesions include oxidised pyrimidines such as thymine glycols (TG) and 5-hydroxycytosine (5-OHC).

   We have studied the repair of some of these lesions in rodent and human male germ cells using cellular extracts, by identifying the relevant proteins and measuring enzymatic activities, and by measuring active repair in live cells. Oxidised purines such as 8-oxoG were efficiently repaired in rat[42] and mouse (unpublished) spermatogenic cells. However, human testicular cells from several individuals showed no repair of these lesions,[42] a result which we found most unexpected. On the other hand, oxidised pyrimidines such as TG were efficiently repaired in both humans and rats. Enzymatic activities and protein levels of the relevant repair enzymes were measured; these were largely present in amounts that were in accordance with the efficiency of repair of oxidised pyrimidines measured in cells. At the level of RNA, a very high amount of Ogg1 mRNA was reported in mouse testis,[43] while in human tissues including the testis, OGG1 mRNA is expressed at similar levels.[44] In conclusion, it appears that, as a consequence of this repair deficiency, human male germ cells may be particularly sensitive to DNA oxidation.

23.6 An Improved Mouse Model for Reproductive Genotoxicity Testing

Mice and rats are by far the preferred species for mutagenicity testing and reproductive toxicity studies. Based on our results suggesting that humans are different from rodents with respect to the repair of important oxidative premutagenic lesions, the development of alternative models should be pursued. For this purpose, we propose the use of a transgenic mouse strain in which the main DNA glycosylase (mOgg1) for removal of 8-oxoG was knocked out.[45] Nuclear testicular extracts show no incision of 8-oxoG,[45] suggesting that the testis has little back-up activity besides mOgg1. We detected no cellular repair of oxidative lesions in the male germ cells prepared from these mice (unpublished observations).

The level of spontaneous 8-oxoG was measured in male germ cells of mOgg1-deficient mice and in mice carrying either both or one mOgg1 allele (Figure 2). It is evident that the spontaneous levels differ markedly; the mOGG1 *null* mouse has at least ten times more 8-oxoG than the heterozygous or homozygous wild type. Furthermore, this level was similar to what was found in human testicular cells, that is in the range 0.75–1.0 8-oxoG per 10^6 dG.[42]

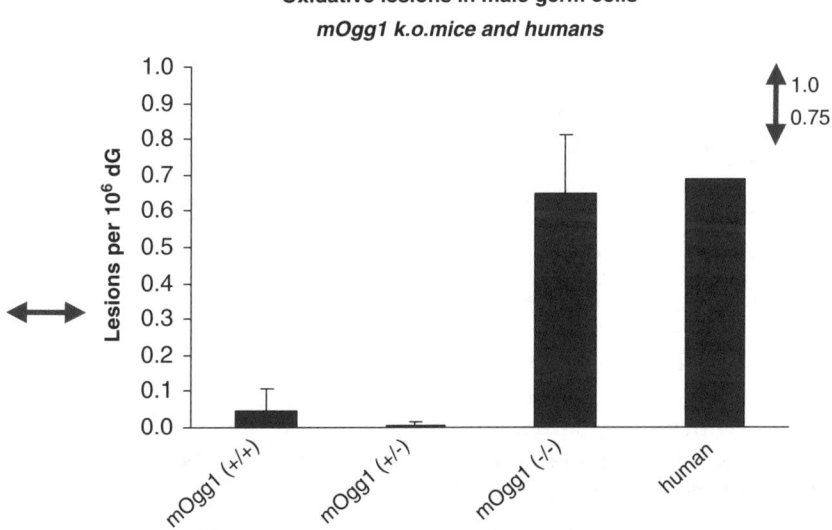

Oxidative lesions in male germ cells
mOgg1 k.o.mice and humans

Figure 2 *Spontaneous levels of oxidative lesions (8-oxoG) in mouse and human male germ cells. 8-oxoG were measured as Fpg-sensitive sites with the comet assay, Fpg being a bacterial enzyme cleaving at some oxidative lesions. Male germ cells were prepared from either homozygous wild-type mice, heterozygous mice, mOgg1 null mice or humans testicular biopsies, as indicated. Right arrows indicate the previously published range of 8-oxoG lesions in human testicular cells;[42] left arrow indicates the median level reported for human lymphocytes in a large interlaboratory validation study[46]*

Measurement of low levels of 8-oxoG in cellular DNA is not straightforward and the most reliable method is still a matter of discussion. The EU project ESCODD recently published the median level of spontaneous 8-oxoG in human peripheral lymphocytes (Figure 2) to be 0.3 per 10^6 dG.[46] It appears that the mOGG1 mouse model is close to humans with respect to its content of 8-oxoG in testicular cells.

We are currently further evaluating the mOgg1-deficient mouse as a model for human male germ cell toxicity testing. BER-related enzymes involved in the later stages of the pathway (DNA ligases, AP-endonucleases, DNA polymerases, Xrcc1[33,47-49]) seem to be present at levels sufficient for functional repair. A further evaluation and characterisation of the mOgg1 mouse model comprises measurements of enzymes required for metabolism of xenobiotics. When mice were treated with B[a]P, Cyp1a1 was induced in both the liver and the testis, whereas – as expected – Cyp1a2 was induced in the liver only (unpublished observations). Important activation pathways for polycyclic aromatic hydrocarbons (PAHs) therefore seem to be present in the repair-deficient mice to the same extent as in wild-type mice. Our laboratory is currently analysing B[a]P-induced mutations in wild-type and mOGG1-deficient mice using the Big Blue®cII system. The mutation signature in testis (cauda sperm) compared to other organs will presumably provide valuable information.

23.7   How do Male Germ Cells Succeed without a Full Set of DNA Repair Systems?

Spermatogenic cells are partly protected from exogenous genotoxicants by the blood–testis barrier and the presence of enzymatic and chemical antioxidant species. During evolution, protection against PAHs and heterocyclic amines generated during cooking of food may have been important for cells lining the GI tracts, as has the repair of UV-induced lesions for dermal cells. Resistance to solar UV-light requires NER and most probably also oxidative lesion repair (it has recently been shown that mOgg1 mice are more prone to UV-induced skin cancers than the wild type[50]). Testicular cells may represent cell types for which NER and oxidative repair is or has been less important, at least in humans, which – unlike rodents – are not selected for high reproduction. This situation may however have changed in the modern society. Norway and Denmark have the highest global incidences of testicular cancer, about 11 cases per 100,000 per year. Together with low sperm counts and increased rates of genital malformations (termed testicular dysgenesis syndrome[51]), these are compelling indications that life-style and chemical exposure represent a novel challenge for the male reproductive system. The identification of possible underlying causes for these endpoints will depend on a thorough mechanistic understanding of all aspects of male-mediated developmental toxicity including DNA repair.

Acknowledgments

This work was supported by the European Community (ENV4-CT95-0204) and the Norwegian Research Council (Project 148703/310). We acknowledge the permission from Elsevier (*Toxicology*) to reproduce data from ref. 4 (Lag et al.), and from Oxford University Press (*Nucleic Acids Research*) to reproduce Fig. 1 in ref. 10.

References

1. T. Lindahl, *Nature*, 1993, **362**, 709.
2. M. Slutsky, J.L. Levin and B.S. Levy, *Int. J. Occup. Environ. Health*, 1999, **5**, 116.
3. T.R. Torkelson, S.E. Sadek, V.K. Rowe, J.K. Kodama, H.H. Anderson, G.S. Loquvam and C.H. Hine, *Toxicol. Appl. Pharmacol.*, 1961, **3**, 545.
4. M. Lag, E.J. Soderlund, G. Brunborg, J.E. Dahl, J.A. Holme, J.G. Omichinski, S.D. Nelson and E. Dybing, *Toxicology*, 1989, **58**, 133.
5. S. Teramoto and Y. Shirasu, *Mutat. Res.*, 1989, **221**, 1.
6. M. Zitzmann, C. Rolf, V. Nordhoff, G. Schrader, M. Rickert-Fohring, P. Gassner, H.M. Behre, R.R. Greb, L. Kiesel and E. Nieschlag, *Fertil. Steril.*, 2003, **79**(3), 1550.
7. J.G. Omichinski, G. Brunborg, J.A. Holme, E.J. Soderlund, S.D. Nelson and E. Dybing, *Mol. Pharmacol.*, 1988, **34**, 74.
8. C. Bjorge, R. Wiger, J.A. Holme, G. Brunborg, T. Scholz, E. Dybing and E.J. Soderlund, *Reprod. Toxicol.*, 1996, **10**, 51.
9. C. Bjorge, G. Brunborg, R. Wiger, J.A. Holme, T. Scholz, E. Dybing and E.J. Soderlund, *Reprod. Toxicol.*, 1996, **10**, 509.
10. J. Jansen, A.K. Olsen, R. Wiger, H. Naegeli, P. de Boer, H.F. van Der, J.A. Holme, G. Brunborg and L. Mullenders, *Nucleic Acids Res.*, 2001, **29**, 1791.
11. D.S. Joshi, J. Yick, D. Murray and M.L. Meistrich, *Radiat. Res.*, 1990, **121**, 274.
12. A.A. van Loon, P.J. Den Boer, G.P. van der Schans, P. Mackenbach, J.A. Grootegoed, R.A. Baan and P.H. Lohman, *Exp. Cell Res.*, 1991, **193**, 303.
13. A.A. van Loon, E. Sonneveld, J. Hoogerbrugge, G.P. van der Schans, J.A. Grootegoed, P.H. Lohman and R.A. Baan, *Mutat. Res.*, 1993, **294**, 139.
14. J.E. Cleaver, *Birth Defects Orig. Artic. Ser.*, 1989, **25**, 61.
15. R.D. Wood, *J. Biol. Chem.*, 1997, **272**, 23465.
16. W.L. de Laat, N.G. Jaspers and J.H. Hoeijmakers, *Genes Dev.*, 1999, **13**, 768.
17. J.R. Mitchell, J.H. Hoeijmakers and L.J. Niedernhofer, *Curr. Opin. Cell Biol.*, 2003, **15**, 232.
18. S. Tornaletti and P.C. Hanawalt, *Biochimie*, 1999, **81**, 139.
19. R.S. Nairn, D.L. Mitchell, G.M. Adair, L.H. Thompson, M.J. Siciliano and R.M. Humphrey, *Mutat. Res.*, 1989, **217**, 193.
20. G. Brunborg, J.A. Holme and J.K. Hongslo, *Mutat. Res.*, 1995, **342**, 157.

21. A. Gospodinov, R. Ivanov, B. Anachkova and G. Russev, *Eur. J. Biochem.*, 2003, **270**, 1000.
22. G. Xu, G. Spivak, D.L. Mitchell, T. Mori, J.R. McCarrey, C.A. McMahan, R.B. Walter, P.C. Hanawalt and C.A. Walter, *Biol. Reprod.*, 2005, **73**, 123.
23. A.K. Olsen, B. Lindeman, R. Wiger, N. Duale and G. Brunborg, *Toxicol. Appl. Pharmacol.*, 2005, **207**, S521.
24. L. Li, C. Peterson and R. Legerski, *Nucleic Acids Res.*, 1996, **24**, 1026.
25. P.J. van der Spek, C.E. Visser, F. Hanaoka, B. Smit, A. Hagemeijer, D. Bootsma and J.H. Hoeijmakers, *Genomics*, 1996, **31**, 20.
26. L. Cheng, Y. Guan, L. Li, R.J. Legerski, J. Einspahr, J. Bangert, D.S. Alberts and Q. Wei, *Cancer Epidemiol. Biomarkers Prev.*, 1999, **8**, 801.
27. M. Shannon, J.E. Lamerdin, L. Richardson, S.L. McCutchen-Maloney, M.H. Hwang, M.A. Handel, L. Stubbs and M.P. Thelen, *Genomics*, 1999, **62**, 427.
28. N.B. Hecht, *Bioessays*, 1998, **20**, 555.
29. K. Steger, *Anat. Embryol. (Berl.)*, 1999, **199**, 471.
30. R.E. Sotomayor and G.A. Sega, *Environ. Mol. Mutagen.*, 2000, **36**, 255.
31. B. Koberle, J.R. Masters, J.A. Hartley and R.D. Wood, *Curr. Biol.*, 1999, **9**, 273.
32. A.K. Olsen, H. Bjortuft, R. Wiger, J. Holme, E. Seeberg, M. Bjoras and G. Brunborg, *Nucleic Acids Res.*, 2001, **29**, 1781.
33. G.W. Intano, C.A. McMahan, R.B. Walter, J.R. McCarrey and C.A. Walter, *Nucleic Acids Res.*, 2001, **29**, 1366.
34. P. Grippo, P. Orlando, G. Lococrondo and R. Geremia, *Prog. Clin. Biol. Res.*, 1982, **85**, 389.
35. B.P. Engelward, M.S. Boosalis, B.J. Chen, Z. Deng, M.J. Siciliano and L.D. Samson, *Carcinogenesis*, 1993, **14**, 175.
36. N.K. Kim, S.H. Lee, T.J. Sohn, R. Roy, S. Mitra, H.M. Chung, J.J. Ko and K.Y. Cha, *Anticancer Res.*, 2000, **20**, 3037.
37. T. Duncan, S.C. Trewick, P. Koivisto, P.A. Bates, T. Lindahl and B. Sedgwick, *Proc. Natl. Acad. Sci. USA*, 2002, **99**, 16660.
38. P.A. Aas, M. Otterlei, P.O. Falnes, C.B. Vagbo, F. Skorpen, M. Akbari, O. Sundheim, M. Bjoras, G. Slupphaug, E. Seeberg and H.E. Krokan, *Nature*, 2003, **421**, 859.
39. M.J. Thompson, S. Abdul-Rahman, T.G. Baker, J.A. Rafferty, G.P. Margison and M.C. Bibby, *J. Reprod. Fertil.*, 2000, **119**, 339.
40. J. Ashby, N.J. Gorelick and M.D. Shelby, *Mutat. Res.*, 1997, **388**, 111.
41. H.M. Shen, S.E. Chia, Z.Y. Ni, A.L. New, B.L. Lee and C.N. Ong, *Reprod. Toxicol.*, 1997, **11**, 675.
42. A.K. Olsen, N. Duale, M. Bjoras, C.T. Larsen, R. Wiger, J.A. Holme, E.C. Seeberg and G. Brunborg, *Nucleic Acids Res.*, 2003, **31**, 1351.
43. T.A. Rosenquist, D.O. Zharkov and A.P. Grollman, *Proc. Natl. Acad. Sci. USA*, 1997, **94**, 7429.
44. K. Nishioka, T. Ohtsubo, H. Oda, T. Fujiwara, D. Kang, K. Sugimachi and Y. Nakabeppu, *Mol. Biol. Cell*, 1999, **10**, 1637.

45. A. Klungland, I. Rosewell, S. Hollenbach, E. Larsen, G. Daly, B. Epe, E. Seeberg, T. Lindahl and D.E. Barnes, *Proc. Natl. Acad. Sci. USA*, 1999, **96**, 13300.
46. C.M. Gedik and A. Collins, *FASEB J.*, 2005, **19**, 82.
47. F. Hirose, Y. Hotta, M. Yamaguchi and A. Matsukage, *Exp. Cell Res.*, 1989, **181**, 169.
48. I. Husain, A.E. Tomkinson, W.A. Burkhart, M.B. Moyer, W. Ramos, Z.B. Mackey, J.M. Besterman and J. Chen, *J. Biol. Chem.*, 1995, **270**, 9683.
49. J. Chen, A.E. Tomkinson, W. Ramos, Z.B. Mackey, S. Danehower, C.A. Walter, R.A. Schultz, J.M. Besterman and I. Husain, *Mol. Cell Biol.*, 1995, **15**, 5412.
50. M. Kunisada, K. Sakumi, Y. Tominaga, A. Budiyanto, M. Ueda, M. Ichihashi, Y. Nakabeppu and C. Nishigori, *Cancer Res.*, 2005, **65**, 6006.
51. N.E. Skakkebaek, E. Rajpert-De Meyts and K.M. Main, *Hum. Reprod.*, 2001, **16**, 972.

CHAPTER 24

3rd International Congress on Male-Mediated Developmental Toxicity: Closing Panel Discussion

JACK BISHOP[a] AND BARBARA F. HALES[b]

[a] National Toxicology Program (NTP), Department of Health and Human Services, Research Triangle Park, NC 27709, USA
[b] McGill University, Montreal, Quebec, Canada

At the close of the last day of the Congress, participants assembled to deliberate the issues discussed over the previous 3 1/4 days and propose future issues to be addressed.

The following action items were identified:

(i) Develop a male-mediated developmental toxicity (MMDT) website with registries of male-mediated developmental toxicants and assisted reproductive therapy (ART) outcomes – Dr. Bernard Robaire

(ii) Write grants for the training of highly qualified personnel in this area, to support student travel and training – All participants

(iii) Write and publish a position paper outlining the most critical research needs in MMDT and making recommendations for future directions aimed at meeting those needs – Dr. Andrew Wyrobek, with Drs. Crow, Anderson, Mulvihill, and Robaire

(iv) Initiate planning for the 4th International Congress on MMDT to be held in 2009 in Research Triangle Park, North Carolina, USA – Dr. Jack Bishop

The participants concurred that, as demonstrated by the presentations of the previous days, MMDT can be produced through many different types of mechanisms, including disorders of chromatin packaging, repeat sequence disorders, chromosomal damage such as large deletions, rearrangements, and insertions, as well as gene mutations (microlesions, alteration of mini- or micro-satellite DNA), and even *via* epigenetic mechanisms such as changes in DNA methylation. It is clear that a variety of models are required to assess these types of damage. There is evidence that a number of highly relevant environmental and therapeutic exposures, including radiation, acrylamide, anticancer drugs, diesel fuel, boron, and endocrine disruptors, have produced male-mediated effects. Data on much more than just reduced fertility and increased malformations will be required to translate animal-model data to human health outcomes. One of the complications is that the health outcomes mediated by an exposure of the male parent may vary, from the more immediate *in utero* death of the conceptus, growth retardation or malformations, to longer-term effects such as increases in postnatal learning deficits, childhood cancer, or other adult-onset diseases.

A broad range of assay endpoints, such as assessment of effects on sperm counts, sperm motility, morphology, and reproductive functionality, as well as chromatin structure including such tests as the Sperm chromatin structure assay (SCSA), the Comet assay, fluorescence *in situ* hybridization (FISH), has been used to detect the damage induced by male exposures. Some more recently developed technologies that include genomic, epigenomic, and proteomic analyses promise to provide assessments that go beyond the morphologically descriptive to the impact on genomic structure and function. There is a need for a review of what each of these assays/approaches contributes to our overall understanding of male-mediated toxicity, their efficacy, and whether and when they do or do not correlate with each other and with specific adverse progeny outcomes.

Other major areas of concern that were discussed include those associated with exposures, such as the dose–response relationship, low-dose effects and acute *vs.* chronic dosing regimens, and critical windows of exposure as a result of the key programing events in germ cell development. One example of a critical window of exposure is the *in utero* exposure of primordial germ cells, which appears to be able to result in epigenetic changes in male germ cells that are transmitted to subsequent generations, as reported by Anway *et al.*[1] (2005) (and in genetic changes as reported by Brinkworth *et al.* at this meeting). Another variable to be considered in assessing MMDT is the repair capacity in oocytes. There is a real need for improved, pregnancy-outcome measures (in animal models and humans) and additional resources for human populations-based epidemiology studies related to male-mediated effects.

A number of technical and logistical issues were discussed. The group indicated the need to develop a website and Dr. Bernard Robaire, McGill University, agreed to lead this effort. It was suggested that registries of male-mediated developmental toxicants and ART outcomes be developed. Grants for the training of highly qualified personnel in this area, to support student travel and training, need to be written. There was also a proposal to form a committee from participants at this Congress to publish a position paper outlining the most critical research needs in MMDT and making recommendations for future directions aimed at meeting those needs. Dr. Andrew Wyrobek was asked to lead this effort and solicit the input of Drs. Crow, Anderson, Mulvihill, and Robaire.

Dr. Jack Bishop noted that a workshop titled "Assessing Human Germ Cell Mutagenesis in the Post Genome Era" was held at The Jackson Laboratory, Bar Harbor Maine, USA, in September 2004, and that a report on this workshop, which is to be published in *Environmental Molecular Mutagenesis* in 2006, will contain recommendations for future research in the field of germ cell mutagenesis. Most of these recommendations should be of interest to researchers in MMDT.

It was decided by the Congress participants that the 4th International Congress on MMDT would be held in 2009 in the Research Triangle Park, North Carolina, USA. The exact date and location are to be determined. Dr. Jack Bishop was asked to coordinate the initial planning for this meeting.

Reference

1. M.D. Anway, A.S. Cupp, M. Uzumcu and M.K. Skinner, *Science*, 2005, **308**, 1466–1469.

Subject Index